Springer Series in Information Sciences 15

Editor: Thomas S. Huang

Springer Series in Information Sciences

Editors: Thomas S. Huang Teuvo Kohonen Manfred R. Schroeder

Managing Editor: H. K. V. Lotsch

William F. Schreiber

Fundamentals of Electronic Imaging Systems

Some Aspects of Image Processing

Third Edition
With 150 Figures

Springer-Verlag

Berlin Heidelberg New York
London Paris Tokyo
Hong Kong Barcelona
Budapest

Professor William F. Schreiber, Ph. D.

Massachusetts Institute of Technology,
Department of Electrical Engineering and Computer Science,
Cambridge, MA 02139, USA

Series Editors:

Professor Thomas S. Huang

Department of Electrical Engineering and Coordinated Science Laboratory,
University of Illinois, Urbana, IL 61801, USA

Professor Teuvo Kohonen

Laboratory of Computer and Information Sciences, Helsinki University of Technology,
SF-02150 Espoo 15, Finland

Professor Dr. Manfred R. Schroeder

Drittes Physikalisches Institut, Universität Göttingen, Bürgerstrasse 42–44,
W-3400 Göttingen, Germany

Managing Editor: Dr.-Ing. Helmut K.V. Lotsch

Springer-Verlag, Tiergartenstrasse 17,
W-6900 Heidelberg, Germany

ISBN 3-540-56018-1 3. Auflage   Springer-Verlag Berlin Heidelberg New York
ISBN 0-387-56018-1 3rd Edition Springer-Verlag New York Berlin Heidelberg

ISBN 3-540-53272-2 2. Auflage   Springer-Verlag Berlin Heidelberg New York
ISBN 0-387-53272-2 2nd Edition Springer-Verlag New York Berlin Heidelberg

Library of Congress Cataloging-in-Publication Data. Schreiber, William F. Fundamentals of electronic imaging systems : some aspects of image processing / William F. Schreiber. – 3rd ed. p. cm. – (Springer series in information sciences ; 15). Includes bibliographical references and index. ISBN 0-387-56018-1 (New York). – ISBN 3-540-56018-1 (Berlin). 1. Image processing. 2. Imaging systems. I. Title. II. Series. TA1632.S33   1993 621.36'7–dc20   92-39695

Preface to the Third Edition

Image processing is a fascinating applications area, not a fundamental science of sufficient generality to warrant studying it for its own sake. In this area, there are many opportunities to apply art and experience, as well as knowledge from a number of sciences and engineering disciplines, to the creation of products and processes for which society has an expressed need. Without this need, work in the field would be sterile, but with it, image processing can readily provide the interested scientist or engineer with a professional lifetime of challenging problems and corresponding rewards. This point of view motivates this book and has influenced the selection and treatment of topics. I have not attempted to be encyclopedic; this service has already been performed by others.[1]

It will be noted that the word "digital" is not in the title of this book. While much of present-day image processing is implemented digitally, this work is not intended for those who think of image processing as a branch of digital signal processing, except, perhaps, to try to change their minds. Image gathering and image display, vital parts of the field with strong effects on image quality, are inherently analog, as are all of the channels and media now used, or likely to be used in the future, to record TV signals and to transmit them to the home. Finally, although the human visual system is certainly made up largely of discrete elements, it is not a digital system by any reasonable definition. In view of the pervasiveness of analog phenomena in the field, it is particularly important for electrical engineers and computer scientists to keep a check on their natural tendency to look for exclusively digital solutions for every problem. This is nowhere more evident than in the current proposals for digital terrestrial broadcasting, which is turning out to be a good deal less simple than many enthusiasts at first believed.

This book originally grew out of courses taught at the Massachusetts Institute of Technology by the author and two of his former students, Thomas S. Huang and Oleh J. Tretiak. Like the courses, it is primarily intended for engineers and scientists who plan to work in the field. It has not proven necessary to spend much time on strictly mathematical or digital issues, since the students already know that material or can learn it easily by themselves.

What is attempted here is to give the reader enough contemporary information about the real world of images and image making that he can effectively apply appropriate science and technology to the solution of practical problems. The

---

[1] W. Pratt: *Digital Image Processing* (Wiley, New York 1991); A. N. Netravali, B. G. Haskell: *Digital Pictures, Representation and Compression* (Plenum, New York 1988)

emphasis is on the transmission and reproduction of images of natural scenes and documents, monocularly viewed, still and moving, in color as well as black and white. The images are intended for viewing by humans, in the normal way that we look at pictures or books, for information, pleasure, or entertainment. The following fundamental questions are dealt with:

- What is an image, and how is it distinguished from an arbitrary function?
- Where does it come from, where does it go to, and what are its fundamental characteristics?
- What are the effects on these characteristics of the laws of physics and the properties of modern imaging devices and systems?
- How do we perceive images? How is the performance of systems affected and how should their design be influenced by the properties of the observer?
- How do we transform effectively between the discrete world of modern signal processing and the continuous worlds of objects and observers?
- To what extent is it possible to take advantage of the properties of scenes to be imaged and viewers to be served to reduce the amount of data required to reconstruct images? How do we evaluate the performance of such compression systems when they invariably alter the objective image parameters? What is the role of statistical compression systems?

Because of the practical bias of much of this work, we have emphasized those aspects of image processing equipment and algorithms that influence picture quality as perceived by the viewer under normal viewing conditions and with the intended kind of subject matter. The choice of topics and their treatment arises out of the personal experience of the author, his students, and his associates at MIT and elsewhere. This experience has included the development of the *Laserphoto* facsimile system and the *Electronic Darkroom* for the Associated Press, the *Autokon* electronic process camera for ECRM, Inc., a real-time two-channel compression system for Sony Corporation, the *Smart Scanner* for Scitex Corporation, a complete prepress data-processing system for a large gravure printer, an interactive color-editing system for Electronics for Imaging, Inc., a number of CRT, vidicon, laser, and CCD scanners, and continuous development of computer-simulation systems for image-processing research. Finally, experience in the Advanced Television Research Program at MIT since 1983 provided the background for the material on TV system design and development reported in Chapter 8 and the Appendix.

It is a chastening experience for an academic to try to get image users, such as printers or television broadcasters, to accept his computer-processed pictures in place of those produced by conventional means. The defects and degradations objected to are not inconsequential. They show that measuring picture quality in a meaningful way is not easy. The effects of the characteristics of imaging components, especially cameras and display devices, on the overall performance of systems are generally ignored in the literature and by many research workers, but are actually very large. Our treatment of this subject is not as detailed as desirable, in part because complete knowledge is lacking, but at least we have tried to give the subject the attention it deserves.

The approach to this subject described here has been readily accepted by MIT students, many of whom are still motivated by practical applications, even though the curriculum, like that at most comparable schools, is primarily theoretical. Exceptionally bright and alert, they really appreciate accounts of first-hand experience, even if (especially if) they do not accord with "theory."

Readers will quickly notice that we have a great deal of respect for the work of traditional image makers who have never taken a university-level image-processing course. We hope the reasons for this will become evident to those who are more scientifically oriented, as they come to realize the very high standards of quality that are used in these fields, and the degree of skill and diligence employed to get the desired results. We have made some attempt to translate the words and concepts used by such artists and craftsmen into language more credible to the scientist, but with limited success.

The first six chapters appeared in the first edition. In the second edition, obvious errors were corrected and Chapters 7 and 8, on color and TV system design, respectively, were added, along with the Appendix. In the third edition, Chapter 8 and the Appendix have been extensively revised and expanded to deal with progress in the race to develop high-definition television broadcasting systems.

The preparation of this book would not have been possible without the help of many students and colleagues who supplied material, read drafts, and offered helpful comments, for all of which I am very grateful. Elisa Meredith and Kathy Cairns served as editorial and administrative assistants, respectively, during the preparation of the first edition. Debra Harring likewise served during the preparation of the third edition. Their labors are much appreciated. Since the inception of the project, I have also had a great deal of encouragement and assistance from Dr. H. Lotsch at Springer-Verlag.

The second edition was completed while serving as Visiting Professor at Ecole Polytechnique Federale de Lausanne. The description of the European HDTV effort benefited greatly from discussions with Murat Kunt, head of the Signal Processing Laboratory at that institution. The section on the Japanese HDTV program was prepared while serving as a member of a panel of the Japanese Technology Evaluation Center of Loyola University. Richard J. Elkus, Jr., chairman of the panel, helped a great deal in understanding the strategies behind the program.

The color chapter developed out of the period that I spent at Technicolor Motion Picture Corporation and out of the gravure development project at MIT. My interest in color began during my relationship with Donald H. Kelly at Technicolor. During the work at MIT on the gravure project, we were most fortunate in having the advice of John A. C. Yule, formerly of Kodak, from whom I received many useful insights. In writing the color chapter, I had the extensive assistance of my former student, Robert R. Buckley, now at Xerox.

In Chapter 8, I have tried to organize what we learned about TV system design in the course of the still-existing project on that subject at MIT. The treatment in both chapters was greatly influenced by the simultaneous production of a videotape version of the entire book at the MIT Center for Advanced Engineering Studies, whose staff was particularly helpful. In the sections on recent developments in Europe, I benefited greatly from the comments of Ralf Schäfer of the Heinrich-Hertz-Institut in Berlin, Gary Tonge of the Independent Broadcasting Commission, Winchester, UK, and Robert Whiskin of Understanding and Solutions, Dunstable, UK. The treatment in both Chapters 7 and 8 was greatly influenced by the simultaneous production of a videotape version of the entire book at the MIT Center for Advanced Engineering Studies, whose staff was particularly helpful. Television, by its nature, lends itself very well to visual demonstrations, many of which are included in the videotape course. It is my belief that using the book and the videotape at the same time enhances the learning experience.

The revisions to the Appendix and to Chapter 8 are intended to bring the account of current HDTV developments up to date and to explore their technical background, respectively. Current proposals to use highly efficient source-coding methods (JPEG and MPEG) and digital channel coding, if actually implemented, will represent the first major practical application of some of the most significant work in image processing, previously confined to laboratories. The near-instantaneous and unqualified acceptance, in the TV broadcasting community, of techniques heretofore regarded with near-universal suspicion, is a remarkable turnabout that is hard to explain. In the view of the author, all parties need to be somewhat more cautious in this enterprise if very unpleasant surprises are to be avoided.

Cambridge, Massachusetts, September 1992   *W. F. Schreiber*

Acknowledgements

Much of the work reported in Chap. 8 was performed at MIT under the Advanced Television Research Program, a research project sponsored by the Center for Advanced Television Studies. The members of CATS have, at various times, included ABC, CBS, NBC, PBS, HBO, Ampex, General Instrument, Harris, Kodak, 3M, Motorola, RCA, Tektronix, and Zenith. A number of colleagues and students contributed to the developments. Most of the images used in Chap. 8 were made by students, including Warren Chou, Ashok Popat, Paul Shen, Adam Tom, and Kambiz Zangi. The help of all these people and organizations is gratefully acknowledged.

The MIT Video Course:
Fundamentals of Image Processing

The present textbook may also be used in conjunction with the new MIT video course *Fundamentals of Image Processing*. In this course Professor Schreiber delineates the latest developments in scanning, recording, coding, storage, enhancement and transmission of images by means of demonstrations and carefully selected examples.

The course materials comprise 23 color video cassettes together with the present text. The book provides essential background reading and is an integral part of the course.

Course materials are available from:

MIT, Center for Advanced Engineering Study
Rm. 9–234, Cambridge
MA 02139–4307
USA

Complete course materials: Order no. 680–1000
Preview tape: Order no. 680–0101
Individual video cassettes are also available.

Contents

1. Introduction

This chapter defines "image" and "image processing" for the purposes of this book. Given these definitions, it becomes evident that exact reproduction is never possible. It is shown that changes in images need not degrade quality, and that it is essential to use perceptually valid quality measures in evaluating processing methods. A generalized image processing system is devised to exhibit the wide range of phenomena associated with imaging apparatus and the correspondingly wide range of knowledge that can fruitfully be brought to bear on the subject. Finally, to make these considerations more explicit, a particular example of a practical imaging system is discussed.

1.1 What is Image Processing?

This book is intended to be written in plain English, in which words have their usual meaning. An example of what we mean by an image is the two-dimensional distribution of light intensity in the focal plane of a camera pointed at a natural scene. Mental "images" will not do; still less a politician's "image" created in the mind of a potential voter by a TV commercial. We do imply visual meaningfulness. Arbitrary two-dimensional functions are not images, nor is a two-dimensional array of numbers, although the latter may represent an image. Processing is the treatment of an image to produce a second image or group of images for some purpose. Making a photographic print from the light distribution in the focal plane is such processing. Recognizing printed English characters in the image of a page is very useful, but is outside the scope of our meaning, as are all types of pattern or character recognition, the work of photointerpreters, and image segmentation or scene analysis except as it may be a step in image coding and reconstruction.

In spite of these restrictions, we have an ample field in which to work. Television, motion pictures, facsimile, and printing together make up an important segment of our economy and one indispensable to education and culture generally. They are all fields that have seen the most modern technology combine with ancient crafts to produce useful and enjoyable products.

Since all these fields have a long history, in the case of printing antedating the discovery of electricity, it is obvious that much image processing can be and routinely is done without reliance on electronics and computers. Clearly the use of modern sophisticated methods, in addition to or in place of traditional techniques, must be justified by some advantage of cost, quality, or perhaps convenience. This applies with special force to the replacement of manual by automated operations. While it is true that, on the average, we all get richer when the cost of producing goods and services is reduced, the mindless replacement of people by machines is not only awkward for suddenly "redundant" workers, some of whom may have devoted a lifetime to acquiring the skills suddenly rendered unnecessary, but the savings may be much less than anticipated unless the entire process is first carefully analyzed. A computer is not an artist, and art is the essence of making good pictures. It often makes much more sense, even economic sense, to design systems to facilitate and augment human capability, rather than to replace it entirely.

Some kinds of image processing make more sense than others. Making a recognizable version of Abraham Lincoln or the Mona Lisa on a line printer may be amusing but hardly useful. Sharpening a chest X-ray by spatial convolution and displaying it on a 512×512 CRT display where it has almost no diagnostic value is a waste of time. Finding the "edges" in an image by repeated correlation operations resulting in a line drawing less representational than can be made by a moderately skilled artist in a few seconds proves very little, although reconstructing a good quality image from compactly coded outlines (even if they are not a pleasing representation) may be very useful indeed. On the other hand, using sophisticated algorithms to produce color separations from which good quality pictures can be printed with inexpensive inks is unquestionably valuable. Coding picture signals so that they can be transmitted with less channel capacity but with subjective quality equal to that obtained by usual methods can be highly advantageous. Taking a corrupted TV image of a planet, received from a space probe, and processing it so that the resulting image is free from defects and that surface features not discernible in the original are made visible is good image processing. Operating on the video signal produced by a television camera so that the image becomes comparable to that from a camera of superior specifications is certainly advantageous. Needless to say, we shall try to concentrate our attention on useful kinds of processing.

1.2 Descriptions of Images

Having distinguished images in our sense from other two-dimensional functions, it remains to describe them in ways that facilitate further processing. In a few special cases, such as test patterns, an algebraic description is possible in which the image intensity can be expressed as a function of spatial and perhaps temporal variables. In the case of graphical images, which we may define as those which can be constructed from a small number of levels, usually two, very compact algebraic descriptions are sometimes possible. More generally, statistical, spectral (e.g., a Fourier transform), or numerical descriptions are required, with the latter particularly valuable, even necessary, for processing by digital machines. We shall devote considerable attention later on to the optimum numerical description of images. The reason for this is that inadequate methods may result in excessive required channel capacity, loss of quality, or both. While only a small amount of capacity is required for a single small monochrome picture, hundreds or even thousands of Megabits per second may be needed to represent high resolution, stereo color motion pictures and perhaps 1000 megabits for the contents of a Sunday rotogravure section. Economy of numerical descriptions of images is one of the most important topics in image processing.

1.3 Sources and Types of Images

Images are often obtained by viewing natural objects with cameras, either optical, mechanical, or electronic. While the image itself is most often a scalar point function of two variables, it may be three-dimensional, if varying in time, and may be a vector, rather than a scalar, if stereo or color. Holographic systems can have three-dimensional images fixed in time. The object may be a previously made image, such as a photograph or document. Many systems have multiple stages in which the image produced by each stage is the object for the following stage.

In virtually all cases, the image does not perfectly represent the object, in that not enough information is present to permit reconstructing the object exactly. Sometimes this matters, but often it does not. In any event, it should be borne in mind that it is in the nature of images merely to approximate the object and very common for the approximation to be gross. "Exact" reproduction is never possible, on account of the quantum nature of the universe, if for no other reason. This is a phenomenon to be dealt with, not a goal to be sought after with its lack of achievement deemed a failure. Likewise the degree of exactness need not be taken as an index of goodness of imaging systems. We often do not know what scale to use to measure such errors. The quality of images, an important subject, must be related to their usefulness for the intended purpose, rather than the degree to which they duplicate the object.

1.4 Processing Images

Since the output of most processes with which we are concerned is another image, both input and output can often have the same type of description, while the process itself must describe the relation between the two images, usually in mathematical, i.e., algorithmic, terms, either algebraic, logical, statistical, or some combination. For example, a picture may be cropped, enlarged, sharpened, subjected to a nonlinear tone-scale transformation, and noise cleaned. Such a process, or sequence of operations, can be described in terms of an algorithm.

We shall always take the point of view that what is characteristic is the algorithm, while of secondary importance is the physical device or system with which the algorithm is implemented.

While many modern image processors, especially those who come to the field from the computer science side, think of image processing largely in terms of computer programs, we shall be willing to evaluate, and perhaps use, any systems that work. These potentially include electronic techniques such as analog and digital hardware, programmable computers, and microprocessors; optical methods, both coherent and incoherent, including conventional optical systems, holographic systems, electro-optical, magneto-optical, and acousto-optical effects; mechanical and photochemical systems, including photography, Xerography and electrostatic systems generally; all kinds of printing processes, both impact and nonimpact; ink jet; and even wood-burning and mosaic tile! The point is that good pictures have been made by all these processes. Some are cheap and repeatable, and describable accurately enough to be both controllable and usable.

1.5 Purposes of Processing

The most common type of processing is reproduction – often at a distance, implying some kind of transmission. Speed and quality in relation to the channel are obvious considerations, and coding an important technique. Storage and retrieval are often involved, with related questions of quality, capacity, cost, access time, etc. The display or recording methods, soft versus hard copy, and the myriad forms of output devices with their individual characteristics must be studied.

Another significant category of processing comprises enhancement and restoration, separate fields or different applications of the same ideas, depending on one's point of view [1.1]. A great deal of attention has been given to this subject by both traditional and modern workers, with the result that a wide variety of very effective methods is available. Recent developments in computer processing foreshadow the early development of cost-effective interactive systems that will enable an editor, i.e., an artist, to manipulate vast amounts of pictorial data easily and quickly to achieve the desired ends.

Although we have specifically excluded pattern recognition and similar techniques from our study, it is appropriate to mention that in certain specialized areas, such as Roman font character recognition, long-standing problems have now been solved to the extent that a number of highly effective machines are commercially available that perform these functions very well. At the present time, the conversion of typed characters to machine code can be done faster, cheaper, and more accurately by machines than by humans. Other types of pattern recognition, such as blood cell classification, chromosome karotyping, and cancer cell detection, are possible but not yet widely used. Interpretation of satellite photos is fairly advanced in some fields, but the goal of completely automated weather forecasting, for example, has not yet been reached.

1.6 Image Quality

Since the images that are the subject of our concern are to be viewed by humans, their quality relates to human perception and judgment. Lacking adequate models of observer reaction, it is generally impossible to calculate precisely, and in some cases even difficult to predict qualitatively, the effect on quality of an alteration in the objective parameters of an image. Finally, since virtually all useful image processing techniques produce such objective changes, subjective testing is required for the evaluation of the comparative merits of different systems. When the quality of a processing system or technique is to be studied, one must, of course, study the quality of the output imagery over the range of inputs and viewing conditions for which the system is designed. It is quite easy to give examples of systems whose rank order depends on the subject matter or the manner in which the output is viewed.

The importance of subjective judgments in television has been recognized for some time. As a result, a great deal of work has been put into contriving experimental setups and statistical methods that give repeatable results. These are admirably summarized in a recent book by *Allnatt* [1.2].

It is not necessary to pretend we know nothing about the observer just because we cannot calculate his reaction exactly. In fact a great deal is known, although many of the classical psychophysical investigations have not measured properties particularly relevant to imaging systems. As the relationship between particular aspects of visual perception and the performance of systems becomes better understood, it has become possible to design new and more useful psychophysical experiments. In some cases, system parameters can be established on the basis of specific measurements.

Some quality factors are self-evident to any technically trained worker. In particular, spatial frequency response can easily be related to resolution and sharpness while temporal frequency response clearly governs motion rendition and flicker. What is not so apparent, especially to scientists and engineers without practical imaging experience, is the great importance of tone rendition. In many if not most cases, it is the most important quality factor. Signal-to-noise ratio (SNR) is another concept familiar to the communication scientist, but neither noise visibility nor the degrading effect of suprathreshold noise follows any simple law [1.3]. Finally, we must deal with a group of factors of everyday overwhelming concern to printers and photographers, but often overlooked by others, namely defects and artifacts. It is almost impossible for laymen to appreciate the degree to which graphic artists abhor defects. In order to achieve the degree of perfection demanded and routinely reproduced by traditional methods, electronic and computer systems require a quality of design and implementation of a very high order. This discipline is not foreign to designers of commercial studio equipment for television and sound broadcasting, but it has to be learned by most other engineers and scientists. The most troublesome factors are geometrical distortion, dirt, streaks and other patterned noise, irregular line or picture element spacing, and nonuniformities of contrast and brightness over the field of view. In color systems, mail-order catalogs for example, color accuracy far beyond that of amateur slide shows or home TV receivers is required, often in spite of rather poor quality paper and ink.

A significant advantage of electronic over traditional systems in the attainment of high quality is the possibility of substantially correcting, in one stage, the distortions introduced in another. An important subject of study is the determination of the theoretical and practical limits to which such correction methods can be carried.

1.7 A Generalized Image Processing System

Figure 1.1 is a block diagram of an image processing system so general that it encompasses a large proportion of systems actually in use. Our purpose in discussing this arrangement is to list the factors that affect the operation of each block and the kinds of knowledge required to analyze and design such systems and their components.

1.7.1 Source/Object

This can be a collection of physical objects or an image produced by another system. It may be self-luminous, or receive light from the surrounding surfaces or from the camera itself in the case of "flying-spot" scanners. It can be fixed or changing, to be viewed in color or achromatically, with one "eye" or two. It can bear a fixed physical relationship to the camera, or there may be relative motion, deliberate or accidental, and change of size and shape with time. The surfaces of the object may be smooth, with various degrees of gloss, irregular, or patterned in such a way as to be described as texture.

It should be noted that the amount and arrangement of the light falling on the object has a profound effect on the quality of the image that is eventually produced. The best results are achieved with diffuse general illumination sufficient to give good signal-to-noise ratio, together with some more highly directional illumination from no more than a few sources to give a three-dimensional effect that photographers call "modeling." A murky or turbulent atmosphere between object and camera deteriorates quality markedly unless some special effect is being sought. The lighting of scenes to produce the desired result is a well developed art [1.4]. In general, it is much harder to produce attractive images in monochrome than in color, especially when the illumination is not optimum or when the atmosphere is not clear.

1.7.2 Optical System

Light emanating from the object impinges on the sensitive surface of the camera, either directly or through some optical system. The latter may be fixed or variable. Simple optical systems constitute a special case of the more general situation in

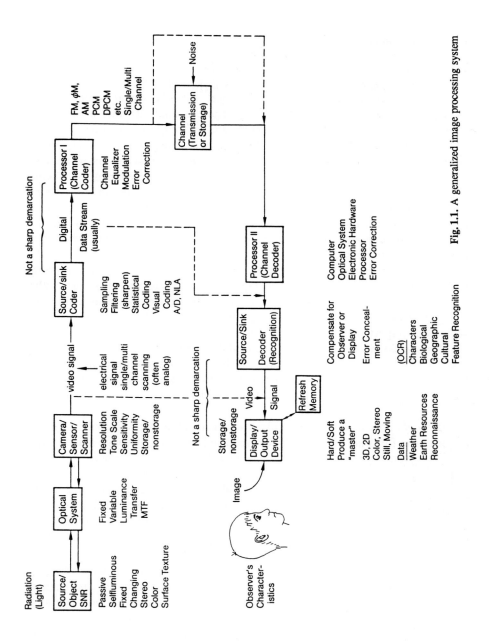

Fig. 1.1. A generalized image processing system

which an output image is a function of an input image. While often almost linear in the sense of system theory, many such systems are shift-varying. Usually, however, they have an effective optical impulse response that varies only slowly across the field. Thus the field can be divided into regions, each of which is shift-invariant but slightly different from its neighbors. Most often both the luminous transfer and the spatial frequency response fall off away from the central axis.

7

1.7.3 The Camera

This is a device that forms an electrical signal from an optical image. Although some systems have been built in which many parallel paths carry information from the camera, generally there is only one, with the signal being obtained by scanning [1.5]. Most scanning methods cause a small aperture to search the image area in a pattern, or raster, of closely and uniformly spaced lines or points, once per image for stills and periodically for moving images. A highly significant difference between classes of cameras is whether the signal at any instant depends only on the light then falling on the corresponding picture element or whether the incident light has been integrated for some time. In nonstorage systems, it is sometimes possible to interchange the light source and light sensor. In this arrangement, called a flying-spot scanner, a small light spot scans over the object, while large-area photodetectors collect a portion of the light reflected or transmitted by the object.

Cameras are characterized by their sensitivity, spatial frequency response, signal-to-noise ratio, and tone-scale reproduction. There are also many other pertinent quality factors common to all imaging devices, such as uniformity and geometrical accuracy. Response speed, or temporal frequency response, is sometimes important.

Cameras, in the sense we have described them, have existed for more than 100 years. There are many important kinds, suitable for almost any need. Interestingly, the very best images are still obtained with mechanical scanners, although the most sensitive cameras, capable of making good TV pictures by candlelight, are electronic.

The commonest storage-type electronic camera is the vidicon. Image orthicons were for years the standard for broadcast purposes, although silicon target vidicons, plumbicons, and saticons have now largely replaced them. Solid-state camera devices consisting of photodiode arrays are beginning to come into use. At the professional level they are comparable in price to tube-type cameras, are more sensitive and give better motion rendition, but have lower resolution. At the consumer level, solid-state color cameras are beginning to be competitive with tube cameras. The image dissector, a very old nonstorage electronic camera, is sometimes used where very high light levels are permissible.

Cameras for still pictures are almost always flying-spot scanners, the Nipkow disk being the oldest device, still in use for some infra-red applications. CRT scanners are sometimes used, but rotating drum devices are most common. Recently flat-bed laser scanners have reached a high level of development and have replaced drum scanners in many graphic arts applications. CCD techniques are coming into use.

1.7.4 Source/Sink Coder

The source/sink encoder is usually understood to be that part of the system that receives the unprocessed video signal from the camera, often in analog form, and processes it to put the information into a form suitable for further use. This may be a very simple step, such as adjusting the linear amplification and band-limiting the signal so that it can be used as the input to a modulator and then transmitted. On the other hand, it may be very complicated, involving analog-to-digital conversion, tone scale transformation, two-dimensional filtering, statistical or visual coding (data compression), sharpening, noise reduction, etc. Source coding means coding that makes use of the characteristics of the signal produced from a particular group of sources by cameras or other devices with particular characteristics. To the extent that the observer's characteristics are taken into account, we can call the operation of the encoder "sink coding." It is the encoder where most of the operations generally thought of as "image processing" are performed. It is the part of the system on which most attention is often focussed, although as we have seen here, it is only one step in a rather long chain from the object to the final result. The output, nowadays, is generally a digital data stream.

1.7.5 Channel Coder

Processor 1, often called the channel coder, prepares the information for transmission or storage. It may be as simple as an amplitude modulator or involve highly sophisticated coding to take advantage of channel characteristics. Error control procedures are often used, since virtually all real channels and storage devices add noise and/or produce errors in the data. In most cases, synchronization signals are added to the picture data, so that data pertaining to the various pictures being transmitted can be sorted out and so that, for each picture element, correct geometrical positioning can be achieved.

1.7.6 Channel

The channel itself may be simply a wire path or a more involved direct communication link such as a coaxial cable, waveguide, or optical fiber system. Electromagnetic wave propagation may be used. In virtually all such situations, sending and receiving apparatus is required that is more complicated if the system is two-way and/or multichannel, or if temporary storage is used. Storage, either momentarily during transmission or for arbitrarily long times as in the case of magnetic recording, adds additional complexity. Channels are normally described as "digital" or "analog," by which terms we refer to the character of the input and output signals. Even though we can describe the performance of digital channels simply in terms of transmission speed and error rate, whereas analog channels typically have many more parameters in their description, all channels are analog in their physical embodiment, and therefore have continuously variable input and output states. In general, dividing a continuous range of states into a discrete number reduces the effective information capacity.[1] It is

---

[1] We use "information" here in its usual, nontechnical meaning. More rigorous definitions are dealt with later.

a general rule that the same physical medium has a lower capacity as a digital link than as an analog link. In order to achieve a more nearly equal capacity in the digital case, it is usually necessary to operate at a large raw error rate and to use extensive error correction schemes.

1.7.7 Channel Decoder

Processor 2, the channel decoder, has the function of recovering, from the channel output, the same data received by the channel coder. Often, some approximation must be accepted on account of non-correctable noise or distortion caused by the channel. Implementation of the channel decoding operations may involve digital or analog hardware, computers (mini or micro) or even optical processing.

1.7.8 Source/Sink Decoder

The decoder sometimes is simply the inverse of the encoder, having the function of recovering a video signal much like that received from the camera. For very compact code descriptions the operation may be nontrivial and may require extensive computation. Another important function, for which we might have provided a separate box, is processing to take account of the characteristics of the display as well as the observer who will view or otherwise use the output. Even in the simplest closed-circuit TV system, the output image is ordinarily viewed under very different conditions of luminance and size from that of the original object, so that some video processing is desirable.

1.7.9 The Display

Displays, or output devices in general, are often classified as "soft" (a temporary image, e.g., on a CRT or optical projector) or "hard" (permanent). Some purists reserve the description of "hard" for printed or nonphotographic copies. As in the case of objects, they may be fixed or moving, color or achromatic, mono or stereo, two-dimensional, or in the case of some holographic systems, three-dimensional. It is very common for the output to limit the quality of the system. In almost all cases, the output device sets the dynamic range of the display luminance. Failure to take this factor into account is the cause for poor image quality in many cases.

Although it is the display that couples most closely with the properties of the observer, the latter must be taken into account at nearly every stage of system design. Fundamental choices of resolution, frame rate (for moving images), and signal-to-noise ratio cannot freely be made just at the output of the system.

The commonest soft-copy display device is the cathode-ray tube, although some progress has been made in developing gas-discharge and other types of arrays, especially for slowly changing images. Storage may be incorporated in the CRT but with the rapidly falling price of digital memory, it is now more common to "refresh" an ordinary CRT. For very large displays, laser flying-spot scanners have sometimes been used. Light-valve devices such as the Eidophor are used for theater projection. Solid-state light valves such as liquid crystals are beginning to come into use for certain applications.

Hard-copy devices are too numerous to list comprehensively. Flying-spot scanners, both CRT and laser, as well as rotating drum, are used in conjunction with photosensitive materials, both conventional and electrostatic. Impact devices are very successful, especially for alphanumerical printing, even at very high speeds. Ink-jet printing has some popularity.

Another category of hard-copy printers involves the making of a "master", i.e., a printing plate or surface. Obviously these are most useful where multiple copies are needed. The three principal printing methods are letterpress, lithography (often called "offset"), and gravure. Making printing surfaces from the output of a digital system is currently a very active field of development. Some printing plates can be made sufficiently sensitive to expose directly to CRTs or laser scanners, but most require an intermediate photographic negative. Electrostatic processes such as Xerography are especially appealing since the photoconductive materials are very sensitive, but very high quality printing is not as yet possible by this process. Laser exposure of offset plates is in use to some extent, while experimental systems for laser exposure of letterpress plates and gravure cylinders have been demonstrated.

1.8 A Simple Television System

In order to make more concrete the idea of a complete image processing system, consider the case of a simple monochrome television arrangement, as shown in Fig. 1.2, capable of providing a remote viewer with an image of live action,

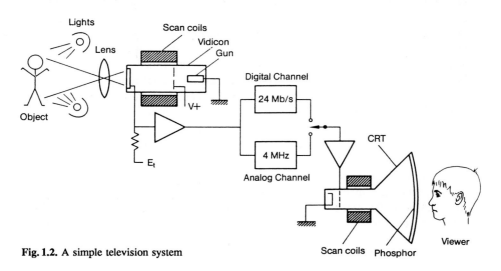

Fig. 1.2. A simple television system

11

subject to the constraint that the channel available to connect the sending and receiving end is either a low error rate 24 megabits/s digital line or a 4 MHz, 40 dB SNR[2] analog link. Let us suppose that the client who wants the system is not technically trained, but can appreciate good pictures. He specifies he wants "the best possible pictures at the lowest cost." The system designer must supply the missing specifications, choose operating parameters, and select the equipment.

A naive designer would complain about the inexact nature of the overall requirements, but an experienced engineer would try to provide the missing specifications himself and proceed with the job. What does he have to know?

1.8.1 Choice of Components

Low cost requires using the commonest input and display devices, a vidicon camera tube and a cathode-ray picture tube. If the quality is to be as high as possible, then it should be limited primarily by the channel, which has been specified. In order to permit the vidicon to produce better pictures than the channel can transmit, adequate illumination is required. The larger the area to be covered, the higher will be the lighting cost. A fixed camera and fixed lens covering the entire field of view is cheaper than a movable camera with a variable lens that can follow the action. A very wide angle will produce small output images, however, so the designer will have to go back to the client to explain the effect of the choice.

1.8.2 Scanning Standards

Having chosen the camera and display device and decided to supply the amount and type of illumination required for good pictures, the remaining choice has to do with selection of the scanning standards [1.6]. This is a problem that has confronted image transmission system designers from the earliest days and has been the subject of much study. Since the light intensity at the photosensitive target of the vidicon is a function of two space variables and one time variable, and the same is true on the phosphor surface of the output picture tube, the problem is to use the channel to "map" the intensity function into the signal to be transmitted in the optimum way to produce an output image of the highest quality. This problem cannot be approached except by careful consideration of the visual capability of the observer, since it is his subjective judgment that is the sole criterion of goodness. Some consideration must also be given to the characteristics of the objects depicted and the use to which the system is to be put.

It is easier to think about this in terms of the digital channel [1.7]. The transmission rate in bits/s is equal to the product of the x, y, and t sampling densities with the number of bits per sample. The choices thus trade off space, time, and tone-scale resolution. If the objects were highly detailed and moved

---

[2] Peak signal/rms noise.

slowly, we would use high spatial and low temporal and tone-scale resolution. If the objects were less detailed but moved rapidly, we would use higher temporal and tone-scale resolution and lower spatial resolution. If any of the resolutions were higher than the observer could perceive, there would be no need to waste channel capacity on that dimension, since that aspect of object characteristics would be imperceptible even if the viewer looked directly at the original scene. Likewise, if the system were for identification or detection of a particular object feature, and it did not matter whether some other feature were well depicted, the choices would be biased.

1.8.3 Sampling in the Time Domain

Special attention must be paid to the sampling along the time axis. Even for objects barely moving, a low sampling rate – say below 20 or so per second – would give rise to flicker on the display. Although the vidicon camera tube integrates the light from frame to frame so that it takes a "time exposure" of the image on its target, most CRTs are brightened only momentarily by the scanning electron beam. Although long persistence phosphors are available, their decay is quasi-exponential. Even if the decay time constant could be freely chosen, if it were long enough to eliminate flicker, it would cause substantial merging of each frame with the next. The use of a storage display tube, or of an external refresh memory in conjunction with a short persistence CRT, would be helpful, but would increase the cost.[3]

For picture tubes of usual brightness and size, 30 frames per second is sufficient to eliminate most of the flicker. (Operation at 25 frames per second, as in Europe, requires a lower maximum luminance than at 30, for equal freedom from flicker.) Even these rates would be insufficient without using interlace, a partly successful exploitation of one of the limitations of human vision.

The perceptibility of flicker depends on the size and brightness of the flickering object – the larger and brighter, the worse the flicker. In order to exploit this fact, it is conventional to choose the relative vertical and horizontal scan frequencies so that alternate scan lines are traced out on each vertical sweep. Thus two sweeps are required for one complete picture. By this means, a lower frame rate can be used with a given brightness, resulting in a higher vertical resolution for the same data rate. If interlace worked perfectly, the vertical resolution would be doubled. In fact, the resolution is increased only marginally, except at very low brightness [1.8] (The resolution of interlaced pictures can be further increased by more elaborate processing than appropriate in this simple system.) There are additional problems with interlace besides the fact that the vertical resolution is less than often thought. In the case of subject matter moving vertically in the picture, for example, the viewer's perception sometimes changes to that of a

---

[3] The latter alternative will probably become quite practical in the near future. Several TV receivers are already on the market that use frame stores to convert the 30-frame/60-field transmitted signal to a 60-frame/60-field display.

picture with half the number of scan lines at twice the frame rate. Interlaced displays are also generally unsatisfactory for small alphanumerics common on video display terminals used in conjunction with word processing systems, on account of severe interline flicker.

Frame rates high enough to avoid flicker usually are adequate for rendition of normal motion, and always would be adequate if the observer happened to fixate his eye in the same manner as the TV camera. Even for rather slow movement of objects with textured or otherwise detailed surfaces, 30 frames per second causes noticeable blurring under these circumstances. This is avoided if the movement is tracked, a process that is natural for humans and difficult and expensive for cameras.

Because of various physical limitations on their operation, most camera tubes integrate only for the field duration, not for the frame duration. For moving objects, this gives a better temporal and poorer spatial resolution than if they integrated for the full frame.

A 30-frame choice leaves 0.8 Mbits/frame, which in a square format could be used for 3 bits/picture element (pel) in a 516×516 format, 4 bits/pel in a 447×447 format, 5 bits/pel in a 400×400 format, etc.[4] It takes as much as 8 bits/pel for the quantization noise to disappear completely, leaving a resolution of only 316×316. This would almost never be the best choice. There is, in fact, no best choice independent of the subject matter, but 4 to 6 bits per pel would be a reasonable compromise.

1.8.4 Explaining Motion Rendition and Flicker

There are several ways by which the diverse phenomena associated with the reproduction of moving objects can be organized and explained. The way most appealing to the mathematically inclined is to regard the light intensity on the target of the camera tube and on the surface of the display tube as scalar point functions of two space variables of limited extent and one time variable of unlimited extent. The rendition of the temporal variations of this function at each (spatial) point in the image is then a classical problem in sampling and reconstruction. Given a certain number of samples per second, it is necessary to band-limit to half the sampling frequency before sampling and transmission. The samples can then be interpolated using a low-pass filter of the same bandwidth to recover the band-limited function and produce a temporally continuous display.

Temporal band-limiting before sampling physically blurs the image of a moving object. At the receiver, technological limitations prohibit true temporal interpolation, which would give a temporally continuous display. In both television and motion pictures, the temporal samples, or frames, are displayed in sequence, and it is the human visual system that converts this flickering image into one that is continuous. In motion pictures, the camera usually operates at 24 frames

---

[4] The resulting 8-, 16-, and 32-step gray scales, as we shall see later, can be converted into stepless noisy (randomized) gray scales that give better quality than the usual quantized representation.

per second (fps), while the projector, using each frame 2 or 3 times, operates at 48 or 72 fps. In existing TV systems, the display is at the same frequency as the camera, although some experiments have been done in which the display rate is higher.

The appearance of the output picture, one would think, should be predictable by measuring the temporal frequency response of the observer and using the ideas of linear system theory. A good measurement of this characteristic has been made by *Kelly* [1.9], shown in Fig. 1.3. Using a wide-field display with a sinusoidal variation imposed on a uniform steady luminance, he measured the percentage modulation for just noticeable flicker as a function of flicker frequency and background luminance, obtaining the results shown here. Kelly's data accurately predict flicker in uniform fields with temporal variations of arbitrary waveform, but do not directly tell us anything about motion rendition. One way to approach that subject is from a psychophysical viewpoint.

A well known aspect of vision called the "phi phenomenon" relates directly to motion rendition. Consider two small light sources that subtend an angular

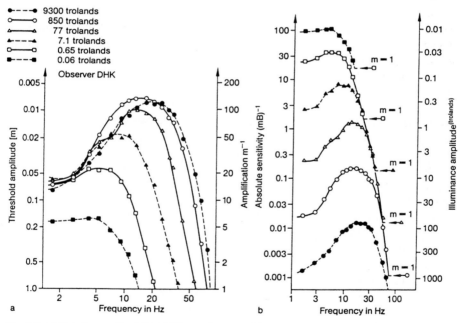

Fig. 1.3. (a) Relative amplitude sensitivity vs. frequency. The "troland" is a unit of retinal illumination. Note that at low frequencies, the curves merge, according to Weber's Law. the curves reach a peak when the adaptation time constant equals one half the period of the stimulus. The final high frequency limitation is undoubtedly neural. (b) Absolute amplitude sensitivity vs. frequency. Note that each curve ends when the modulation amplitude equals the average illumination since light intensity cannot be negative. The asymptote of the curves on the high frequency side is consistent with the CFF being proportional to the log of the amplitude of the flicker. Along this curve, there is no adaptation to the variation – the perception depends on the average illumination only

15

separation of about one degree at the observer. If they are turned on for one millisecond each, less than ten milliseconds apart, they appear to flash simultaneously, while if the time interval is more than one second we see first one and then the other. If the interval is about ten milliseconds, the light seems to *move* from one point to the other. Thomas Edison is sometimes (erroneously) credited with turning this psychological curiosity to practical use by demonstrating that a sequence of slightly displaced images could, under proper conditions, give a convincing illusion of continuous motion.

Referring to Kelly's data, and leaving aside the obvious nonlinearity of the results, the most striking feature is that in a significant range of low frequencies the eye acts as a differentiator while for a band of high frequencies it is more properly described as an integrator. All successful systems of motion rendition employ sampling rates (16 or 24 fps for movies, 25 or 30 fps for TV) in the differentiation range, while all good systems that avoid flicker, even though the display is actually discontinuous, use field rates in the integration range. Quite obviously, the successive flashes of the display must be integrated to avoid flicker. Equally obviously, if one reproduced the entire frequency range to which the observer is sensitive, the output ought to look like the input.

In many systems the original image is not sufficiently bandlimited by the camera to satisfy the conditions of the sampling theorem, resulting in temporal aliasing. In a motion picture camera using a very short exposure per frame, objects in motion produce multiple images. Continuously rotating objects may seem to move forward or backward at strange speeds. If this result is achieved deliberately we call it the stroboscopic effect, but if accidental it is highly annoying.

In cameras, either photographic or electronic, which integrate the image for a full frame, precise band-limiting does not take place, since the transform of the rectangular temporal characteristic is not an ideal low-pass filter. Likewise the display is not an ideal interpolator. Since frame-rate reduction is an obvious way to conserve bandwidth, several studies have been carried out to see if adequate motion rendition could be accomplished at very low sampling rates using improved filtering. *Baldwin* [1.10] averaged movie frames and showed that a rate of 12 frames per second was adequate in some cases. *Cunningham* [1.11] stored enough frames so that he could use fairly accurate low-pass filtering and interpolation and showed that even lower rates were good enough for close-up pictures of subjects speaking. Such experiments require what were once thought of as huge computation facilities, although much less equipment would be needed if recursive filtering could give the desired frequency response.

In very low frame-rate systems, band-limiting appropriate to the frame rate may well give images that are so blurred as to be useless. In such cases where good motion rendition is not possible, it may be preferable to use short enough exposures to obtain a sequence of unblurred images and to display them either with a quick change from one to the next or with a dissolve of a fraction of a second.

To the thoughtful reader, this short discussion may have presented more questions than answers. Indeed, the present state of knowledge of the subject is

16

far from complete. The spectral approach is dealt with in more detail in Chap. 2 while the psychophysical viewpoint is discussed in Chap. 3. A recent discussion is found in [1.8].

1.8.5 Sampling in the Space Domain

We have so far not discussed the sampling and quantizing operations necessary to convert the spatially continuous image at the camera tube target into a set of numbers. However, we have noted that the temporal sampling is preceded by periodic integration in the camera tube, while the inverse operation at the display tube involves an exponentially decaying phosphor, the equivalent of a one-pole low-pass interpolation filter. The equivalent in the space domain of the temporal integration performed by the vidicon is to divide the target into small squares and integrate (or average) the light intensity over each square before analog-to-digital conversion. Spatial interpolation in the display involves the shape of the CRT electron beam and depends on whether the beam is moving during illumination of each pel. Some integration, both spatially and temporally, may be done by the observer.

These simple-minded procedures are special cases of what is probably the single most important element in digital image processing. There is as yet no completely satisfactory theory capable of prescribing optimum image sampling or reconstruction, and practice is rather haphazard. "Solutions" given by the sampling theorem of linear systems theory can easily be shown not to be optimum. Furthermore, only under conditions of wasteful oversampling does it not make any difference how these operations are performed [1.12].

1.8.6 Analog Transmission

The analog case is no easier to understand, since it normally involves sampling in one space dimension and on the time axis, as compared to the digital channel, which requires sampling on three axes. Although it is possible, in principle, to exchange tone scale resolution with spatial resolution as in the digital case, this is almost never done. Thus the tone-scale resolution is set by the signal-to-noise ratio of the channel. The remaining decisions on scanning standards trade off spatial versus temporal resolution. There is some uncertainty in equating vertical and horizontal resolution, since sampling occurs only in one direction. This situation has been dealt with in the literature [1.13] by the use of the "Kell Factor," but the phenomenon is more complex than realized by most authors who use this term. A good deal of light is shed on this topic in a recent paper by *Hsu* [1.14].

In fact, the comparison between horizontal and vertical resolution is impossible without prescribing the precise sampling and interpolation methods to be used. In practice, TV broadcasting standards [1.15] have generally provided the possibility of more resolution in the vertical direction in the hope that, since the analog channel was more susceptible to improvement by filtering, methods would eventually be found to increase the horizontal sharpness without increasing the bandwidth. In practice, much of the higher vertical resolution is lost for reasons connected with the physics of camera tube operation as well as perceptual matters related to the use of interlace [1.8]. Recent work on more sophisticated receivers using frame stores has shown that it is possible to attain nearly the theoretical vertical resolution. The situation thus may reverse itself with changes in technology. In any event, one does not make too great an error by assuming, for purposes of comparison, that the resolution of the analog channel in the horizontal direction is equivalent to two samples per cycle of bandwidth and that each such sample is equivalent to one TV line in the vertical direction [1.16].

With these assumptions, the 4 MHz channel is equivalent to 8 million samples per second, or 267,000 pels per frame – a 516×516 pel image for a 30 frame system. The 40 dB SNR is roughly equivalent to a properly randomized quantization into 29 levels, or 4.85 bits/sample. In both the analog and digital systems, we would observe the noise in the displayed images to be much more noticeable in the dark than in the light areas if the video signal were proportional to light intensity. For reasons associated much more with the technology of camera and display tubes rather than with psychophysics, in current practice, the video signal is corrected at the transmitter for the nonlinear characteristic of the picture tube. This nonlinear distortion, which is called gamma correction, fortuitously produces a nearly uniform noise sensitivity throughout the tone scale.

1.9 Lessons for the System Designer

Our aim in discussing this simple system has been to show the wide variety of knowledge needed for adequate design of imaging systems, and the inadequacy of an approach limited to the tenets of digital signal processing. We clearly need to know a great deal about visual perception, about the use of lighting for aesthetic purposes as well as how to calculate the amount of light needed for a given noise performance, about the properties of photoelectric materials useful in camera tubes and how to characterize camera tube performance, about the best methods of converting between continuous images and their digital representations, about coding for transmission through various kinds of channels, and about display devices – their physical properties and mathematical characterizations. Finally, we need to sum up this diversified knowledge so that we can learn to use it intelligently in system design.

2. Light and Optical Imaging Systems

In this chapter, we shall discuss some elementary concepts in optics that are necessary to deal with the most common optical portions of imaging systems. Significant limitations are placed on the performance of such systems due to the amount of light from the object collected in the image, and the quantum and wave natures of electromagnetic energy. Practical electronic and optical phenomena that produce additional limitations are also introduced. The Fourier treatment of optical images and systems, which has proven so useful in both analysis and design, is also developed. The introduction of the scanning principle permits the calculation of the spectrum of video signals, whose shape is shown to be limited by the size of the objects depicted.

2.1 Light Sources

Many important light sources, such as the sun or common light bulbs, radiate electromagnetic energy because they are hot. Although a single wavelength is associated with the energy (photon flux) due to each specific atomic transition, incandescent bodies have so many different transitions and therefore so many different wavelengths that the emission has an essentially continuous spectrum, or distribution of energy, at different wavelengths. The efficiencies of both radiation and absorption are affected in the same way by physical surface properties. Thus a perfect radiator is also one that absorbs all, and reflects none, of incident radiation. A good absorber is a black-walled cavity with a small entrance hole. This is also, therefore, a nearly perfect radiator. Such an arrangement is used as a standard for luminous (i.e., visible) radiation, as we shall see below. In this section, we are dealing with radiation as measured in physical units such as watts.

The emission of perfect radiators, also called "black bodies", is described by Planck's law, shown graphically in Figs. 2.1 and 2.2.

$$M_T(\lambda) = \frac{2c^2h}{\lambda^5 \exp\left(\dfrac{ch}{\lambda kT} - 1\right)} \qquad \text{power/(unit area} \times \text{unit wavelength)}$$

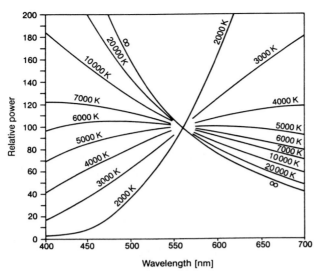

Fig. 2.1. Relative power distribution of a black body [2.2]

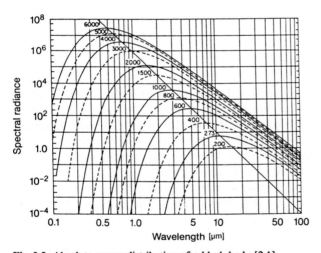

Fig. 2.2. Absolute power distribution of a black body [2.1]

$$\lambda : \text{Wavelength of light}$$

$$c : \text{Velocity of light}$$

$$k = 6.625 \times 10^{-34} \, W : \text{Boltzmann's constant}$$

$$h : \text{Planck's constant}$$

$$T : \text{Absolute temperature, Kelvin [K]}$$

The total radiation as a function of temperature can be found by integrating Planck's law over wavelength, yielding the Stefan-Boltzmann Law.

20

$$M(T) = \frac{2\pi^5 k^4}{15 h^3 c^2} T^4 \quad \text{[power/unit area]} .$$

The wavelength of peak radiation can be found by differentiating Planck's law with respect to wavelength and setting the result equal to zero. This gives the Wien displacement law,

$$\lambda_{max} = \frac{2.88 \times 10^{-3}}{T} \quad \text{[m]} .$$

These laws give results in close agreement with our everyday experience of hot bodies. As the temperature is raised, invisible but sensible infrared (heat) radiation is emitted. As the temperature is further increased, the radiation increases rapidly and the body begins to glow – first red, then orange, yellow, white, and finally blue-white. We sometimes speak of colors in terms of the temperature K, ("color temperature") of a black body of the same color. Since incandescent black bodies only assume certain colors, (never green or purple, for example) color temperature can be misleading when applied to other sources, such as phosphors.

The sun is approximately 6000 K, at which temperature the maximum radiation is at 480 nm. A substantial fraction of sunlight is scattered in passing through the earth's atmosphere even on a perfectly clear day. Since short waves are scattered more than long, and since multiple scattering produces a uniform flux density with equal intensity in all directions, the sun appears yellowish when viewed through the atmosphere, while the sky, which is illuminated by the scattered light only, appears bluish. A perfect, nonabsorbing scattering medium sends half the scattered light into each hemisphere. Thus the earth, when viewed from a long distance, is surrounded by a blue haze, while the total spectral composition of the light reaching the earth's surface is deficient in blue light, compared to the sun's emission, by half that which is scattered, as shown in Fig. 2.3.

If there were no atmosphere, the sky would be black as it is on the moon. When additional substances are present besides air, or if the path through the atmosphere is longer, the scattering is increased beyond normal. The sky and the sun may then take on many more vivid colors as they do sometimes at sunset or sunrise. Particularly striking effects are produced when particulate debris from large fires or volcanic eruptions is present in the upper atmosphere.

Presumably as a result of evolution, the spectral composition of the sum of yellowish sunlight and bluish skylight reaching the earth's surface, very nearly equal energy per unit wavelength over the visible spectrum, is perceived as "white" or neutral. Actually we have a rather large tolerance for unequal distributions. As long as they do not have pronounced peaks and valleys, good color perception is possible. Since it is not practical to operate incandescent lamps at high enough temperatures to give white light, we have long since become accustomed, and sometimes prefer, to look at faces and other colored objects with 2800 K light, which is distinctly yellow-orange.

Fluorescent lamps and TV tubes can easily be made more neutral, as a result of which they appear quite blue alongside incandescent light. Color perception,

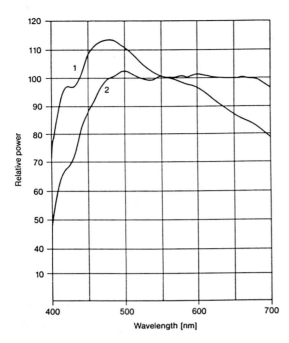

Fig. 2.3. The variation of solar radiation with wavelength: (1) Relative distribution of energy in sunlight above the earth's atmosphere, (2) Relative distribution of energy in noon sunlight at Washington. [2.2]

for reasons that we shall discuss later, is often seriously deficient using fluorescent light, even though it looks more like daylight.

In outdoor scenes, a portion of surface area is illuminated by direct (yellowish) sunlight. A larger portion, including most of the directly illuminated area, receives (bluish) skylight. Another variable portion receives reflected light from nearby surfaces. We are frequently unaware of the effect of the nonuniformity of the color of the illumination on direct viewing. Color slides, however, in which the effect is increased by high contrast and film deficiencies, are sometimes disappointing for this reason.

The appearance of scenes is greatly influenced by the directional properties of the light sources as well as the surface properties of the objects. The particular combination of diffuse and direct illumination found outdoors on clear days works very well. Professionally lighted scenes are treated in a similar manner, except that the directional "key" lighting is usually relatively less, compared to the diffuse general illumination, in order to reduce the dynamic range of the scene so as to fit the output process.

2.2 Photometry

Two completely separate systems for numerically describing the amount of light in optical systems are in use. Radiometric quantities, in units of energy, power, power density, etc., are usually used where the sensors are physical devices, whereas photometric quantities, in units of light flux, flux density, etc., are often used where human perception is of greatest importance. The basic photometric unit is the lumen, or quantity of light flux. It has the dimensions of power, and is therefore conserved in a nondissipative medium.

If light from an incandescent source is spread out by a dispersive prism into a spectrum of continuous wavelengths (the power and wavelength at each point in the spectrum can readily be physically measured), we find, as did Newton, that when the familiar rainbow is seen, white and purple are missing, and that wavelengths outside the relatively narrow band from 400 to 700 nm are invisible.

Persons with normal color vision have similar ability to distinguish light of the different wavelengths. By convention, we have all learned to use the same color names. Of course, there is no way of knowing that we all have the same color sensations.

If the energy in the visible band is roughly equal energy per unit wavelength, then the recombined spectrum looks white. If red and blue spectrum light are added together, we obtain the missing purple. We shall come back to the description of color later, but for the present purpose we need a system by which we can describe the brightness of colored lights independent of their color, and means to relate the intensity of the perception to some radiometric measurement.

The connection between photometric and radiometric units is given by the luminosity curve of Fig. 2.4: the relative sensitivity of the "standard observer" (standardized by international agreement) to light of various wavelengths together with a scale factor that tells how much power is in a lumen of light at a particular wavelength. The lumen itself is defined by international agreement to be 1/60 of the light flux emitted per cm^2 by thorium oxide at 2042 K, the temperature of melting platinum. If the surface brightness of the thorium oxide

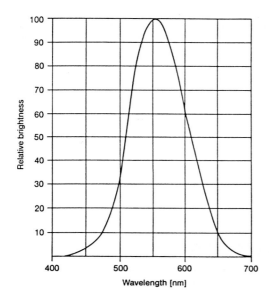

Fig. 2.4. The luminosity curve: The visibility curve for a normal eye. Specifically, this curve indicates the relative brightness of equal energies of spectrum colors as a function of wavelength. The maximum visibility occurs at 555 nanometers (nm) and the visibility becomes essentially zero at 400 and 700 nm. [2.2]

is matched in a flicker photometer[1] with monochromatic green light, it is found that 1 lumen is equivalent to 1.61 mW 550 nm. Since this is the wavelength of peak visual sensitivity at normal levels of illumination, more power is needed to match one lumen at other wavelengths. For light of more than one wavelength, the lumen equivalent is found by summing the lumen equivalent of each spectral component. The lumen equivalent of "white" light is not meaningful without stating the spectral composition of the light, since many different spectral compositions may look white. In fact, using photometric units for light of peculiar colors is often very misleading.

With the lumen defined and standardized, other quantities can be defined as convenient. The *illuminance* of a surface is the incident flux density, which can be expressed in several different units:

1 phot: 1 lumen/cm^2
1 lux or meter-candle: 1 lumen/m^2
1 ft-candle: 1 lumen/ft^2.

The *luminance* of a surface is the flux emitted per steradian per unit area of the emitting surface, as shown in Fig. 2.5. It is nearly always a function of θ, the angle between the normal and the direction of emission. A perfect Lambertian source is defined as one in which the luminance is proportional to the cosine of the angle of emission. Such a surface appears equally bright (the word brightness is reserved for appearance) from all directions, since the emitting area is foreshortened by the viewing angle so as precisely to cancel the cosine factor. The units of luminance are the foot-lambert, equivalent to $1/\pi$ lumens per steradian per ft^2 in the normal direction, and the lambert, equivalent to $1/\pi$

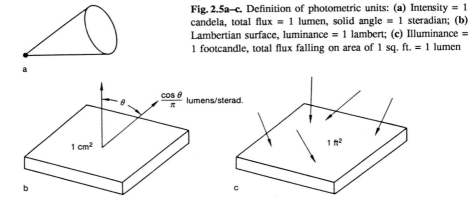

Fig. 2.5a–c. Definition of photometric units: (a) Intensity = 1 candela, total flux = 1 lumen, solid angle = 1 steradian; (b) Lambertian surface, luminance = 1 lambert; (c) Illuminance = 1 footcandle, total flux falling on area of 1 sq. ft. = 1 lumen

[1] A flicker photometer is a device for presenting two patches of light in rapid succession in the same position in the visual field. A frequency of succession can be chosen at which the color flicker disappears but the brightness flicker remains. Adjustment of the relative intensity of the patches to eliminate the brightness flicker establishes the photometric equivalence of the two lights.

lumens per steradian per cm^2, in the normal direction. One lambert = 929 ft-lamberts.

When a perfectly diffusing surface with a reflectivity of unity has an illuminance of one phot (one lumen/cm^2) the total emitted flux is, of course, one lumen/cm^2. Regardless of the direction of illumination, such a surface has a luminance of one lambert, emitting $(1/\pi)\cos\theta$ lumens per steradian cm^2. Real surfaces have reflectivities less than unity, generally a function of wavelength, and usually have at least some specular (mirror-like) as well as some diffuse reflection, since it is very difficult to make either a purely diffuse or purely specular surface.

The quantity luminance can be used to describe the light emission from either a self-luminous or passive illuminated surface. In either case, if the emitting area is small compared to the distance at which the light is measured, the effect is as if the light were emitted from a point source. In such cases, the source can be described by its *intensity*, a one candela source producing one lumen per steradian and therefore an illuminance of one foot-candle on a normal surface one foot distant.[2] An isotropic source produces 4π lumens per candela. Since most sources are nonisotropic, we often speak of the intensity in a certain direction, rather than $1/4\pi$ times the total lumen output.

2.3. Luminous Transfer of Simple Optical Systems

Because the amount of light incident on the target of a camera tube is such a vital element in image quality, it is important to know how to calculate its value, given a certain illumination on the object together with a certain optical system.

Consider the simple case shown in Fig. 2.6 of a single lens imaging a diffuse reflecting surface of luminance L onto an image plane. For object points near the axis and for a lens diameter d, small with respect to the object distance p, the lens is uniformly illuminated with L/π lumens per steradian per ft^2 of object. The solid angle subtended by the lens at the object is

Object (L) Image (I)

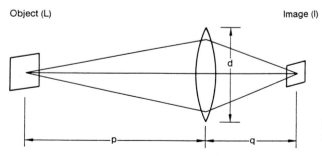

Fig. 2.6. Luminous transfer of a simple optical system

---

[2] The standard source was once a candle burning whale oil at a certain rate!

$$\frac{\pi d^2}{4} \frac{1}{p^2} \ .$$

The total flux incident on the lens is

$$\frac{L}{\pi} \frac{\pi d^2}{4} \frac{1}{p^2} \quad \text{lumens per ft}^2 \text{ of object .}$$

This flux is concentrated in an image area $(q/p)^2$ per ft^2 of object. Hence, the image illuminance is

$$I = \frac{L d^2}{4 q^2} \quad [\text{lumens/ft}^2] \ .$$

This expression can be put into a more useful form in terms of the magnification, $m = q/p$, and the "f/number", $f = F/d$, where F is the focal length. Using the relationship

$$\frac{1}{F} = \frac{1}{p} + \frac{1}{q} \ ,$$

we obtain

$$I = \frac{L}{4 f^2 (1 + m)^2} \quad [\text{lumens/ft}^2] \ .$$

In accordance with the second law of thermodynamics, since f cannot be much less than unity for geometrical reasons, the image illuminance is always less than (usually very much less than) the object luminance. Furthermore, the fraction of emitted light flux collected by typical optical systems is very small.

The image illuminance is independent of focal length. Therefore all photographic cameras of the same f/number require the same shutter speed to record pictures. However, in electronic cameras of a fixed number of picture elements per frame, longer focal length lenses (but the same f/number) appropriate to larger formats receive more flux per picture element. Clearly the light-gathering power of a lens is directly proportional to its area. Astronomical telescopes are governed by some of the same factors as electronic cameras, which is one of the reasons they are also made as large as is practical.

In normal camera use, $m \ll 1$, so that the illuminance is independent of magnification. However, as the camera is moved closer to make a larger image, the illuminance decreases significantly. At unity magnification, the light is decreased by a factor of four, requiring four times longer exposure or two lens stops increase in aperture. (One "stop" corresponds to a factor of two in exposure.) When $m \gg 1$, as in a slide or film projector, the image illuminance is inversely proportional to the image area, as would be expected since the same amount of flux is distributed over the larger area.

2.4 Some Nonideal Behavior of Simple Optical Systems

2.4.1 The Cosine^4 Law and Vignetting

The derivation above of the luminous transfer of a simple optical system makes the implicit assumption that both the image and object subtend small angles at the lens. In a real case, even with an ideal thin lens, the illumination in the image plane will be found to decrease away from the optical axis at least as the fourth power of the cosine of the angle. In practical cases, with thick lenses, the illumination will fall off even more rapidly due to "vignetting." This effect is caused when, because of size limitations of some of the lens elements, some light entering the entrance pupil fails to emerge from the exit pupil. Vignetting, if present, always depends on the diaphragm setting, generally disappearing at small openings.

A simple derivation of the $\cos^4$ law, based on Fig. 2.7, is as follows: If the object is a Lambertian source with luminance L, the flux density in the direction of the lens is

$$\frac{L}{\pi} \cos \theta \text{ lumens per steradian per ft}^2 \text{ of object .}$$

The solid angle subtended by the lens aperture at the object is

$$\Omega = \frac{A \cos \theta}{r^2} \,, \quad r = \frac{p}{\cos \theta} \,, \quad \Omega = \frac{A \cos^3 \theta}{p^2} \,.$$

Thus the total flux incident on the lens per ft^2 of object is

$$\Omega \frac{L}{\pi} \cos \theta = \frac{LA}{\pi p^2} \cos^4 \theta \text{ lumens per ft}^2 \text{ of object .}$$

But

$$A = \frac{\pi d^2}{4} \,.$$

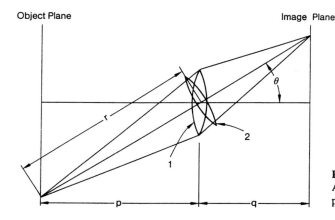

Object Plane   **Image Plane**

Fig. 2.7. The cosine^4 law: (1) A is lens area; (2) A $\cos \theta$ is projected lens area

27

Then using the same geometrical relationships as previously, we obtain

$$I = \frac{L \cos^4 \theta}{4 f^2 (1 + m)^2} \quad [\text{lumens}/\text{ft}^2] \ .$$

2.4.2 Aberrations

The previous derivations assumed ideal thin lenses and further assumed that all the light emitted by a point on the object plane, and which fell on the entrance aperture of the lens, emerged from the exit aperture and fell on a single point in the image plane. (It was noted that lens transmission and vignetting may attenuate, but not redirect, the rays of light.) In fact, even in lenses whose surfaces are perfectly spherical, this assumption is only an approximation. Rays striking the outer portions of a simple lens converge at a closer point than the more central rays. This is called spherical aberration. Light of different colors may be focussed in different planes, an effect called chromatic aberration. Astigmatism, a different focal length in the vertical and horizontal planes, occurs when rays strike the lens off center. Another aberration known as coma produces comet-shaped focal spots when rays strike the lens diagonally. These aberrations can be markedly reduced by using multiple elements of various indices of refraction and different dispersion, properly spaced, but still with spherical surfaces. Other aberrations are produced or primary aberrations may not be completely corrected, if errors or imperfections of surface ("the figure") occur. In all such cases, the "focal point" of a bundle of rays is defined as the point where the flux density is highest – the "circle of least confusion." In general, the set of focal points corresponding to all the points in the object plane does not form a perfectly flat image plane. Often this situation is expressed in terms of the set of points in object space which gives best focus for a plane image, in which case we call it "curvature of field" [2.3].

2.4.3 Diffraction

Modern computer methods of calculating lens parameters and modern methods of measuring radius of curvature during polishing, especially holographic techniques, make it possible to produce nearly perfect optical performance, even for large apertures and wide fields, up to the limit set by the wave nature of light. It is much easier to obtain diffraction-limited performance for moderate focal lengths, small apertures, and narrow fields. Custom-made diffraction-limited lenses for more demanding applications may cost several tens of thousands of dollars. Nearly diffraction-limited performance is obtained in some mass-produced $f/9$ process lenses for as little as $300.

There are several ways of understanding diffraction, which ultimately requires the solution of the wave equation for the given boundary conditions, but perhaps the easiest to visualize is Huygen's principle [2.4]. Each point on a wave front acts like the origin of a secondary spherical wavelet. Knowing the phase of the

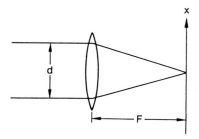
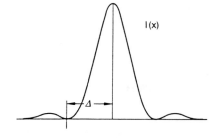

Fig. 2.8. Diffraction: $\Delta = 1.22\lambda F/d = 1.22\lambda(f/\text{number})$

incident wave at each point in an aperture, for example, permits the calculation of the resultant electromagnetic field strength at any point beyond the aperture simply by summing up the contributions from each point in the aperture, taking account of the propagation time (equivalent phase shift) from aperture point to distant point. The most striking effects are produced with monochromatic light. The situation is different if the incident light is coherent, as from a laser, or incoherent, as from an incandescent source. In the coherent case the amplitudes add, while in the incoherent case the powers add.

For monochromatic incoherent light, if a circular aperture of diameter d is uniformly illuminated by a plane wave, a diffraction pattern is produced, as shown in Fig. 2.8, consisting of a central bright spot surrounded by alternating light and dark rings. The distance, Δ, from the center to the first dark ring is Rayleigh's criterion of resolution, frequently used in rating telescopes. Another important case is that of a laser scanner, in which the aperture illumination is Gaussian rather than uniform. This produces a Gaussian[3] profile in the image plane, whose "width" is inversely proportional to the "width" of the aperture illumination. For example, if

$$I(r) = A \exp\left[-\frac{1}{2}\left(\frac{r}{r_0}\right)^2\right] \quad \text{is the aperture illumination}$$

and

$$I(s) = B \exp\left[-\frac{1}{2}\left(\frac{s}{s_0}\right)^2\right] \quad \text{is the image illumination },$$

where r and s are the radial distances in the respective planes, then

$$s_0 = \left(\frac{4\lambda F}{\pi r_0}\right),$$

where (F/r_0) is a kind of f/number.

---

[3] In general, the intensity function in the image plane, properly scaled, is the Fourier transform of the intensity function in the lens aperture [2.5, 6].

29

In both of these cases, and also in general, diffraction defocussing is proportional to wavelength times f/number. Since the aberrations of lenses usually increase with larger apertures, i.e., lower f/numbers, most lenses have an optimum aperture, larger openings resulting in worse aberrations and smaller openings in increased diffraction. Camera lenses usually perform best about two stops smaller than wide open. It is thus not true, as often thought, that maximum sharpness occurs at the smallest opening, though it is true that smaller openings give larger depths of field.

Another point to be noticed in the diffraction relations is that since the absolute size of the focussed spot depends only on f/number, using a larger format with the same f/number and same angle of coverage improves the resolution as well as collecting more light. For electronic imaging systems with a fixed number of picture elements per frame, it is the lens diameter that governs both resolution and light gathering performance. Doubling the lens diameter quadruples the collected light and halves the spot size for a given focal length and image size. Doubling the focal length but leaving the lens diameter unchanged causes the same amount of light to be collected per picture element, but doubles both the spot diameter and the picture element width, resulting in equal performance.

Diffraction also makes it increasingly difficult to produce imagery of a given resolution in terms of picture elements (pels) per scan line or picture width, in smaller and smaller formats. Lenses designed for making integrated circuit masks, for example, are very expensive, as they must achieve diffraction-limited performance at very low f/numbers.

2.5 Fourier Optics and the Modulation Transfer Function

2.5.1 The Linearity of Optical Systems

The imperfect optical system discussed in the previous section is linear as that word is used in linear system theory. If the input is considered to be the sum of a large number of small light sources in the object plane, the output then consists of a large number of somewhat defocussed spots in the image plane [2.7]. For incoherent light, the total illumination at each point is the sum of that due to all the contributing spots in the neighborhood. Taken as a whole, therefore, a specific linear combination of two objects produces the same linear combination of the two images corresponding to the objects taken one at a time; that is, if

$$O_1 \rightarrow I_1 \quad \text{and} \quad O_2 \rightarrow I_2 \quad \text{and}$$

$$a_1 O_1 + a_2 O_2 \rightarrow a_1 I_1 + a_2 I_2 \,,$$

then the system is linear.

This fundamental property of optical systems permits the entire panoply of techniques from linear system theory to be applied. The image of a point source is called the point-spread function (PSF), and takes the role of the impulse response. Since the PSF usually is not uniform across the field, optical systems are not shift-invariant, but they are often considered approximately so for the sake of simplifying calculations. A better approximation is to use several PSFs at various angles or distances from the optical axis.

The Fourier transform of the PSF is the spatial frequency response of the system, usually called the optical transfer function (OTF); its magnitude is usually called the modulation transfer function (MTF). The reason for this nomenclature is that since negative light is physically unrealizable, frequency response must be measured by superimposing a sinusoidal variation onto (i.e., modulating) a constant level. The change in percentage modulation from input to output is a measure of the frequency response.

The use of MTF to characterize systems implies thinking of images in terms of their spectra, i.e., the Fourier transform of their description in terms of spatial variation. While we do not ordinarily think of images in this way—i.e., as the sums of sine waves—it is as valid and useful as the transform methods long used by communications engineers in describing temporal signals and filters. In fact, Fourier's original work [2.8] was applied to heat flow problems in which spatial variations are of primary importance.

Because of the immense power of transform methods in analyzing systems, it is tempting to apply such methods to cases of doubtful validity. For example, we often hear about the spatial frequency response of photographic film or of the human eye, both of which are grossly nonlinear. It is important in such cases to state exactly what is meant by frequency response and to what conditions of a particular physical system the analysis applies.

Although the mathematics of linear system theory is precisely the same in optics as in electric circuits, the physical systems are different and the variables in question—flux density on the one hand and current or voltage on the other—have different constraints. Light cannot be negative, for example. Causality in the time domain has no correspondence in optics since the temporal "before" and "after" are replaced by "left", "right," "up", and "down" in the space domain. Time is one-dimensional, space two- or three-dimensional. Many optical systems are circularly symmetrical, in which case polar coordinates are convenient. The Fourier transform is then replaced by the Fourier-Bessel or Hankel transform, in which sine waves are replaced by Bessel functions.

2.5.2 The Spectrum of Images

Consider the fixed image, $L(x, y)$, of Fig. 2.9, $2a$ units wide and $2b$ units high. To clarify the analysis, we first calculate the spectrum where the luminance varies horizontally only, so that $L(x, y) = L(x)$. We can expand $L(x)$ in a Fourier series for $-a < x < a$:

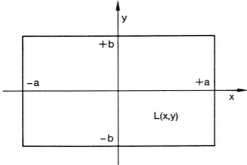

Fig. 2.9. An image

$$L(x) = \sum_{k=-\infty}^{\infty} A_k \exp(j\omega_k x) \, ,$$

where

$$A_k = \frac{1}{2a} \int_{-a}^{a} L(x) \exp(-j\omega_k x) dx \, ,$$

where

$$\omega_k = k\pi/a \, .$$

The fundamental spatial frequency ($k = 1$) is one cycle, or 2π radians per picture width. The spectrum thus consists of a zeroth term, the average image luminance, plus an infinity of harmonics of the fundamental. There is no obvious upper limit to the spectrum.

It is important to note in what sense the series represents the image. The coefficients A_k are chosen so as to minimize the mean square error in the $2a$ range of the image. All real images are physically constrained to meet the mathematical conditions that cause the rms error to die out steadily as the number of terms increases. However, in the neighborhood of sharp transitions in $L(x)$, the truncated series invariably rings as shown, with a ringing frequency equal to that of the highest retained harmonic. The sum of the infinite series exhibits the Gibbs phenomenon as shown in Fig. 2.10. The visual effect of this performance is unsatisfactory if the spatial frequency of ringing is within the range of high visibility.

The sum of the series is periodic with period $2a$. It properly represents the original image between $-a$ and $+a$, but repeats the image endlessly over all other values of x.

If we now take account of the vertical luminance variations, a two-dimensional Fourier series is required. Thus,

$$L(x, y) = \sum_{k=-\infty}^{\infty} \sum_{l=-\infty}^{\infty} A_{kl} \, \exp[j(\omega_{xk}x + \omega_{yl}y)] \qquad \begin{matrix} -a < x < a \\ -b < y < b \end{matrix} \, ,$$

where

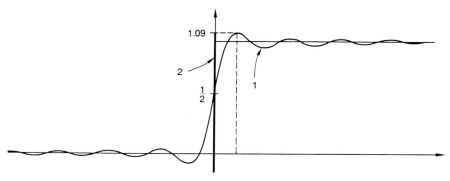

Fig. 2.10. The Gibbs phenomenon: (1) sum of the finite series; (2) sum of the infinite series

$$\omega_{xk} = \frac{\pi}{a}k \; ; \quad \omega_{yl} = \frac{\pi}{b}l \quad \text{and}$$

$$A_{kl} = \frac{1}{4ab} \int_{-b}^{b} \int_{-a}^{a} L(x,y) \, \exp\left[-j\left(\frac{\pi kx}{a} + \frac{\pi ly}{b}\right)\right] dx \, dy \; .$$

The same remarks about convergence apply to the two-dimensional case. The series is doubly periodic so that its sum is a postage stamp-like double infinity of copies of the original image. The spectrum is likewise two-dimensional and theoretically infinite. In actuality the spectrum is limited and, within the bandwidth of significant harmonics, their amplitude generally decreases monotonically with frequency.

There are several ways to approach the overall shape of the spectrum. One direct method is to recall that virtually all images are formed by optical systems[4] and that the latter are characterized by impulse responses (PSFs) that amount to small defocussed spots. In the one-dimensional case the output image $I(x)$ can be calculated from the convolution:

$$I(x) = \int_{-\infty}^{\infty} L(\xi)H(x - \xi)d\xi$$

or

$$I(x) = \int_{-\infty}^{\infty} L(x - \xi)H(\xi)d\xi \; ,$$

where $H(x)$ is the impulse response, here considered shift-invariant. There is an unavoidable, but usually unimportant, edge anomaly here, since at the edge of the image this expression incorrectly convolves $H(x)$ with the opposite edge of the image, due to the array of duplicate images represented by the series.[5] Since

---

[4] Optical systems may enlarge or reduce, a function not directly achievable in temporal filters. We assume in this analysis that the unit of length in the image plane is scaled by the magnification.

[5] This is the equivalent to the so-called "circular convolution" of discrete Fourier transforms.

33

for practical cases the spatial extent of $H(x)$ is small compared to the image, and since some anomaly must take place at edges anyway, we shall disregard it here.

We now substitute in the previous equation the series expansion for L, giving

$$I(x) = \int_{-\infty}^{\infty} \sum_{k=-\infty}^{\infty} A_k \, \exp[j\omega_k(x - \xi)]H(\xi)d\xi \, .$$

Reversing the order of integration and summation[6]

$$I(x) = \sum_{k=-\infty}^{\infty} A_k B_k \, \exp(j\omega_k x) \, ,$$

where

$$B_k = \int_{-\infty}^{\infty} H(\xi) \, \exp(-j\omega_k\xi)d\xi$$

is the Fourier transform of the impulse response, evaluated at $\omega = \omega_k$.

A typical impulse response is the Gaussian:

$$H(x) = \frac{1}{\sigma\sqrt{2\pi}} \, \exp\left[-\left(\frac{x}{\sigma}\right)^2\right] \, , \quad \text{whose Fourier transform} \, ,$$

$$F[h(x)] = \exp\left[-\frac{1}{2}(\omega\sigma)^2\right] \, , \quad \text{and}$$

$$B_k = \exp\left[-\frac{1}{2}\left(\frac{\pi}{a}k\sigma\right)^2\right] \, .$$

Thus $I(x)$ is represented by a Fourier series with terms of the same spatial frequencies as $L(x)$ but with each coefficient multiplied by a sample of the Fourier transform of the PSF. In the typical case of a blurry spot such as the Gaussian, the higher components are substantially decreased. For spatial wavelengths equal to σ, the attenuation is about 100. The PSF thus sets an effective limit to the bandwidth. Roughly speaking, the number of significant harmonics is equal to the number of resolvable picture elements, multiplied by a factor between 1/2 and 2.

In the separable Gaussian two-dimensional case, the PSF is

$$H(x,y) = \frac{1}{2\pi\sigma_x\sigma_y} \, \exp\left\{-\frac{1}{2}\left[\left(\frac{x}{\sigma_x}\right)^2 + \left(\frac{y}{\sigma_y}\right)^2\right]\right\} \, ,$$

and the equivalent final image is

---

[6] $L(x)$ and $H(x)$ are physical quantities. There is rarely a question of convergence to trouble such an inversion.

$$I(x,y) = \sum_{k=-\infty}^{\infty} \sum_{l=-\infty}^{\infty} A_{kl} B_{kl} \, \exp[j(\omega_{xk}x + \omega_{yl}y)] \,,$$

where

$$B_{kl} = \exp\left\{ -\frac{1}{2} \left[\left(\frac{\pi}{a}k\sigma_x\right)^2 + \left(\frac{\pi}{b}l\sigma y\right)^2 \right] \right\}$$

B_{kl} is the Fourier transform of the two-dimensional impulse response, evaluated at $\omega_x = \omega_{xk}$ and $\omega_y = \omega_{yl}$. Just as in the one-dimensional case, the higher harmonics are decreased in amplitude by the effect of the impulse response so as to set an effective limit to the spectrum. This limit is an area of the ω_x, ω_y plane: for PSFs with circular symmetry, the passband is a circular region. Since wide impulse responses have narrow spectra and vice versa, elliptical PSFs correspond to elliptical passbands, oriented at right angles in the space and frequency domains.

Another common impulse response is the circular cylinder, corresponding to a defocussed optical system (aberrations and diffraction ignored) in which the image plane is removed from the true focal plane, as shown in Fig. 2.11. The transform is found to be

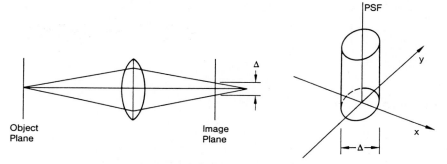

Fig. 2.11. Defocussed optical system and its point-spread function

$$B(\omega_x, \omega_y) = 2\pi\Delta \frac{J_1(\omega_r \Delta)}{\omega_r} \;; \quad \omega_r = \sqrt{\omega_x^2 + \omega_y^2} \,,$$

which is circularly symmetrical in the frequency plane with alternating annular zones of negative and positive values.

Negative Fourier coefficients are equivalent to phase inversion, as dramatically shown in Fig. 2.12, where the imagery is negated for certain spatial frequencies. When the optical system aperture is not circular (four- and five-sided apertures are not uncommon in today's cost-cutting environment) the effect of convolving the object with the impulse response is sometimes quite obvious.

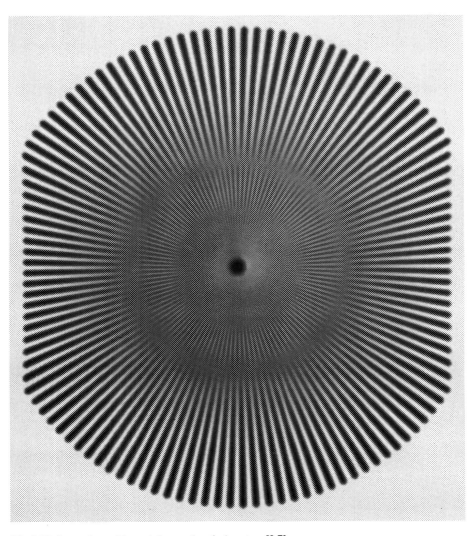

Fig. 2.12. Image formed by a defocussed optical system [2.7]

2.5.3 The Aperture Effect

There are a number of practically important cases in which systems are most directly characterized by the aperture function rather than the impulse response. One example is a flying-spot scanner where the scan lines are closely spaced compared to the beam diameter. Another is a CRT display. The entire class of passive mechanical scanners such as the old Nipkow disk, the Viking (Mars Lander) camera, or many satellite cameras such as Landsat, with scan density finer than aperture size, are of this type. The relationship of the aperture description

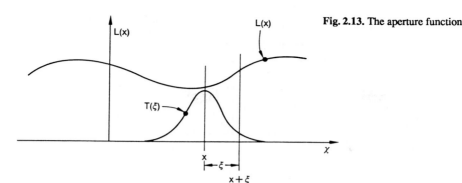

Fig. 2.13. The aperture function

to the impulse response is readily shown by the one-dimensional example of Fig. 2.13.

Consider an aperture whose transmission is $T(x)$ scanning an image $L(x)$. When the aperture is at x, the total light sensed is

$$I(x) = \int_{-\infty}^{\infty} T(\nu)L(x+\nu)d\nu \ .$$

Let $x + \nu = \xi$

$$I(x) = \int_{-\infty}^{\infty} T(\xi - x)L(\xi)d\xi \ .$$

Now if $T(\nu) = H(-\nu)$, we have

$$I(x) = \int_{-\infty}^{\infty} H(x - \xi)L(\xi)d\xi \ ,$$

which is the same as we obtained previously. Therefore the aperture transmittance is the reflection of the impulse response with respect to the origin, the "memory function" of older literature.

Care is required in evaluating the aperture function in some situations. Flying-spot scanners are linear, but electronic camera tubes are not. The effective aperture of the latter is therefore not simply the cross-sectional current density of the scanning beam, but a varying, oddly shaped function that depends on signal level.

If the aperture intensity function is circularly symmetrical, so is the point spread function. The two functions (of r) are related by the Fourier-Bessel or Hankel transform. A short table of such transforms is found in [2.9].

2.5.4 Temporal Video Signals

The previous analysis makes it very easy to find the temporal spectrum of a video signal generated by scanning an image. In the one-dimensional case, we had

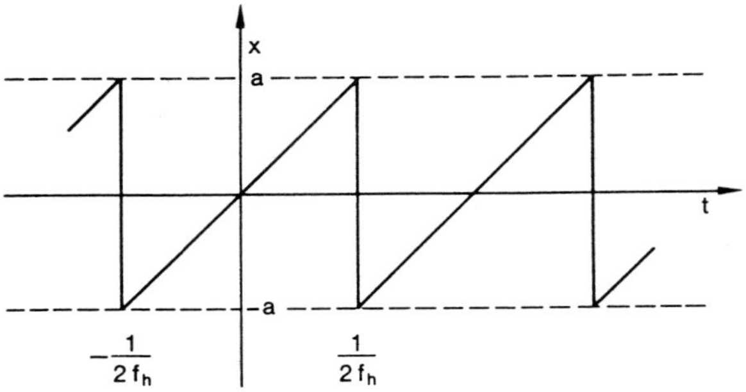

Fig. 2.14. The waveform of repetitive scanning

$$I(x) = \sum_{k=-\infty}^{\infty} A_k B_k \, \exp(j\omega_k x) \, ,$$

in which the series represents a periodic image. Thus, repetitive scanning as shown in Fig. 2.14 is equivalent to unidirectional scanning through the periodic image with $x = 2af_{\mathrm{h}}t$. Thus

$$I(t) = \sum_{=-\infty}^{\infty} A_k B_k \, \exp(j2\pi k f_{\mathrm{h}}t) \, ,$$

and the spectrum is seen to comprise a dc term plus harmonics of the line scanning frequency [2.10–12].

In the two-dimensional case, periodic scanning of the image vertically and horizontally with frequencies f_{v} and f_{h}, respectively, is equivalent to unidirectional scanning through the postage stamp-like array of identical images given by the series but with $x = 2af_{\mathrm{h}}t$ and $y = 2bf_{\mathrm{v}}t$. Substituting above we have

$$I(t) = \sum_{k=-\infty}^{\infty} \sum_{l=-\infty}^{\infty} A_{kl} B_{kl} \, \exp[j2\pi(k f_{\mathrm{h}} + l f_{\mathrm{v}})t] \, .$$

Each spatial frequency $k/2a$, $l/2b$ is mapped into a temporal frequency $k f_{\mathrm{h}} + l f_{\mathrm{v}}$. Thus the temporal spectrum consists of a dc term plus harmonics of the horizontal (line) scanning frequency. Each of these terms is surrounded by a cluster of components consisting of harmonics of the vertical (frame) scanning frequency. The number of each kind of harmonics relates to the number of resolvable elements in each direction.

If $f_{\mathrm{h}} = n f_{\mathrm{v}}$, where n is an integer, the raster repeats after n scan lines, and the vertical spectral components surrounding the various horizontal harmonics coincide as shown in Fig. 2.15.

If f_{h} and f_{v} are mutually prime, the raster never repeats and the spectrum is actually continuous. If $f_{\mathrm{h}} = (n/m)f_{\mathrm{v}}$, where n/m is a reduced rational number, the full raster has n scan lines and repeats only after m vertical periods, a scheme called mth order interlace. The spectral components

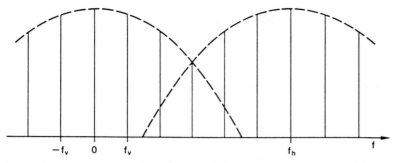

Fig. 2.15. The spectrum when the horizontal frequency is an integral multiple of the vertical frequency

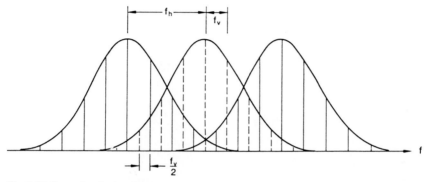

Fig. 2.16. Second-order interlace

clustered about each harmonic of f_h are now interlaced, the lowest theoretical component occurring at f_v/m. For example, in the NTSC television standard, $f_v = 60\,\text{Hz}$, $f_h = 15,750\,\text{Hz}$, and $n/m = 525/2$. Such second-order interlace is shown in Fig. 2.16. There are 525 lines per full frame, traced out in 1/30 second. In practice, in order to have any substantial power below 60 Hz, the image would have to have some significant average difference between odd and even lines. The tendency of most camera tubes to erase the target completely on each vertical scan obliterates such differences as may exist except where vertical motion is present in the image.

If the vertical spot size is small compared to the line spacing and if the image contains significant vertical variation, the cluster of sidebands surrounding each harmonic of f_h overlaps those of the next higher and lower horizontal harmonic. Each such temporal frequency component can then be interpreted as a number of different linear combinations of f_h and f_v, each combination representing a unique pattern of parallel stripes with a certain spacing and orientation. The result of this "aliasing" is moiré patterns and very poor reproduction of sharp edges and fine lines, as shown in Fig. 2.17. The control of this phenomenon will be discussed later in connection with sampling and interpolation filters.

Fig. 2.17. An example of aliasing, including moiré patterns, produced by analog scanning in the vertical direction with aperture much smaller than the line spacing. Notice the difference in rendition between the horizontal and vertical resolution wedges

2.5.5 The Correlation Function and the Shape of the Spectrum

Further insight into the expected shape of the spectrum, either spatial or temporal, can be gained by considering the distribution of types and size of objects usually depicted. From this information the autocorrelation function can be calculated, and from the latter the general shape of the spectrum can be inferred [2.13].

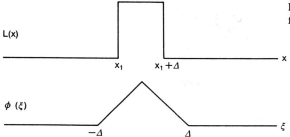

Fig. 2.18. A pulse and its correlation function

Sensible images must depict objects that are mostly much larger than the smallest resolvable spot, or else the image would look like sand on a beach. For the present purpose, let us suppose that the (one-dimensional) image consists of a number of uniform patches of random width and intensity. The autocorrelation function $\phi(\xi)$ is defined as

$$\phi(\xi) = \frac{1}{2a} \int_{-a}^{a} L(x)L(x - \xi)dx .$$

For the pulse in Fig. 2.18, $\phi(\xi)$ is a triangle of base 2Δ. The sum of a large number of triangles with various heights and widths can be assumed without serious error as

$$\phi(\xi) = k \ \exp(-k_1|\xi|) .$$

Actual measurement confirms the generally exponential behavior of the autocorrelation function [2.10].

We define the power spectrum of a function $L(x)$ as:

$$\Phi(\omega) = A(\omega)A^*(\omega) ,$$

where $A(\omega)$ is the Fourier transform of $L(x)$. The Wiener-Khinchine theorem shows that

$$\Phi(\omega) = \int_{-\infty}^{\infty} \phi(\xi) \ \exp(-j\omega\xi)d\xi ,$$

i.e., the autocorrelation function and the power spectrum are a Fourier-transform pair. For the symmetrical exponential above,

$$\Phi(\omega) = \frac{2k_1}{(k_1^2 + \omega^2)} .$$

Thus we see that amplitude of the power spectrum begins to decrease at $\omega = k_1$ and, at high spatial frequencies, falls as $1/\omega^2$. Since k_1 depends on the typical object size, the break point is higher for images with copious fine detail.

In a two-dimensional image, a similar exponential autocorrelation function can be expected in both directions, so that the power spectrum decreases as

41

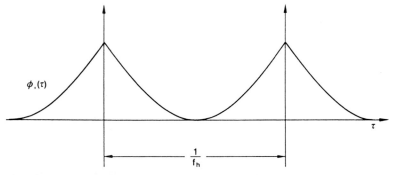

Fig. 2.19. Autocorrelation function of a video signal in the absence of vertical detail

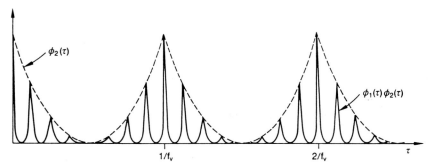

Fig. 2.20. Autocorrelation of a video signal from a still image

$(1/\omega_x)^2$ and $(1/\omega_y)^2$ at high frequency. For time-varying images, a similar auto-correlation function can also be expected, not because of the size of the objects, but because the image can change only slowly in time if motion is to be well depicted. Thus, in the time direction also, the spectral intensity decreases rapidly with frequency.

Autocorrelation analysis can also be used to predict typical spectra of temporal video signals resulting from scanning. The autocorrelation in the horizontal direction is comparable to that in the one-dimensional spatial case. Repeated scanning of the source line would result in a periodic version of the one-dimensional spatial function, shown in Fig. 2.19. Because of the comparable statistical relationship in the vertical direction, progressive scanning of adjacent lines in a still picture yields an autocorrelation function shown in Fig. 2.20 which is the product of $\phi_1(\tau)$ with a second, similar function, $\phi_2(\tau)$, whose period is the frame duration.

When we consider the effect of motion, we find that the function of Fig. 2.20 must be multiplied by a third factor of the form $\exp(-K|\tau|)$ to represent the slow change from one frame to the next. Then the final result is as shown in Fig. 2.21. Note that for usual pictures, ϕ_2 and ϕ_1 are substantially zero in between their periodic cusps.

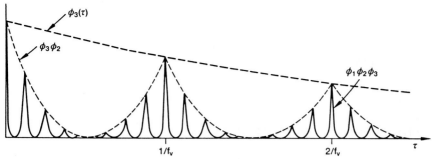

Fig. 2.21. The autocorrelation of a video signal from a moving image

To find the corresponding power spectrum, we convolve the transform of the three separate factors. Since ϕ_1 and ϕ_2 are periodic, each gives rise to a periodic line spectrum, one consisting of harmonics of f_h and the other of harmonics of f_v. Their convolutions comprise harmonics of f_h, each surrounded by a cluster of harmonics of f_v, as we showed above. Now, however, the overall spectral shape is seen to depend on the typical size of objects depicted. For large objects, with a broad correlation function, the overall spectrum as well as the sidebands around each harmonic of f_h, decay rapidly and vice versa.

To account for the effect of time variation, we note that $\phi_3(\tau)$ is pulse-like, rather than periodic. Its transform is also a pulse, wide for rapid variation and narrow for slowly changing images. When this spectrum is convolved with the double line spectrum, each infinitesimal line is replaced by a finite pulse.

In mathematical form,

$$\phi_1(\tau) = \sum_{k=-\infty}^{\infty} \exp(-K_h|\tau - kT_h|) \ .$$

The typical term has a transform

$$\frac{K_h}{K_h^2 + \omega^2} \ \exp(-j\omega_k T_h) \ ;$$

thus

$$\Phi_1(\omega) = \frac{K_h}{K_h^2 + \omega^2} \sum_{k=-\infty}^{\infty} \exp(j\omega_k T_h)$$

$$= \frac{1}{T_h} \frac{K_h}{K_h^2 + \omega^2} \sum_{k=-\infty}^{\infty} u_0(f - kf_h) \ ,$$

where u_0 is the unit impulse, and where $f_h = 1/T_h$. Similarly,

$$\Phi_2(\omega) = \frac{1}{T_v} \frac{K_v}{K_v^2 + \omega^2} \sum_{l=-\infty}^{\infty} u_0(f - lf_v) \ .$$

43

Convolving these spectra and discarding constant factors yields

$$\Phi_{1,2}(\omega) = \frac{1}{(K_v^2 + \omega^2)(K_h^2 + \omega^2)} \sum_{k,l=-\infty}^{\infty} u_0(f - kf_h - lf_v),$$

$$\Phi_3(\omega) = \frac{K_t}{K_t^2 + \omega^2}.$$

Convolving this expression with $\Phi_{2,3}(\omega)$ gives

$$\Phi(\omega) = \sum_{k,l=-\infty}^{\infty} \frac{1}{[K_t^2 + (\omega - \omega_{kl})^2](K_h^2 + \omega^2)(K_v^2 + \omega^2)},$$

where $\omega_{k,l} = 2\pi(kf_h + lf_v)$. Again, this is a result similar to that obtained previously. However, we now have the factors K_t, K_h, and K_v related to characteristics of the object, which have been shown to control the spectral shape.

2.5.6 Concluding Remarks about the Video Spectrum

The previous analysis shows that if the image depicts mostly objects much larger than the sample spacing, and, in the case of moving objects, if good motion continuity obtains, then the spectrum falls off rapidly at higher frequencies, the vertical sidebands fall off rapidly around each horizontal harmonic, and the temporal sidebands fall off rapidly around each vertical harmonic. This is not to say that this high frequency power can be further reduced or even discarded without loss. Even though there are few small objects or sharp edges in the image and thus little high frequency power, it is on the faithful reproduction of such detail that the impression of sharpness depends. Along the time axis, the result of inadequate reproduction of the corresponding spectral components is excessive blurring of moving objects.

A notable feature of both spatial and temporal spectra of images, besides the fall-off at high frequencies, is the relative emptiness of the space between harmonics of the fundamental frequencies. This has led to numerous proposals and some attempts at compression methods based on using the empty space. Little success has attended these efforts, with one exception. That is the use of spectral interleaving to transmit both chrominance components of the NTSC color TV signal on one subcarrier and the placement of that subcarrier within the luminance band, its harmonics interleaved with luminance harmonics.

A complete discussion of the NTSC system is not appropriate at this point. It suffices to say that the two chrominance components modulate, in quadrature (i.e., one modulates a sine wave and the other a cosine wave), a subcarrier which is an odd multiple of one-half the horizontal scanning frequency. Thus, modulation is straightforward. The problems occur in demodulation. The complete separation of luminance and chrominance components requires three-dimensional filtering [2.11]. Even two-dimensional filtering is generally done only in professional equipment, not in the vast majority of TV receivers, while three-dimensional filtering requires a frame store. Inadequate filtering produces "cross-color" and "cross-luminance" effects that substantially reduce image quality. These effects were not noticed very much in the early years of color transmission, since they were masked by the characteristics of picture tubes and other physical components of the system. However, with the improved components of modern times, and perhaps also with the heightened sensitivity of ever more demanding professional viewers, these defects have become increasingly bothersome. As a result, most of the recent proposals for improvements in TV broadcasting [2.12] abandon the spectral overlap of chrominance and luminance information that was once considered such a technological triumph.

2.6 Quantum Phenomena and Related Noise Sources

2.6.1 The Quantum Nature of Light

Many experiments can be carried out that demonstrate that for very small flux densities, light appears to consist of discrete quanta called photons. For example, if a diffraction experiment is set up and the pattern recorded on photographic film, for very brief exposures and/or very low light levels, the pattern is very "noisy", i.e. it consists of a number of discrete points. With a detector sufficiently sensitive to count single photons, it is seen that each is "diffracted" randomly but that with a sufficient number, the expected pattern is built up.

Further experiments show that $E = h\nu$ is the energy per photon, where $\nu = c/\lambda$. For $\lambda = 555$ nm, the wavelength of green light of maximum luminosity, we find

$$E = (6.625 \times 10^{-34} \times 2.998 \times 10^8 \text{m/s})/(555 \times 10^{-9}\text{m})$$
$$= 3.58 \times 10^{-19} \text{joules/photon} ,$$

or

$$1 W = 2.80 \times 10^{18} \text{ photons/s} .$$

We have previously noted that for green light, 1 lumen is equivalent to 1.61 mW. Therefore, 1 lumen of green light is equivalent to

$$1 \text{ lumen at } 555 \text{ nm} = \frac{0.00161\text{W}}{3.58 \times 10^{-19}\text{Ws/photon}} = 4.50 \times 10^{15} \text{photons/s} .$$

It is not useful to speak of such an equivalent for "white" light without specifying exactly what we mean, but the figure 1.36×10^{16} photons/s is often seen in the literature.

One result that can easily be calculated from these numbers is the sensitivity of certain photodetectors. Photoemissive surfaces such as are used in photomultiplier tubes and image orthicon camera tubes, as well as reverse-biased silicon photodiodes, operate by capturing photons, producing unit electronic charge per captured photon. The photocurrent then is

$$I[\text{A}] = L[\text{lumens}] \times \frac{\text{photons}}{\text{s lumen}} \times (\text{q.e.}) \times 1.602 \times 10^{-19} \, \text{C/electron}$$

or

$$I[\text{A}] = L[\text{W}] \times \frac{\text{photons}}{\text{Ws}} (\text{q.e.}) \times 1.602 \times 10^{-19} \text{C/electron}$$

where the quantum efficiency (q.e.) is the fraction of photons captured.

At 555 nm with 100 % q.e., the conversion factor is $1.602 \times 10^{-19} \times 2.80 \times 10^{18}$ or 0.448 A/W. The conversion factor is higher for longer wavelengths since there are more photons per watt. In photometric units, the conversion factor is $1.602 \times 10^{-19} \times 4.5 \times 10^{15}$ or 721 μA/lumen at 555 nm. For other wavelengths, taking account of the luminosity curve, the factor rises sharply at both ends of the visible spectrum, since more power is required to evoke visual response, as shown in Fig. 2.22.

Silicon photodiodes have quantum efficiencies of 60–80 %, relatively uniform across the visible spectrum, but falling off in the infrared, so as to peak at about 1 μm (micrometer). Photoemissive surfaces may have a q.e. of as much as 25 %, but generally peak in the 300–450 nm region, since a minimum energy/photon is required to overcome the potential barrier at a metal surface. Since camera tubes other than those with silicon targets have a q.e. that varies a great deal with wavelength, these curves must first be multiplied by the spectral sensitivity in order to be used in calculations.

Whenever a sensitivity is stated to be higher than those given here for 100 % q.e., some multiplication has taken place. This must be taken into account in noise calculations, since the signal-to-noise ratio depends on the unmultiplied photoelectron flow.

The photon rate is numerically so high, even for rather small levels of luminous flux, that it is at first difficult to realize that the discrete nature of light could ever be a problem. Yet it does in fact set a fundamental performance limitation to all imaging systems, although in most cases other noise sources, also related to the quantum nature of the universe, are more significant.

2.6.2 Shot Noise

If a stream of randomly emitted particles is carefully observed, it will be found that the number emitted in any fixed interval varies around some average value. This variation constitutes an ac "noise" superimposed on the dc (average) value. In Appendix 2.8, it is shown that the probability of getting K particles in τ seconds

$$P(K, \tau) = \frac{(a\tau)^K}{K!} \exp(-a\tau) \,,$$

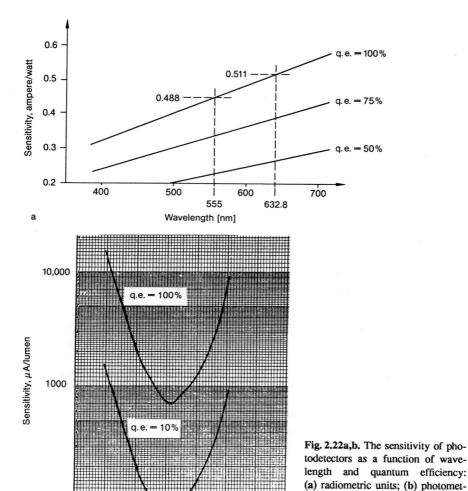

a

b

Fig. 2.22a,b. The sensitivity of photodetectors as a function of wavelength and quantum efficiency: **(a)** radiometric units; **(b)** photometric units

which is the Poisson distribution where a is the average number per second. The rms deviation

$$\sqrt{(K - \bar{K})^2} = \sqrt{\bar{K}} \ .$$

Shot noise is random and "white", that is, uniformly distributed over an essentially infinite bandwidth. For a particular observation interval τ, the "signal" is $a\tau = \bar{K}$, and the "noise" is $\sqrt{\bar{K}}$. Thus

$$\text{SNR} = \frac{\bar{K}}{\sqrt{\bar{K}}} = \sqrt{\bar{K}} \ .$$

As the observation interval decreases, i.e., as the observation bandwidth F increases, the SNR worsens. Assuming

$$F = 1/2\tau, \quad \text{SNR} = \sqrt{\frac{a}{2F}} \quad \text{and the rms noise flux is } \sqrt{2aF}.$$

Thus a flux of 8×10^6 particles/s has a SNR of unity when $F = 4 \times 10^6$ Hz.

In the case of current consisting of randomly emitted electrons (photoelectrons or current through a reverse-biased p-n junction),

$$a = I/e \text{ with } e = 1.602 \times 10^{-19} \text{C/electron},$$

$$\text{SNR} = \sqrt{I/2eF},$$

$$I_n = \sqrt{2eIF} \text{ A: noise current}.$$

For unity SNR at $F = 4$ MHz
$$I = 2eF = 1.28 \times 10^{-12} \text{ A}$$
For 100:1 (40 dB) SNR
$$I = 12.8 \times 10^{-9} \text{ amps}.$$

These seem like very small currents and particle fluxes. Nevertheless, they are sometimes encountered and can limit system operation significantly. To illustrate this, we now calculate the illumination required on an object to obtain 40 dB SNR in the case of a perfect (shot-noise limited) camera tube with 25% quantum efficiency, $f/4.5$ lens, $1/2 \times 3/8''$ target, and typical object reflectivity, ϱ, of 20%.

$$I_s = I_T A \times 1.36 \times 10^{16} \frac{\text{photons/s}}{\text{lumen}} \times 0.25 \times 1.602 \times 10^{-19}$$

$$= 12.8 \times 10^{-9}$$

$$I_0 = \frac{4f^2}{\varrho} I_T = \frac{4f^2}{\varrho} \times \frac{12.8 \times 10^{-9}}{A \times 1.36 \times 10^{16} \times 0.25 \times 1.602 \times 10^{-19}}$$

$$= 7.31 \text{ ft-candles of "white" light}.$$

Thus if we had such a near-perfect camera, it would be possible to get excellent TV signals with fairly low illumination levels. In actual practice, at least ten times as much light is needed and often much more.

2.6.3 Other Noise Sources

To be more realistic in predicting the performance of electronic cameras, we must include other noise sources besides the shot noise that inevitably accompanies the signal. Two additional sources always present in electronic camera systems are amplifier noise and "Johnson" noise of the resistor used to develop a signal voltage from the signal current that is the normal camera output.

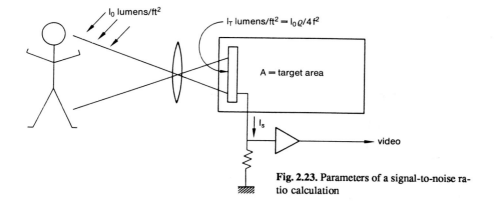

Fig. 2.23. Parameters of a signal-to-noise ratio calculation

In Fig. 2.23, the essential parameters required for a realistic calculation of noise performance are shown. A typical camera circuit is shown in Fig. 2.24a. The signal current is passed through the load resistor R and the resulting signal amplified. Because of the presence of the capacitance (due to both camera tube and amplifier input circuit), the resulting signal falls off at high frequencies. As we shall see, it is best not to choose R low enough to obtain the desired bandwidth. Instead, a rather large value of R is used, and the amplifier frequency characteristic designed to give uniform overall response within the band of interest. Figure 2.24b shows a superior circuit, which has the same noise performance but automatically equalizes the response.

For purposes of this analysis, we assume that all three noise sources are flat and random, so that they can be characterized by a uniform "power" spectral density which, when integrated over the passband F, gives the mean square noise current or voltage. The shot noise in the signal has the value

$$i_{ns}^2 = 2eI_sF$$

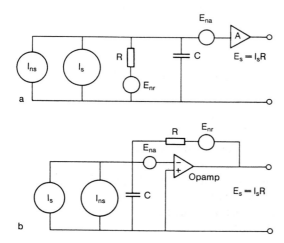

Fig. 2.24a,b. SNR calculation for a photon-limited camera

49

and thus has the spectral density

$$G_{ns} = 2eI_s \ .$$

The Johnson noise in series with the resistor has the value

$$\overline{e_{nR}^2} = 4KTRF \ ,$$

and thus the spectral density

$$G_{nR} = 4KTR \ .$$

If we assume that the amplifier noise is equivalent to the shot noise in a dc current I_0, and that the signal component of this current is $g_m e_{in}$, where g_m is the transconductance of the amplifier, then this is equivalent to an input noise

$$e_{na}^2 = \frac{2eI_0 F}{g_m^2} \ ,$$

which therefore has the spectral density

$$G_{na} = \frac{2eI_0}{g_m^2} \ .$$

Putting all three noise components into the circuit and solving for the output noise spectral density, we have

$$G_{no} = \frac{2eI_s R^2}{1 + (\omega^2 R^2 C^2 / A^2)} + \frac{4KTR}{1 + (\omega^2 R^2 C^2 / A^2)}$$
$$+ \frac{2eI_0}{g_m^2} \frac{1 + \omega^2 R^2 C^2}{1 + (\omega^2 R^2 C^2 / A^2)} \ .$$

Several points can be gleaned from this expression. At high frequencies, the amplifier noise predominates. At low frequencies, the Johnson noise contribution can be minimized by making R large. When

$$2eI_s R^2 = 4KTR \ ,$$

the two components are equal, and under those circumstances,

$$I_s R = 2KT/e = 50 \text{ mV at } T = 290 \text{ K} \ .$$

Thus a useful rule of thumb in design is that the load resistor should be chosen large enough so that the (low frequency) output signal is at least 50 mV.

Amplifier noise cannot be so easily minimized. Instead, the designer concentrates on getting as large an optical signal as possible, on minimizing the capacitance, and on obtaining a high g_m^2 / I_0 ratio.

In Fig. 2.25 we have plotted these various spectra. Note that $1/RC$ would have been the signal bandwidth without the feedback amplifier; this is increased

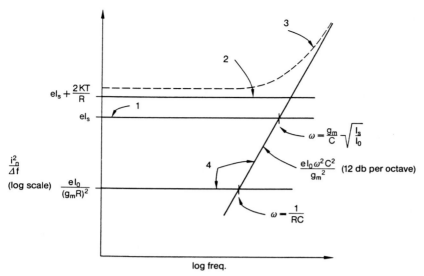

Fig. 2.25. Output noise (current^2) spectrum: (1) noise due to input shot effect; (2) noise due to shot effect plus resistor noise; (3) total noise; (4) amplifier noise asymptotes

by A with the amplifier. However, the feedback does not similarly control the noise – it has a $1/RC$ corner frequency. The shot noise and Johnson noise have a bandwidth of A/RC at the output, as does the signal, due to the equalizing effect of the feedback.

A perhaps more useful way to show the noise performance is by means of the Operating Spot Noise Figure (OSNF). This is simply the ratio of the narrow band input SNR to output SNR. To calculate this we note that the input signal-to-noise ratio spectral density is $G_s/2eI_s$ where G_s is the input signal spectral density, while the output signal-to-noise ratio spectral density is

$$\frac{G_s}{G_{no}}\frac{R^2}{1+(\omega^2 R^2 C^2/A^2)} .$$

Thus, OSNF is

$$1 + \frac{2KT}{e}\frac{1}{I_s R} + \frac{I_0/I_s}{g_m^2 R^2}(1 + \omega^2 R^2 C^2) .$$

This is shown in Fig. 2.26 for a practical case of a vidicon system in which

$$I: 1\,\mu A\,,\ g: 0.0125 A/V\,,\ I: 5mA\,,\ \text{and}\ C: 30pf\,.$$

The curve is shown for $R = 200,000\,\Omega$, where the shot noise is much larger than the Johnson noise, for $R = 50,000\,\Omega$ where the two noises are equal, and for $R = 1330\,\Omega$, which would give a bandwidth of 4 MHz without feedback. It can readily be seen that, within the band of interest, the noise performance can be greatly improved by correct choice of resistance.

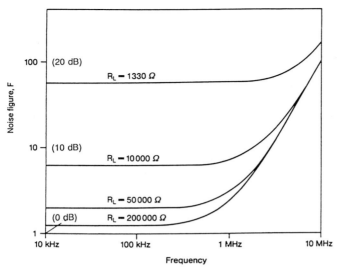

Fig. 2.26. Operating spot noise figure (OSNF) of a vidicon-preamp combination, expressed as a power ratio, F

2.7 Lessons for the System Designer

In this chapter, we have discussed physical limitations on the performance of imaging systems due to the quantum and wave nature of light. Practical problems in achieving this limiting performance have also been discussed, with emphasis on methods by which actual performance can be calculated in typical situations. Fourier methods have been introduced and applied, first to optical systems and then, by extension, to scanned television and facsimile systems. The spectral description of images and imaging systems is an alternative to their description in terms of sample values and impulse responses, respectively. It is most important for the system designer to become fully comfortable with this alternative description, as it is very helpful in understanding the operation of many important processes, especially in coding. In addition, it often makes it easier to relate the properties of the observer to the parameters of systems.

The main message of this book deals with the perceptual aspects of imaging systems design, because that seems to the author to be the most important (and most often neglected) group of issues in this field. However, the physical and mathematical ideas introduced here have proven very useful in the analysis and design of imaging systems, although they almost never give final answers to anything. Experience alone will provide the research worker with the most effective balance of mathematical, physical, and perceptual techniques.

2.8 Appendix: Statistical Considerations

Shot Noise. If a stream of randomly emitted particles of average value a particles per second is carefully observed, it will be found that the actual number emitted or observed in a fixed interval fluctuates about some average value. This fluctuation constitutes an ac noise signal superimposed on the dc, or average value. The magnitude of the ac noise depends on the bandwidth of the measurement and decreases, in relation to the dc signal, as a increases. This can be seen qualitatively by considering the stream to consist of a train of very narrow pulses:

The ac energy obviously is distributed through a very wide bandwidth, so that a wide-band measurement will give a higher value than a narrow-band measurement. Likewise, for a fixed bandwidth, adding more pulses will cause the ac *power* to increase in direct proportion to the increase in average value a (since random-phase currents add in the power domain) while the dc power increases as a^2.

This fluctuation of low-intensity currents or light is called shot noise. It is of great importance in image transmission systems because it represents an inherent, natural barrier to noiseless operation of all light-sensitive devices, including camera tubes and the human eye. For this reason, it is of interest to calculate the statistical properties of shot noise and use these properties to discuss the sensitivity and SNR of imaging devices.

The Poisson Distribution. We need to know the probability of getting K particles in a period of τ seconds, if the average particle current is a per second. To do this, we divide τ into a large number, n, of intervals. The probability of getting a particles in such a small interval is $a\tau/n$ and if $a\tau/n$ is small enough, the probability of more than one is negligible. The probability of getting exactly K particles in τ seconds is thus

$$P_n(K, \tau) = \frac{n!}{K!(n-k)!} \left(\frac{a\tau}{n} \right)^K \left(\frac{1-a\tau}{n} \right)^{n-K}$$

This is the well known binomial distribution, where the coefficient is the number of different ways K events can occur in n trials, where $(a\tau/n)^K$ is the probability that a particle occurs in K particular intervals and $(1 - a\tau/n)^{n-K}$ is the probability that *no* particle occurs in the other $n - K$ intervals. To facilitate going to the limit of $n \to \infty$, the expression may be rewritten:

$$P_n(K, \tau) = \frac{\overbrace{n(n-1)(n-2)\ldots(n-K+1)}^{\to n^K}}{K!}$$

$$\times \frac{(a\tau)^K}{n^K} \underbrace{\left[\left(1 - \frac{a\tau}{n}\right)\frac{n}{a\tau}\right]^{a\tau}}_{\to \exp^{-a\tau}} \underbrace{\left[1 - \frac{a\tau}{n}\right]^{-K}}_{\to 1}.$$

$$P(K, \tau) = \lim_{n\to\infty} P_n(K, \tau) = \frac{(a\tau)^K}{K!} \exp(-a\tau),$$

which is the Poisson distribution.

In order to calculate the value of ac noise we must know the mean square deviation of K. To find this directly by summing $K^2 P(K, \tau)$ over K is quite difficult, and other methods are used. Eventually we find

$$(K - \bar{K})^2 = a\tau = \bar{K}$$

In other words, the mean square deviation is numerically equal to the average number of particles in the time τ.

SNR in a Photon Stream. Consider a weak light, consisting of an average of a photons/s. In τ seconds, the mean square fluctuation is $(a\tau$ photons$)^2$. This corresponds to an ac photon flux of

$$\frac{\sqrt{a\tau}}{\tau} = \sqrt{\frac{a}{\tau}}$$

photons/s. Thus, the SNR

$$\frac{S}{N} = \frac{a}{\sqrt{a/\tau}} = \sqrt{a\tau}$$

The observation interval τ is obviously of great importance. For an observation bandwidth F,

$$\tau = \frac{1}{2F} \ .$$

Thus, $S/N = a/2F$, and the rms noise flux is $2aF$ photons/s.

For example, with a flux of 8×10^6 photons/s. with an observation bandwidth of 4×10^6 Hz,

$$\frac{S}{N} = \frac{a}{2F} = \frac{8 \times 10^6}{2 \times 4 \times 10^6} = 1 \ .$$

SNR in an Electron Stream. For an electron current of I amperes, $a = I/e$, e being the electronic charge. Thus

54

$$\frac{S}{N} = \sqrt{I/2eF}$$

The noise current itself is $\sqrt{2aF} = \sqrt{2IF/e}$ electrons/s, or $\sqrt{I_n} = \sqrt{2eIF}$ amperes.

This result, which was discussed by Schottky, implies that the noise power is uniformly distributed over the band, a situation consistent with the assumption that the current is made up of very narrow pulses, one for each electron.

It is of interest to calculate the current at which unity signal/noise level prevails for 4 MHz bandwidth. This would be the minimum useful signal current in a shot-noise limited camera tube (or any other camera device) for conventional TV:

$$1 = \sqrt{I/(2eF)} \, ,$$
$$I = 2eF \, ,$$
$$I = 2 \times 1.6 \times 10^{-19} \times 4 \times 10^6 \, ,$$
$$I = 1.28 \times 10^{-12} \text{A} \, .$$

Noisy Multiplication of Noisy Currents. Frequently in imaging systems, a noisy photon or electron current undergoes a (more or less) noisy multiplication. It is then necessary to calculate the noise in the output stream. We distinguish two cases.

Photoemission. When incoherent light strikes a photoemitter, experiment demonstrates that each photon releases one photoelectron with probability p, where p is called the quantum efficiency, the maximum value of which is unity. Practical photoemitters have p's of 0.01–0.25,[7] depending on the spectral region and materials involved. This amounts to taking the incoming photon stream, which may be thought of as a train of very narrow pulses with random emission time, and eliminating some of them, again at random. The resulting photoelectron stream is statistically indistinguishable from a noisy pulse train governed by the Poisson distribution with the same average current, which was *not* the result of photoemission. Therefore, we can find the SNR directly from the dc value of photoelectric current. For example, in the situation given above, where a flux of 8×10^6 photons/s was found to have unity SNR when observed with 4 MHz bandwidth, if the photons were incident on a cathode with 5 per cent quantum efficiency, then $1/0.05 \times 8 \times 10^6 = 1.6 \times 10^8$ photons/s would be needed to have unity SNR in the photoelectric current, and 1.6×10^{12} photons/s for 100:1 SNR.

Secondary Electron Multiplication. When a weak current is increased by secondary emission, experiments have shown that what actually happens is that each

---

[7] Silicon photodiodes can have p's of 0.75 or more.

primary electron produces exactly m electrons, each of which escapes or not with probability p, so that the secondary emission ratio

$$\delta = mp .$$

Thus, the output current is a series of pulses coincident in time with the input pulses. The output pulse amplitude fluctuates about an average value which is δ times the input amplitude. It will clearly have a lower SNR than the primary current unless δ is so large that the fluctuations in δ are relatively very small.

The probability of getting just N secondaries for n primaries is

$$P(N/n) = \frac{(nm)!}{(nm - N)!N!} p^N (1 - p)^{nm-N} ,$$

whereas the probability of getting n primaries in a certain period is

$$P(n) = \frac{(\bar{n})^n e^{-\bar{n}}}{n!} ,$$

where $\bar{n}$ is the average number of primary electrons in the period. The total probability of getting N secondaries in a certain period is

$$P(N) = \sum_n P(N/n)P(n) .$$

Although this series could conceivably be summed in closed form, it is much easier to assume the period long enough so that N is much larger than one. In that case, both distributions become Gaussian (by using Stirling's formula for the factorial).

$$P(N/n) = \frac{1}{\sqrt{2\pi n\delta}} \exp \left[-\frac{(N - n\delta)^2}{2n\delta} \right] ,$$

$$P(n) = \frac{1}{\sqrt{2\pi\bar{n}}} \exp \left[-\frac{(n - \bar{n})^2}{2\bar{n}} \right] ,$$

and

$$P(N) = \int_0^\infty P(N/n)P(n)dn .$$

At this point, we may note that in such cases, mean square deviations add linearly or else we may actually carry out the integration (by completion of the square in the exponent) and find that

$$P(N) = \frac{1}{2\pi\bar{n}(\delta + \delta^2)} \exp \left[-\frac{(N - \bar{n}\delta)^2}{2n(\delta + \delta^2)} \right] \quad \text{with} \quad \bar{n}\delta = \bar{N} .$$

Thus the mean square fluctuation in N, the secondary electrons emitted in a certain time,

$$(N - \bar{N})^2 = \bar{n}\delta + \bar{n}\delta^2 = \bar{N} + \bar{n}\delta^2 .$$

The first term is the shot noise to be expected in N electrons, while the second term is δ^2 times the shot noise in the primary current. The ac shot noise current for a dc current of I amperes is

$$I_n = \sqrt{2eIF} .$$

We have, for the case of secondary emission,

$$I_n^2 = 2eI_{sec}F + 2e\delta^2 I_{pri}F ,$$

where $I_{sec} = \delta I_{pri}$, so that we can also write

$$I_n^2 = 2eI_{sec}F(\delta + 1) .$$

Since if the multiplication were noiseless we would have only the multiplied primary noise, the mean square noise current is increased by $1/\delta$, so that, for example, if $\delta = 10$, I_n is increased only by 1%. For s stages of secondary emission

$$I_n^2 = [\delta^s + \delta^{s-1} + \ldots \delta + 1]2eFI_{out}$$

$$I_n^2 = 2eFI_{out} \sum_{j=1}^{s} \delta^j = 2eFI_{pri}\delta^s \sum_{j=1}^{s} \delta^j .$$

Of this noise, a portion is due to amplified noise in the primary current, which is

$$\delta^{2s} I_{n,pri}^2 = \delta^{2s}2eFI_{pri} .$$

Thus the ratio of the actual noise power output to the value obtainable with noiseless multiplication is

$$\frac{2eFI_{pri}\delta^s \sum_{j=1}^{s} \delta^j}{2eFI_{pri}\delta^{2s}} = \delta^{-s} \sum_{j=1}^{s} \delta^j ,$$

which can be regarded as the "noise figure" of the multiplier. For a typical case of a ten-stage multiplier with a per-stage gain of 4, we have

$$4^{-10} \sum_{j=1}^{10} 4^j = 4^{-10} \frac{4^{11} - 1}{3} = 1.33 ,$$

so that the noise power is increased by 33 % or 1.24 dB.

3. Perception of Images

The purpose of this chapter is to introduce the reader to those aspects of human vision that are pertinent to the understanding and design of man-made image processing systems. As far as possible, we shall concern ourselves with psychophysics, i.e., the response of observers to visual stimuli, and shall avoid, wherever possible, inferring response from physiology, i.e., the structure of the visual apparatus. We shall even more assiduously avoid inferring structure from psychophysical data. In a few cases where the connection between psychophysics and physiology is well understood, it may be discussed to simplify understanding or just because it is interesting.

The reason for this approach is that, in fact, vision is so complicated that we are very far from having good physical explanations for many phenomena. I believe that it is much better to design on the basis of a psychological result that is well known than on the basis of speculation about physiological structure. This is particularly true for color reproduction but holds as well for noise visibility.

A limitation of the psychophysical approach is that in most measurements the observer is limited to a yes/no response. It is easy to get matching or threshold data in this way, but difficult to predict the appearance of, or subjective reaction to, perturbations in images at suprathreshold levels.

3.1 Seeing in the Dark

Everyone knows that our eyes become more sensitive after some time in the dark. Although some of this adaptation is due to a change in the size of the pupil (about $10:1$ in area), a process that takes no more than a second, the larger part of the adaptive process has been identified with chemical changes in the photosensitive constituents of the retinal receptors [3.1, 2]. There are two distinct time constants in the adaptation, so that the first stage is substantially complete after about ten minutes[1] and the final stage after about 30 min. It has been established that these two time constants are associated with two different kinds of receptors. The less sensitive cone cells come into play under photopic (normal bright light) viewing conditions, and are responsible both for color vision and high visual acuity in

---
[1] A substantial portion of this response occurs very quickly. As a result, different portions of the retina can be at different states of adaptation (and sensitivity) at the same time.

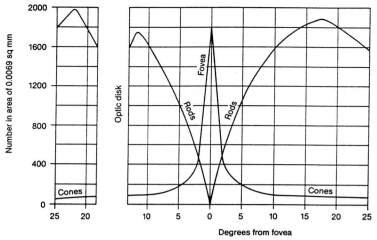

Fig. 3.1. The variation of receptor density with distance from the fovea. The cone density drops rapidly with distance while the rod density rises [3.1]

the center of the field. The more sensitive rod cells are color blind,[2] have the longer time constant, and are used in night vision (scotopic viewing conditions.)

Figure 3.1 shows the distribution of rods and cones across the retina. The central fovea, only a few degrees across, consists exclusively of cone cells. It is only in this area that we have truly sharp vision. Thus the eye is constantly roving over the important part of the field of view in order to gain accurate information about a greater area than can be seen sharply at one time. The density of rods increases away from the fovea, reaching a maximum about 20 degrees from the center. Thereafter the density of cells continues to decrease, although some visual sensitivity is found more than 90 degrees from the center.

The luminosity curves of the two types of cells also differ, as shown in Fig. 3.2, the cones having peak sensitivity at 556 nm and the rods at 510 nm. (This shift in peak sensitivity is called the Purkinje effect.) These curves coincide very closely with the absorbance curves of the photochemical substances, taking into account the spectral transmittance of the ocular media. Thus rhodopsin, the photosensitive substance of the rods, is sometimes called visual purple. In the case of rod vision, the relative insensitivity to red light ($\lambda > 620$ nm) makes it possible to maintain a high state of dark adaptation under room light if deep red goggles are worn.

---

[2] Some authors believe the rods play some role in color vision.

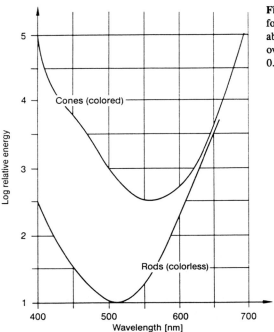

Fig. 3.2. Threshold spectral sensitivity for rods and cones. The actual increment above threshold for color vision varies over the retina. In the parafovea, it is 0.1 to 1.0 log units [3.1]

3.1.1 Visual Thresholds

It is instructive to measure the minimum amount of light that can be seen when the eye is fully dark adapted. It was found by *Hecht* [3.3] that for maximum sensitivity the light must strike the retina about 20 degrees from the fovea, within a time interval of less than 1 ms and within an angle of less than 10′ of arc. In such cases, a response occurs when an average of 5 to 14 receptors are each stimulated by a single photon. This result is interesting for a number of reasons. In the first place, this is a very small amount of light, making the performance of the eye comparable to that of the most sensitive man-made light detectors. In the second place, it indicates that, at least under these conditions, a group of nearby receptors can cooperate. Finally, this very significant experiment points up the statistical nature of the detection process. For such small stimuli, since the light is produced by a Poisson process, the actual number of photons varies from trial to trial. If one measures the percentage of trials in which the stimulus is detected as a function of target luminance, the slope of this curve at the 50% detection point can be used to determine the average number of photons taking part in the decision.

More significant for ordinary viewing is the detection of a difference in luminance between a small area and a uniformly illuminated background as shown in Fig. 3.3. Provided that both the target and background luminance are so low that no significant portion of the photosensitive material is excited, one would expect the ability of each receptor to sense a photon to remain at its completely dark-

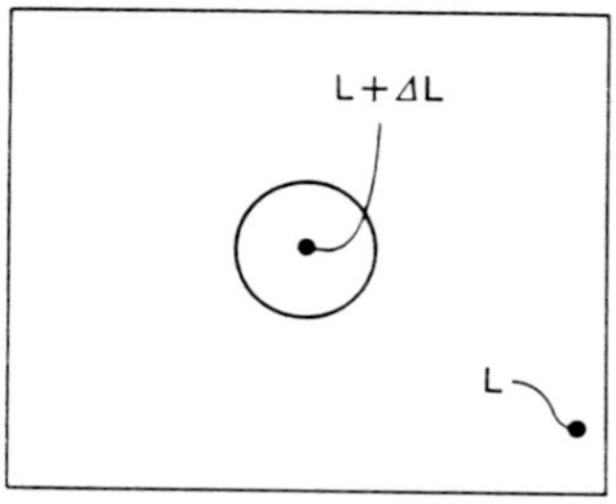

Fig. 3.3. Contrast sensitivity target

adapted level. In that case, the target is detected when its luminance sufficiently exceeds the expected statistical fluctuation in the luminance of the background. We define the contrast

$$C = (\Delta L)/L \ .$$

Thus,

$$C_{min} = (\Delta L_{min})/L = (\Delta N_{min})/N \ ,$$

where N is the number of photons emitted during the detection interval and ΔN_{min} is the least detectable deviation in N. Since ΔN is the rms deviation in N due to the Poisson process, then we assume that for detection

$$\Delta N_{min} = k\sqrt{N} \ ,$$

where k is a number near unity. Thus,

$$C_{min} = k\sqrt{N})/N = k/\sqrt{N} \ ,$$
$$N = k'/C_{min}^2 \ ,$$

but

$$L = (k''N)/A \ ,$$

where A is the area of the spot. Combining these relations [3.4], we have

$$ALC^2 = K$$

for threshold vision. This expression is well borne out by experiment. For example, it predicts that $\Delta L_{min}/L = K'/\sqrt{L}$. On a log-log plot, this gives a slope of $-1/2$. Between 0.0001 millilamberts and 0.01 millilambert, a typical experiment [3.5] gives a slope of -0.42. It also predicts visual acuity at low luminance levels, since the minimum detectable area is K/LC^2. Since the spatial frequency is proportional to the reciprocal of the square root of the area, the maximum detectable spatial frequency is $= K''\sqrt{L}$. In fact acuity does rise as the square root of luminance up to the photopic range, where it levels off. This expression is a generalization of Ricco's Law, which states that AL is a constant, other things being equal. This is just another way of saying that for small spots, the

61

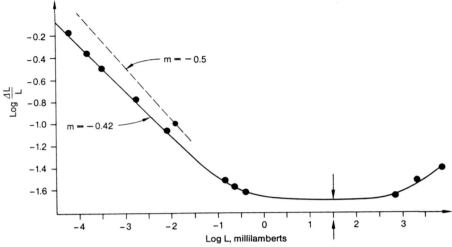

Fig. 3.4. Contrast sensitivity: arrows indicate the Weber-Fechner fraction [3.6]

visibility depends on the total amount of light emitted, and not the size or shape of the spot. The law obviously does not hold for large spots.

As the background luminance rises, more and more of the visual pigment becomes bleached and the sensitivity drops. From about 0.1 millilambert to 800 millilamberts, $\Delta L/L$ is within a factor of 2 of its minimum value. At higher brightness, saturation sets in and the relative sensitivity again falls as shown in Fig. 3.4. It should be noted that this curve is obtained only if the viewer is allowed to adapt thoroughly to the background luminance.

The minimum contrast in the range of normal seeing is called the Weber-Fechner fraction [3.1]. It varies from 1% to 3%, depending on the exact size and shape of the spot.

The constancy of the Weber fraction over so large a brightness range makes tempting the proposal of a simple model for the process, shown in Fig. 3.5. Let the luminance, L, be the input to a nonlinear amplifier and S, the sensation, be the output. If ΔS is the minimum detectable change in S and is assumed to be constant, then

$$\Delta S = G\Delta L = k \text{ (a constant) },$$

where G is the incremental gain of the amplifier. To be consistent with the experimental data showing that $\Delta L/L$ is constant, G must be inversely proportional to L. Thus

Fig. 3.5. A model

$$G = k/L .$$

Then

$$\Delta S = k(\Delta L)/L .$$

We can now integrate this expression, giving

$$S = k \log L + \text{constant} ,$$

which is, of course, only valid in the luminance range where $\Delta L/L$ is constant. This is the justification for the statement that the visual system is logarithmic.

It is also possible to find a relationship between S and L over the entire range of L, and not only where $\Delta L/L$ is constant. Let

$$\Delta L/L = f(L)$$

be the experimental data, such as in Fig. 3.4. Then

$$\Delta S = G\Delta L = k = (dS/dL)\Delta L = (dS/dL)Lf(L) ,$$

$$S = \int k/[Lf(L)]dL .$$

As before, this leads to the logarithmic relationship if $f(L)$ is constant. In the scotopic region where

$$f(L) = k''/(\sqrt{L}) , \qquad S = \int k''/(\sqrt{L})dL = k'''\sqrt{L} .$$

Experimental data such as that of Fig. 3.4 leads to a curve like that of Fig. 3.6, where the straight section corresponds to the region of constant Fechner fraction.

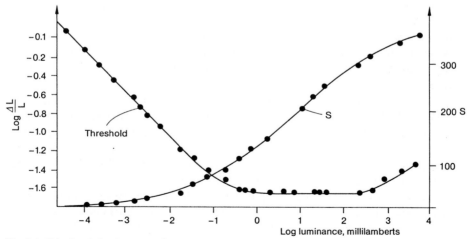

Fig. 3.6. Calculated absolute sensation

At higher and lower luminance the slope is less. The significance of this will be explored in the next section. In any event, it is worth noting that experiments designed to measure absolute sensation are less satisfactory than those dealing with thresholds, and cannot be as directly verified. Conclusions about experimental results can generally be drawn from the $\Delta L/L$ curves. Nevertheless, it is sometimes convenient to think of visual response in absolute terms in order to understand tone reproduction more readily.

3.2 Contrast Sensitivity

Sensitivity of the observer to small changes in luminance is an important factor in the design of imaging systems, since it determines noise visibility and establishes the accuracy with which luminance must be represented.

In the previously discussed experiments, the ability of the dark-adapted observer to distinguish a small spot slightly brighter than the surround was found to be a statistical phenomenon. As the light level rises into the photopic region, a significant portion of the photosensitive pigment becomes bleached and the sensitivity drops, remaining nearly constant over the entire range of normal light levels. This phenomenon is deceptive, however, since in order to achieve the enormous dynamic range implied by Fig. 3.4, the experiment must be carried out slowly enough for the eye to become adapted, i.e., for the pigment bleaching level to become stabilized, to each background luminance.

The conditions under which pictures are viewed are of course considerably more complex than those of Fig. 3.4. Additional experiments are needed, not all of which have been done. Thus the following account, although it seems to explain the main phenomena adequately, is somewhat speculative at this time.

In the previous experiment, the retina must have been adapted very nearly to L_o. When viewing pictures, unless the eye is fixated at one point for some time, the adaptation level must be some weighted average luminance of the picture and the surround against which it is viewed, perhaps modified by recent previous stimuli. The problem, then, is to find the least noticeable luminance difference between adjacent small patches, neither of which is at the adaptation level.

This situation, admittedly grossly simplifed, is represented by the arrangement of Fig. 3.7. L_0 is the adaptation level or near it, and represents the weighted average luminance of the picture. L represents the luminance at some point in the image, and ΔL is the just noticeable difference in L. Under these conditions it is found that $\Delta L/L$ is constant over a much smaller range, and that the center of the range depends on L_0, as shown in Fig. 3.8.

A help in understanding this phenomenon is to notice what a small central patch looks like, as a function of the surround luminance. It can be changed from perceptually "white" to perceptually "black" by making the surround sufficiently (about 100 times) darker or lighter, respectively. The threshold goes to infinity

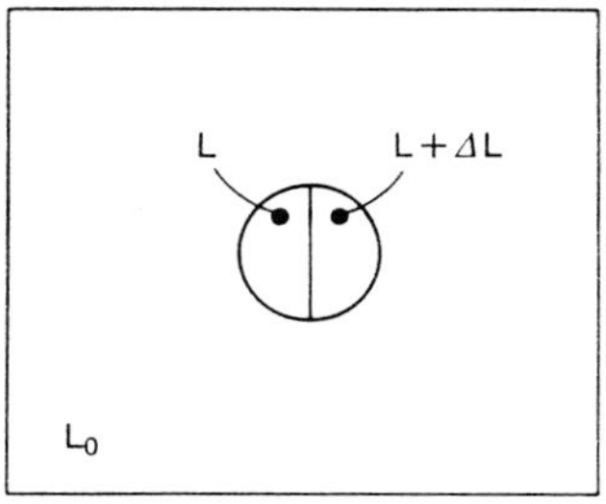

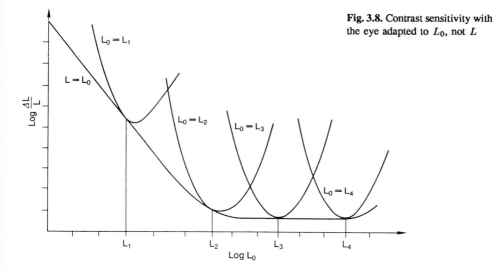

Fig. 3.7. A better contrast target

Fig. 3.8. Contrast sensitivity with the eye adapted to L_0, not L

in both cases, and clearly must be minimum somewhere in between. Numerous experiments suggest that the minimum always occurs when $L = L_0$. Since the entire field is then equal to L_0, that must be the adaptation level. There are grounds to believe that both spatial and temporal contrast sensitivity, for all shapes of time or space perturbations, are maximum for luminance equal to the adaptation level.

The curves shown in Fig. 3.8 are idealized, since they assume that the adaptation level is controlled by the surround and is unaffected by the luminance of the central patch. This assumption is probably true for patch luminances darker than the surround. As the patch becomes increasingly brighter, it begins to affect the adaptation level. When it is sufficiently brighter it begins to look like a light source rather than an illuminated surface. (This never occurs in reflection prints.) It is not known at present whether the transition to the appearance of a light source occurs exactly at the point where the adaptation level changes. The appearance of such stimuli was extensively studied by *Evans* [3.7] in a series of experiments going back to the late 1950s.

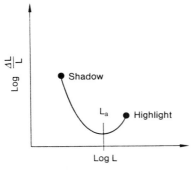 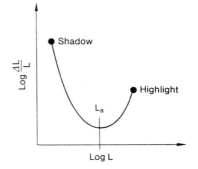

Fig. 3.9. Dynamic range of a reflection print   **Fig. 3.10.** Dynamic range of a transparency

The exact "law" that governs the threshold operates only over the dynamic range of the image and also depends on which aspect of the total stimulus actually controls the state of adaptation. In the case of reflection prints viewed in normal room light, it is quite difficult to achieve satisfyingly "black" blacks and "white" whites. This is because the physical dynamic range is barely two log units and because the adaptation level is not much below the highlight luminance. Transparencies projected by good equipment in a dark room or theater are much better because the physical dynamic range is higher and because the image itself controls the adaptation level, the balance of the room being dark. The adaptation level therefore is much lower than in the case of a reflection print of equal luminance viewed in a bright room. Thus the image highlights of the transparency seem much brighter than the corresponding areas of the print, while the shadows actually are, and seem, much darker. This situation is related to adaption-dependent thresholds in Figs. 3.9, 10.

The curves above have significance because from them we can predict the observer's contrast sensitivity over the tone range of the viewed image, and on this basis decide on the optimum distribution of quantization levels to minimize the appearance of quantization noise.[3] For example, if the situation of Fig. 3.4 prevailed and if the Weber fraction were really constant, then the size of quantization steps should be made proportional to luminance. This could be accomplished with a linear quantizer preceded by a logarithmic compressor and followed by an exponential expander. However, the result of such a process is that the quantization steps are much more noticeable in the highlights than the shadows – a result readily confirmed by viewing a photographic reflective step wedge of uniform density increments. The steps show the highest contrast between 0 and 0.4 to 0.5 density units, becoming progressively less noticeable at higher densities, essentially vanishing at about 1% reflectance ($D = 2.0$). Linear quantization of

---

[3] Other noises, such as shot noise, additive noise, or multiplicative noise, may not be equally visible over the tone scale since it may not be possible to control the scale at the point where the noise occurs.

the luminance signal, on the other hand, gives higher contrast in the dark tones. Clearly the best result is in between these two extremes.

3.2.1 Lightness Curves

The previous phenomena have been studied independently by psychologists and others who were interested in applying the results to practical problems. An example is in the standardization of inks and other colorants, where it is desired to cover the necessary gamut (range) of color and brightness with the smallest number of different sample colors. In this case, successive elements of the sequence should be equally different from each other. In a typical experiment, a subject is given, say, 100 small grey cards, ranging from black to white, and asked to select 10 spanning the range in equal steps. If the sensation associated with a tone is taken to be the number of steps from black, then the curve of step number versus sample luminance is the S vs L curve of Fig. 3.6.

The most extensive series of experiments are those associated with the Munsell system [3.8], which is widely used to identify paints, inks, and other surface colors. Munsell "value" corresponds to lightness of neutral colors [3.9]. Numerous other scales have been proposed, such as power law (square root), logarithmic, etc. A modified log curve has been found to work well on TV displays [3.10]. Several such curves are shown for comparison in Fig. 3.11 [3.11].

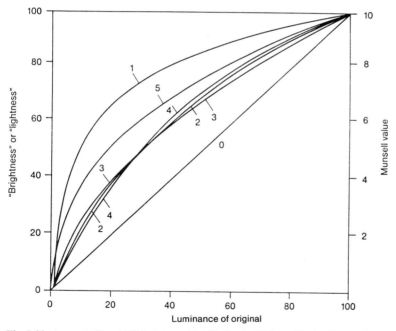

Fig. 3.11. A comparison of lightness scales: (*0*) linear, (*1*) logarithmic, (*2*) modified log, (*3*) power law, (*4*) bilinear, (*5*) Munsell

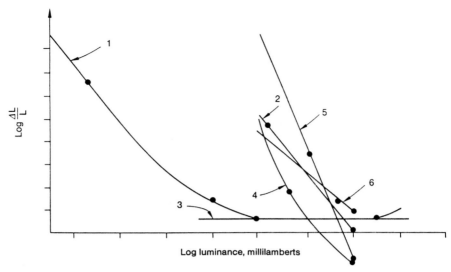

Fig. 3.12. Proposed lightness scales superimposed on experimental data: (*1*) Koenig and Brodhun, (*2*) square root, (*3*) log, (*4*) modified log, (*5*) linear, (*6*) Munsell

Further insight into the meaning of the lightness curves can be gained by applying the methods of Sect. 3.1, where a mathematical relationship was found between these curves and contrast sensitivity data. Since

$$\Delta S = \frac{dS}{dL} \Delta L = k \ .$$

Then

$$\Delta L = \frac{k}{dS/dL} \ , \qquad \frac{\Delta L}{L} = \frac{k}{L(dS/dL)} \ .$$

Therefore, given the lightness curve, we can readily calculate an associated contrast sensitivity. Figure 3.12 shows such curves calculated for the Munsell, log, and square root relationships superimposed on the data of Fig. 3.4. The vertical position is arbitrary. A peak luminance of 100 ft-lamberts, comparable to a bright TV picture tube, is assumed.

Two conclusions can be drawn from the results of this calculation. The $\Delta L/L$ curve derived from the log lightness law is a horizontal line, and we know from numerous experiments that this "law" is untrue, particularly for luminances well below the adapting luminance. All the other curves show an operating dynamic range, at least on the dark side, much more like those of Fig. 3.8 than that of Fig. 3.4, lending strength to the view that the actual dynamic range on the low side is rather small.

None of the curves flatten out and rise again. However, our conclusion from this is that in all these experiments the adaptation level must have been *higher* than the peak luminance. The Munsell curve is "best" in this case. Since the sample chips are placed on a white background in the Munsell experiment, we may assume that the adaptation level is nearly white.

Reflection prints are generally viewed under conditions where they are no brighter than the surround. Projected transparencies, however, are usually viewed in a dark room so that the transparency is so much brighter than the surround that the eye adapts to the image itself. We are therefore led to speculate that the usual lightness curves do not apply to the latter case. If this is so then we may expect that a careful experiment, if extended into the range where peak luminance is substantially *above* the adaptive level, would show an inflection point like that of Fig. 3.6.

An important practical result that grows out of these considerations is that the relative visibility of noise throughout the luminance range of an image depends on the tone scale at the point where noise is added, as shown in Fig. 3.13. For minimum noise visibility at a given SNR, the noise should be equally visible throughout the tone scale. If it is not, the visibility of the most obvious noise can always be reduced by redistribution, using a different tone scale before adding the noise. The best proof that the modified log scale is nearly perceptually uniform is provided by Fig. 3.13b, which has the most uniform noise visibility.

3.3 The Effect of Shape

The minimum contrast of a light pattern against a background (detail if we want to see it, noise if we don't) depends on its shape. Sharp-edged spots are easier to see than blurred spots, and the latter are most visible when they subtend 1/2 to 1/10 degree at the eye. In the case of white noise in a TV display at a signal-to-noise ratio of 44 dB (160 to 1), 10% of the observers are bothered by the noise and 50% notice it. In this case, the rms value (standard deviation) of the noise is about 1/3 of the Fechner fraction as measured with a small sharp-edged target. At this level the Fechner fraction is exceeded by the actual noise only 0.3% of the time.

The application of linear systems theory to shape perception has the potential to unify and simplify the understanding of a wide variety of visual phenomena, just as in the analysis of optical systems. However, the pronounced nonlinearity exhibited by contrast sensitivity is a major obstacle. Some attempts to bypass this problem involve multistage models of vision, some stages linear but frequency-dependent and some static but nonlinear. Others have appealed to a piecewise linear analysis, pointing out that for small signals[4] even grossly nonlinear systems can be treated as incrementally linear [3.12].

Alternatively, aspects of shape sensitivity can be measured directly. A wide variety of patterns is in use, including Snellen charts, Landolt "*C*" rings, resolution wedges, the Air Force resolution chart, variable spatial frequency gratings,

---

[4] Obviously, at threshold, *something* must be small.

70

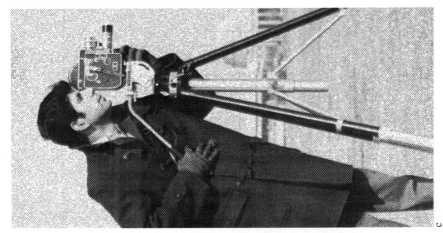

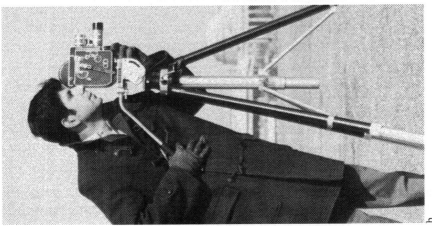

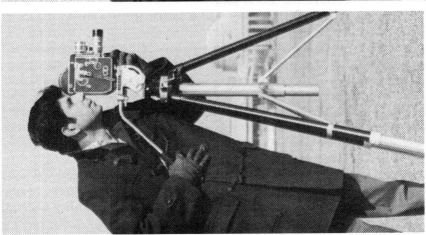

Fig. 3.13a–c. For caption see opposite page

etc. All are useful, but do not promote generalization to arbitrary shape. The most useful studies are those that have attempted to measure spatial frequency response in spite of the existing nonlinearities.

Lowry and *DePalma* [3.13] measured spatial transient response by a matching method and used the ratio of Fourier transforms of response and stimulus as the MTF. They also used sine wave and square wave grating patterns. *Kelly* [3.14] used circularly symmetrical Bessel functions, instructing his subjects to fixate at the center of the pattern. *Mitchell* [3.15] used two-dimensional filtered random noise patterns to avoid the fixation difficulty. He also used an electrophysiological measurement to quantify suprathreshold responses.

All these measurements have practical and theoretical difficulties. Nevertheless, there is a remarkable consensus in the results, indicating that the spatial frequency response peaks at about 5 cycles/degree. For lower frequencies the eye acts as a differentiator, being responsive mainly to edges and gradients, while at high spatial frequencies it integrates, blurring fine detail together. Tests made with variable-frequency square waves do not show as much fall-off in response at low frequencies, indicating that large targets are made visible primarily by their edges.

These results suggest an explanation for an otherwise puzzling phenomenon of vision. In the case of images composed of discrete elements such as TV with visible scanning lines, half-tone pictures, mosaic tile images, or even oil paintings with broad brush strokes, viewing from too close causes the imagery to disappear, leaving only the structure visible. We see better by squinting or moving back.

As shown in Fig. 3.14, the spatial frequency of the structure is twice that of the image bandwidth. If the former is at a frequency of higher visibility, it becomes predominant. Only when the sampling frequency is located over the hump of the frequency response curve does the eye tend to integrate, i.e., smooth the samples.

The very nonuniform frequency response implies substantial waveform distortion. A common example is a step wedge such as in Fig. 3.15 in which the illusion that the steps are nonuniform is so strong that the neighboring steps must be covered to demonstrate otherwise.

On the other hand, substantial actual distortion, particularly of sharp edge transitions, is sometimes misinterpreted. For example, the three edge transitions of Fig. 3.16 are all seen as undistorted edges of varying degrees of *contrast* as well as sharpness. In Fig. 3.17, Strip A is seen as brighter than Strip B, while in Fig. 3.18, although Mach bands are visible, the As are seen as lighter than the Bs. This effect is so strong that no amount of willpower can overcome it. (Viewing

---

Fig. 3.13a–c. Tone scale and noise visibility: Since contrast sensitivity varies with local luminance, the visibility of additive noise depends on the tone scale at the point where noise is added. Here we compare the effect of adding noise in (a) the linear (luminance) domain, (b) the quasilog (lightness) domain, and (c) the log (density) domain. In this case, since the shadows have more detail than the highlights, noise is less visible in the shadows, due to masking, in all three pictures

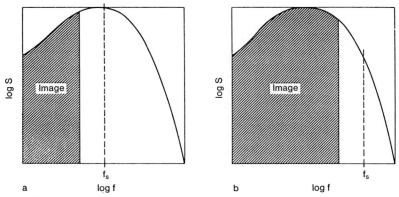

Fig. 3.14a,b. Sensitivity as a function of spatial frequency

Fig. 3.15. An illusion (the steps are physically uniform)

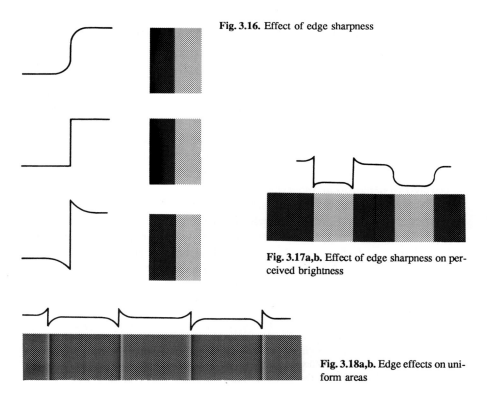

Fig. 3.16. Effect of edge sharpness

Fig. 3.17a,b. Effect of edge sharpness on perceived brightness

Fig. 3.18a,b. Edge effects on uniform areas

through holes so as to cover the boundaries does overcome the illusion.) It is most likely related to the long-known phenomenon of simultaneous contrast, in which the brightness of a patch is greatly affected by the luminance of the surround [3.15].

Accentuation of spatial high frequencies, so as to cause symmetrical overshoot at edges, increases apparent contrast as well as apparent sharpness, both generally desirable. It is therefore one of the commonest methods of image enhancement, having been achieved by a wide variety of methods, including photographic unsharp masking, computer processing [3.16], the Herschel effect [3.17], analog or digital hardware [3.18], and optical filtering [3.19].

Regardless of the technique used, however, a most interesting result is that the quality can often be improved beyond that of the "original" by enhancing the high frequencies to an extent substantially greater than the loss in the transmission process [3.20]. My own conclusion from numerous experiences over a long period of time is that most observers, given the choice, will increase the high frequency contrast of a picture to the point where some defects or artifacts appear. In the absence of other defects a limitation will eventually set in due to the appearance of overshoots at high-contrast edges. The overshoots, sometimes called Mach bands, look like light and dark lines parallel to edges. (They sometimes appear in normal vision [3.2].) A goal of edge enhancement systems is to produce the maximum possible sharpening with the minimum amount of visible defects or overshoot.

In sharpening images, as in minimizing noise visibility, it is important to consider the tone scale of the image, as it is desirable to have the sharpening effect uniform throughout the luminance range. In Fig. 3.19, it is shown that the most uniform sharpening occurs when the signal that is sharpened is in the lightness domain. The effects are rather subtle. In general the log image shows excessive overshoot on the light side of transitions, while the linear image shows them on the dark side. The lightness image has the more nearly uniform effect.

3.3.1 Relationship Between Sharpness and Contrast

A somewhat surprising result of edge sharpening experiments is that the perceived overall contrast of an image is increased when the intensity of spatial high frequencies is increased [3.21]. The perceived difference in brightness of an area in an image from that of its surround increases with the sharpness of the transition. The effect is so strong that it is sometimes possible to make an area that is actually lighter than its surround appear darker by proper manipulation of the transition. The phenomenon, which is another aspect of simultaneous contrast [3.5], is demonstrated in Fig. 3.20, in which the apparent brightness of the central patch is changed by controlling the sharpness of its outline.

A striking example of the ability to increase the apparent contrast, in color as well as brightness, between an object and its surround, can be produced by outlining objects with thin black lines. Even when viewed from a distance at which the lines themselves cannot be directly detected, very large differences in appearance are produced.

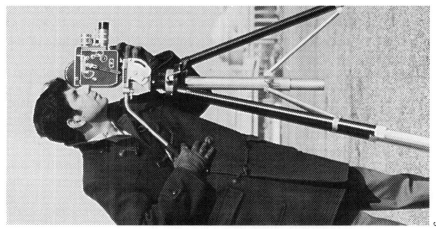

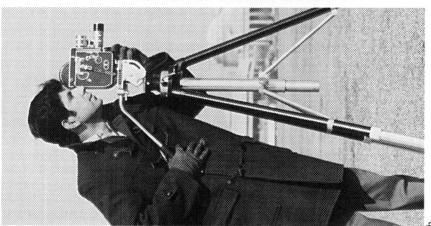

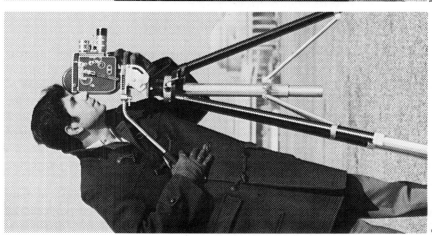

Fig. 3.19a–c. For caption see opposite page

74

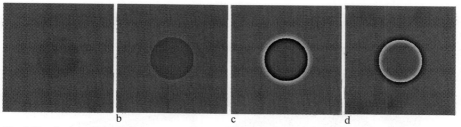

Fig. 3.20a–d. An effect of sharpness on contrast. The central patch is about 10% darker than the surround. The original is at (a), while in (b) and (c), the edge is sharpened moderately and greatly, respectively. In (d), the edge gradient is reversed

Another example of the relationship between apparent contrast and edge rendition occurred in the mid-50s when an edge enhancement process was introduced by the Technicolor Corporation in making color motion picture prints from color negatives. If color positive film is exposed in a projection printer using high f/number optics, the nearly collimated light is refracted by the edge of the relief image in the negative sufficiently to produce very noticeable overshoots on the prints. This greatly improves the apparent sharpness of the images projected in the theater. It so increases the apparent contrast that the prints must deliberately be made with less actual contrast. This has the advantage that the highlight and shadow portions of the image are rendered further away from the "toe" and "shoulder" of the print film D-log E curve, thus actually increasing their incremental contrast [3.22].

The converse effect also takes place. Sharpness depends on contrast as is shown in Fig. 3.21. Pictures of low contrast, especially if they contain no areas of perceptual black and white, appear less sharp, and vice versa. It is interesting to note that nonprofessional observers generally do not distinguish between contrast and sharpness. Instead, they often refer to pictures as clear, distinct, or "snappy," when both aspects are satisfactory. The terms "fuzzy," "muddy," unclear, or blurry, are used nearly interchangeably for unsatisfactory pictures. This interchangeability is well demonstrated by the typical overly high "contrast" setting found on home TV receivers, where apparently viewers attempt to compensate for low sharpness with high contrast.

Experience shows that clarity of seeing depends on both contrast and sharpness. What nontechnical observers notice is that the image is poor, rather than the objective characteristics that make it so.

Fig. 3.19a–c. Tone scale and sharpening: Since contrast sensitivity varies with local luminance, the effect of sharpening depends on the tone scale of the signal at the point of sharpening. Here we compare sharpening in (a) the linear (luminance) domain, (b) the quasilog (lightness) domain, and (c) the log (density) domain

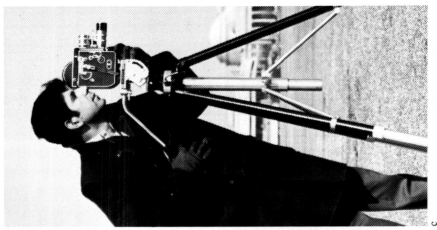

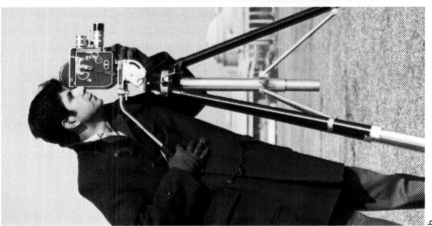

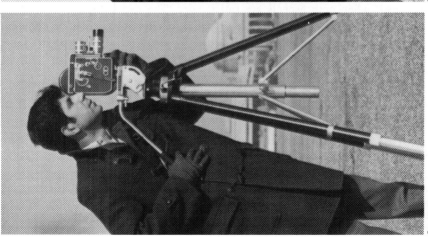

Fig. 3.21a–c. For caption see opposite page

3.4 Masking

Response to visual stimuli, and, in particular, the response of retinal receptors, is markedly influenced by the the presence of other visual stimuli in the immediate spatiotemporal neighborhood. This is not a small effect; in many cases it is the most important aspect of visual perception. Like other characteristics of the sense organs, it is probably a result of evolution, since in most cases it helps us to see better, by separating what is important from what is likely to impede the extraction of knowledge from a scene.

We have already discussed the class of such effects in which incremental sensitivity is increased by the presence of similar stimuli. These are the adaptive phenomena and permit us to see under very widely varying levels of illuminations. Another class of phenomena, called masking, results in reduced sensitivity due to the presence of conditioning stimuli.

An example of masking is the reduced sensitivity to spatial gratings seen shortly after, or even just before, exposure to a grating of similar spatial frequency [3.23]. Likewise, exposure to a temporal sinusoid reduces the response to temporal sinusoids of similar frequency [3.24]. An important example is reduced noise sensitivity in image areas containing a great deal of "activity" in the form of sharp edges or fine detail [3.25]. The resulting reduction of the visibility of random noise in detailed or "busy" areas relative to noise visibility in more-or-less blank areas, as much as 25 dB in some cases [3.26], makes practical many coding schemes, such as DPCM, which confine most of their coding errors to areas of high detail. This phenomenon is also the main reason why SNR, even weighted according to the variation of noise visibility with frequency, is a very poor indicator of image quality. Simple images require a much higher SNR for the same quality than complicated ones.

A similar phenomenon is the masking of detail in a new scene by the presence of the previous scene, a phenomenon first pointed out by *Seyler* [3.27]. Repeating this experiment at MIT by motion picture methods, we found that a new scene could be defocussed radically, without visible effect, if the camera were brought back into focus, after the scene change, with a time constant of about 0.5 s. Glenn has shown that it takes about 0.2 s to perceive higher spatial frequencies in newly revealed areas [3.28]. This effect is of great benefit in temporal differential transmission systems, since it allows new scenes to be built up over a period much longer than one frame.

Fig. 3.21a–c. The effect of contrast on sharpness: These pictures are identical except for a nonlinear transformation that has altered the contrast in an obvious manner. Which seems sharpest?

3.5 Temporal Phenomena

3.5.1 Frequency Response and Flicker

Even though the previous sections have assumed images that were static
time, it is clear that the temporal aspects of the visual response affect
sensitivity and spatial frequency response measurements as well. Not or
adaptation (in time, of course) play such an important role, but the volu
involuntary motion of images on the retina produces time-varying stimul
retinal receptors in the neighborhood of spatial image gradients. This l
tor is not merely incidental. Should the retinal image be stabilized in
by external means, in spite of random ocular tremor, vision disappears
[3.29] in a few seconds! Perception returns if the image is suddenly m
if the luminance level is changed. Even today (1986) the role of eye mov
in normal vision is not completely understood. However, the dramatic p
ena mentioned make it perfectly clear that the visual system is extraor
sensitive to change of any kind, i.e., it often acts as a differentiator.

Classical psychophysics concentrated on flicker measurements. Thus t
(critical flicker or fusion frequency) has been frequently measured as a f
of luminance, duty cycle, and angle of view, most often with sharp-edged
stimuli. The failure of such measurements to lead to a clear understan
flicker led system analysts to propose the use of a stimulus as purely te
and of as simple a waveform as possible. In 1953, *Kelly* [3.12] performed a ser
of experiments using a 65 degree wide "edgeless" uniform field, in which the
stimulus consisted of an average luminance level on which was superimposed
a sinusoidal variation. The threshold modulation was measured as a function
of frequency and of average luminance for a number of observers, for colored
as well as white light. The results are shown in Fig. 1.3 where the threshold
modulation is plotted both relatively and absolutely. An artificial pupil was used
to eliminate the effect of the temporal response of the iris.

The most striking feature of both sets of curves is that for a significant band
of frequencies, the eye is a differentiator, and that for another higher band of
frequencies, it acts like an integrator. Fusion of a sequence of separate luminous
stimuli requires that they occur at a high enough frequency to be integrated,
which, the experiment shows, depends on the luminance level; the higher the
luminance, the higher the repetition frequency must be to avoid flicker. All
commercial motion picture systems, including TV, use flicker rates in this region
– 48 or 72 Hz for films and 50 or 60 Hz for TV. Motion rendition, on the
other hand, presumably requires good response to the entire band of temporal
frequencies to which we are significantly sensitive. Much of this information is
in the differentiation region, which is invariably lower than the flicker frequency.

For frequencies below about 5 Hz, all the curves for photopic light levels
coincide on the relative plot. This is the region of validity of the Weber-Fechner
law. The eye is adapting to the modulation waveform in the course of each cycle

and is therefore sensitive to the fractional modulation. An arrangement in which the gain of an amplifier depends on the input level, but responds to changes with some time constant, can be considered a feedback system. When the response time constant is about half a cycle of the stimulus, the sensitivity changes fall in phase with the input and *raise* the response, which clearly happens in this case. This implies a rather short adaptation time constant – perhaps 0.1 s. If this were the only mechanism involved, the response should remain constant at higher frequencies, but in fact experiments on animals have shown that a high frequency limit exists due to the maximum rate of firing of neurons; hence the overall shape of the curve. Note that the curve at scotopic levels shows little rise in the range where the other curves are differentiators, no doubt because little adaptation takes place at such low light levels. Differentiation is also not shown if sharp-edged targets are used, rather than the edgeless wide field. In that case, the edges probably induce higher frequency stimuli in some receptors because of ocular tremor.

The plot of absolute threshold modulation is a little harder to understand. Note that each curve ends at 100% modulation, at which point the alternating component of stimulus is equal for all curves. These endpoints form a curve that is asymptotic to the line: frequency is proportional to log modulation. This is the same result as the classical CFF experiments. At these frequencies, no adaptation takes place in each cycle, so that the sensed brightness depends on the average luminance only. The state of adaptation evidently governs the flicker sensitivity.

What these curves do not show is that flicker is more noticeable at the periphery of the visual field than in the more central regions, and that the apparent flicker frequency is generally lower than the actual flicker rate. Some additional phenomena concerning flicker were discussed previously in connection with TV interlace systems.

The similarity of the temporal frequency response curves to those previously given for spatial frequency response is quite suggestive. Some attempts have been made to measure a generalized response as a function of spatiotemporal stimulus frequency, but have so far not led to any formulations of special value in the design of image processing systems.

3.5.2 Motion Rendition

Many Americans believe that it was Edison[5] who discovered that the illusion of motion could be produced by viewing a rapid sequence of slightly different images. This is an extension of the "phi motion" of psychology, in which the successive flashing of two small lights with the appropriate time and space separation, makes it appear that the light moves from the first position to the second [3.31]. Should the angular jump be too large or the interval too long between

---

[5] Actually, the phenomenon itself is much older; many others contributed to the invention of motion pictures [3.30].

successive images, or "frames," the motion effect is discontinuous and in some cases can even be retrograde. We have all seen wheels or other moving objects standing still or sometimes moving backward. This stroboscopic effect, which is often very useful, is an example of temporal aliasing. Like other kinds of aliasing, it is but one possible defect that should be traded off against others for optimum image quality. The smoothness of motion is directly related to filling the gaps between successive positions. The degree of temporal and/or spatial band-limiting that can absolutely preclude temporal aliasing has the effect of blurring moving objects [3.32]. Especially in low frame-rate systems, it may be preferable to show a sequence of sharp still images rather than a continuously moving image so blurred as to be useless.

Careful observation shows that motion is generally smoother in TV than in motion pictures. This is because the TV system actually comprises 60 pictures per second, as compared to 24 for film. In addition, most TV cameras integrate for the full 1/60 s while all motion picture cameras use exposure times of less than (and sometimes very much less than) 1/24 s.

Objects moving across the retina while our eyes are fixated elsewhere are blurred by the temporal upper frequency limit of the HVS. The same thing happens in TV cameras, which is perfectly all right unless the observer happens to be tracking the object in question, in which case the TV (or motion picture) representation will be disappointing. In fact, there is no simple way the TV camera can satisfy the entire audience when the scene contains two or more important moving objects.

A fairly plausible explanation of the perceived smooth motion that sometimes results from the presentation of a sequence of distinct, sharp, displaced images can be devised on the basis of spectral considerations [3.33]. Using the nomenclature of Chap. 2, consider an image, $v(x, y)$, fixed in time, which has a two-dimensional Fourier spectrum $V(\omega_x, \omega_y)$. The three-dimensional transform of this fixed image is an impulse "sheet" on the plane $\omega_t = 0$, and is zero elsewhere. If, however, the image is moving in the positive x direction with a velocity of v units of length per second, the three-dimensional spectrum can be shown to have its nonzero values along the plane $\omega_t = -v\omega_x$.

A series of sharp images that are displaced in time can be thought of as the product of the continuously moving image just mentioned with a temporal sampling signal consisting of an impulse train with periodicity T seconds, where $1/T$ is the frame rate. The resulting spectrum is the convolution of the spectrum of the continuously moving image with that of the sampling pulse train. This action produces replicas of the impulse-sheet spectrum displaced in temporal frequency by multiples of $2\pi/T$ rad/s. These spectra are shown in Fig. 3.22.

Obviously, if we were to perceive only the baseband spectrum, the effect would be exactly as if we had looked at the smoothly moving object. In the sampled case, which applies to motion pictures and television, the perception must be the same as if we saw the original if and only if the sidebands of the baseband signal are not seen. This can come about if (a) they are outside the range of human spatiotemporal frequency response, or (b) if the sampled

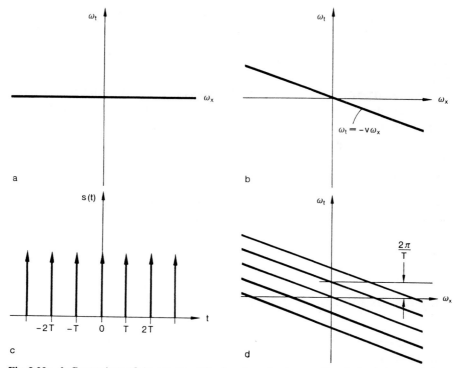

Fig. 3.22a–d. Comparison of the spectra of fixed and moving images. For clarity, it is assumed that the image has no y variation: (**a**) spectrum of a fixed image, (**b**) spectrum of a moving image, (**c**) temporal sampling pattern, (**d**) spectrum of the sampled moving image

signal is passed through a three-dimensional filter that removes the sidebands that otherwise would be perceptible. Case (a) is the situation in which the frame rate is sufficiently high with respect to the angular velocity of motion. Case (b) is one where, without the filter, the motion would be more or less discontinuous, and with a rectangular filter the sharpness of the moving objects would be more or less reduced. It is conceivable, however, that a nonrectangular, nonseparable filter that just separated the baseband spectrum from the sidebands could preserve the sharpness of moving objects without at the same time reducing their sharpness.

The example just given, in which the entire image is in translation, corresponds to camera motion, or "panning." In many cases, the image consists of a stationary background with a number of objects in the foreground moving with different velocities. Obviously, filtering for optimum reproduction of all these objects is quite complicated, but perhaps very rewarding.

Another important case is when the viewer is tracking an object moving in the scene. Here, the "stationary" background is in motion with respect to the retina, while the "moving" portion is more or less stationary. Smooth rendition of the background in this case requires the same conditions that produce smooth rendition of moving objects when the gaze is stationary. As far as the foreground is concerned, an important question is just how long it takes to lock onto an object in motion and how accurately we can immobilize the image of a moving object on the retina. This aspect of visual performance directly determines the required sharpness of reproduction of objects in motion relative to those that are stationary.

What emerges from this brief discussion is that motion rendition is more complicated than it seems. With standard systems, lacking sophisticated adaptive filters, there is no practical frame rate high enough to give flawless motion rendition for very rapid motion, especially when the viewer tracks the moving objects. In addition to the theoretical aspects of motion rendition, what is achieved in particular cases may depend in an important way on details of the equipment, such as integration time of the camera, and lag, or unwanted carryover of information from one frame to the next. In general we would expect TV motion rendition to be much smoother than that of motion pictures because of the higher frame rate and 100% exposure duty cycle, but in many cases, lag causes excessive blurring of moving objects.

3.6 Lessons for the System Designer

This brief review was intended to help system designers create imaging systems that are in consonance with, if they do not actually take advantage of, the main features of human visual perception. It should be obvious that we have not presented every piece of information the designer needs. This is partly the result of limited time and space (and the limited energy of the author) but, in significant ways, because of lack of knowledge. Visual research has not in the past been carried out primarily to provide information for system designers. We therefore do not have all the facts we would like. For example, the interaction between spatial and temporal frequency response is not known very well, nor are the limits of the applicability of the frequency response concept itself, in view of the pronounced nonlinearity of the human visual system. Visual acuity in the tracking mode seems not to have been adequately studied. The phenomenon of masking, which is of obvious importance to coding systems, is understood only sketchily, as are the various aspects of motion perception. Nevertheless, enough is known to design systems with much more uniform noise visibility, a key factor in using channel capacity efficiently. The general principles for trading off sharpness against aliasing are known, and the relationship between sharpness and contrast is understood, albeit qualitatively.

Unfortunately, we do not know enough to establish objective criteria for quality, and this limits the designer in many ways. It means that, at the design stage, there is always some uncertainty as to how well a proposed system will work. Subjective testing is therefore required to quantify system performance, and this must be done with typical subject matter, typical audiences, and typical viewing conditions.

A subject that merits additional study, and that must be based on the subject matter of this chapter, is enhancement of still and moving pictures. Although edge sharpening is commonplace in graphic arts and "aperture correction" often used in television, the usual methods are far from taking advantage of current knowledge of perception. For maximum effectiveness, additional perceptual studies are needed, especially related to masking and to motion rendition and related temporal phenomena.

4. Sampling, Interpolation, and Quantization

This chapter deals with the conversion between real (i.e., spatially and temporally continuous) images and their discrete representation. The mathematics of the conversion processes is discussed and is found to be straightforward. The introduction of perceptual considerations makes the subject more complex, but also more rewarding. It is shown that substantial improvements in image quality, for a given amount of digital data, are possible using perceptual, rather than mathematical, criteria, particularly in the choice of presampling filters and postsampling interpolators.Similarly, perceptual considerations can be brought to bear on the design of quantizers. Appropriate placement of quantization levels and the use of randomization techniques can significantly reduce the visibility of quantization noise.

4.1 Introduction

The discrete representation of continuous images is required whenever digital transmission or processing of natural pictures is carried out. While two- or three-dimensional arrays of numbers are sometimes called images, and some of the tools of image processing can be applied to such arrays, in the vast majority of practical applications both input and output are spatially continuous. While it is also true that it is possible to look at – even enjoy – discrete optical arrays, this is generally done only in the context of an art form such as mosaic tile work or, in more mundane situations, where we tolerate the structure better than the loss of sharpness that accompanies attempts to obliterate the structure.

In previous chapters, we discussed in a general way the operations necessary to represent a spatially continuous image by a set of numbers so that processing can be carried out digitally. It was pointed out that all TV and facsimile systems that employ scanning do, in fact, sample the image in the vertical direction, whether they are considered to be analog or digital. We now shall explore these operations in more depth.

In the most general sense, any discrete representation of an image can be thought of as the result of sampling and quantization. In cases where there is no reason to economize on the amount of data in the representation, many methods prove equally satisfactory. Usually, however, there is reason to try to obtain the best possible picture quality with the least amount of data.This is also the main objective of coding systems. As we shall see, there is no "natural" way to discretize an image. There are only better and worse ways. The difference between a "true" coding system and a well engineered but otherwise straightforward digitizer is therefore not qualitative, but quantitative. These are not fundamentally different kinds of systems, but are simply at opposite ends of a scale that measures cost, complexity, and effectiveness.

A "solution" to this problem is provided by the sampling theorem [4.1], which states that a signal with a bandwidth of B cycles per unit length can be "exactly" reconstructed from $2B$ samples per unit length. This can readily be extended to serve as the theoretical justification for discrete image representation. "Exact" in this case means in the Fourier sense, i.e., the method minimizes the rms error between input and output. As a practical matter, signals cannot be precisely confined to the bandwidth B. The rms error is then simply the total energy in the original signal at frequencies above B.

4.2 The Sampling Theorem

A straightforward proof of the sampling theorem will be given shortly, but its reasonableness may be seen from the following considerations: The spectrum of the product of two signals is $1/2\pi$ times the convolution of their separate spectra. Consider any sampling function $v_s(t)$, periodic in T, generally but not necessarily a train of narrow pulses. It may be expanded in a Fourier series.

$$v_s(t) = \sum_{k=-\infty}^{\infty} V_{sk} \exp\left(jk\frac{2\pi}{T}t\right) .$$

The sampling function therefore has a spectrum consisting of lines at harmonics of the sampling frequency $1/T$. If we now multiply a message function $v_m(t)$ that has a spectrum $V_m(\omega)$ by a sampling wave, $v_s(t)$,[1] the spectrum of the sampled message, v, is the convolution of the two spectra. That is, if

$$v(t) = v_m(t)v_s(t) , \quad \text{then} \quad 2\pi V(\omega) = V_m(\omega)^* V_s(\omega) .$$

From Figs. 4.1–3, it is evident that the final spectrum consists of a multiplicity of message spectra, centered on the harmonics of the sampling wave. Evidently the message can be recovered without error by filtering out any one band. If the band about zero frequency is taken, the signal is recovered directly, while if the band about any other harmonic is selected, the message appears as double sideband amplitude modulation.[2]

---

[1] Sampling waves are usually assumed to be impulses for purposes of analysis, although they are always pulses of nonzero width in real cases. For the sake of a simpler discussion, some complexities of sampling with such pulses are delayed until consideration of Fig. 4.4.

[2] Since $v(x) > 0$, there is always a dc component in an image and therefore a carrier at each harmonic of the sampling frequency.

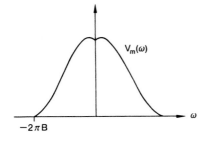

Fig. 4.1. Message spectrum

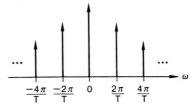

Fig. 4.2. Sampling spectrum

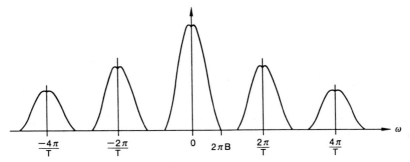

Fig. 4.3. Spectrum of sampled message

To avoid confusion (aliasing) the various spectral bands must not overlap and therefore

$$\frac{2\pi}{T} > 2 \times 2\pi B , \quad \text{or} \quad \frac{1}{T} = f_s > 2B .$$

We have not yet really proved anything, since this demonstration is related to a particular method of demodulation. Now consider the general function of limited spectrum, following *Oliver* et al. [4.2]:

$$v(t) = \int_{-2\pi B}^{2\pi B} V(\omega) e^{j\omega t} \frac{d\omega}{2\pi} .$$

Within the band, $V(\omega)$ can be represented by a Fourier series so that

$$v(t) = \int_{-2\pi B}^{2\pi B} \sum_{n=-\infty}^{\infty} a_n \exp\left(j\frac{n\omega}{2B}\right) e^{j\omega t} \frac{d\omega}{2\pi} ,$$

$$v(t) = \sum_{n=-\infty}^{\infty} 2B a_n \frac{\sin 2\pi B(t + n/2B)}{2\pi B(t + n/2B)} .$$

Hence the function $v(t)$ is equal to the sum of a train of sinc pulses centered at the sampling times, each pulse having an amplitude $2B a_n$. Now

86

$$a_n = \frac{1}{4\pi B} \int_{-2\pi B}^{2\pi B} V(\omega) \exp\left(-j\frac{n\omega}{2B}\right) d\omega \;,$$

which by comparison with the inverse transform of $V(\omega)$ is simply

$$a_n = \frac{1}{2B} v\left(-\frac{n}{2B}\right) \;.$$

Substituting above and letting $m = -n$, we have

$$v(t) = \sum_{m=-\infty}^{\infty} v\left(\frac{m}{2B}\right) \frac{\sin 2\pi B(t - m/2B)}{2\pi B(t - m/2B)}$$

$$= \sum_{m=-\infty}^{\infty} v\left(\frac{m}{2B}\right) \operatorname{sinc}(2Bt - m) \;.$$

Thus the amplitude of each sinc pulse is simply the value of the function at the sampling point.

Noting that sinc $(2Bt)$ is the impulse response of an ideal low-pass filter of bandwidth B, we see that a band-limited function can be thought of as a train of sinc pulses. Each such pulse is zero at the center of all the others. Hence the value of the sum of the pulse train at each sample point is independent of the value of all the other samples.

The "original" function can be reconstructed either by creating a train of suitable sinc pulses or by passing the signal, sampled with an impulse train, through an ideal low-pass filter. This latter procedure is confirmed by examination of the spectrum of the sampled signal.

This proof of the sampling theorem not only gives insight as to why a continuous function can be represented by samples, but also furnishes a procedure for sampling and reconstruction (Fig. 4.4). The signal is first band-limited, then multiplied by an impulse train $v_s(t)$.

$$v_s(t) = \sum_{m=-\infty}^{\infty} u_0(t - (m/2B))$$

$$v_s(t)v_2(t) = \sum_{m=-\infty}^{\infty} v_2(m/2B)u_0(t - (m/2B)) \;.$$

Since the impulse response of the filter F_2 is given by

$$h(t) = \operatorname{sinc}(2Bt)$$

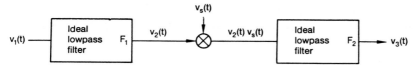

Fig. 4.4. Sampling and reconstruction

the output is

$$v_3(t) = \sum_{m=-\infty}^{\infty} v_2(m/2B)\text{sinc}(2Bt - m) \,.$$

Thus $v_3(t)$ is the band-limited version of $v_1(t)$.

An alternative procedure for obtaining the sample values, $v_2(m/2B)$, of the band-limited function directly from the wideband signal $v_1(t)$ is as follows:

$$v_2(m/2B) = \int_{-\infty}^{\infty} v_1(m/2B - \tau)h(\tau)d\tau \,.$$

This is often easier. One simply convolves v_1 with the impulse response of F_1, and evaluates the result at the sampling points. This is equivalent to using a physical aperture of nonzero extent. Note that each sample value thus depends on the image intensity values over a nonnegligible area around the pel. In a digital processing system, the sample values, suitably quantized, form the numerical representation of the image.

The difficulty with the "solution" provided by the sampling theorem is that the ringing (Gibbs phenomenon) associated with the sharp bandwidth limitation is totally unacceptable. Every sharp luminance transition is surrounded by ringing characteristic of the step response of the ideal low-pass filter. Of course, if the original picture has no sharp transitions there is no ringing, but in that case, we have the intuitive feeling that the original bandwidth must have been smaller and therefore it should have been possible to use fewer samples per unit area.

The scheme of Fig. 4.4 can be varied by using different presampling and interpolation filters. In general, filters that avoid ringing produce images of lower sharpness and resolution. Practitioners of discrete image transmission have evolved a variety of techniques to deal with the problem [4.3]. It has been found that the sampling structure can be reduced in visibility using a variety of interpolation filters, including sample-and-hold, somewhat smoothed sample-and-hold, linear interpolation, raised cosine impulse response, etc. All trade off sampling structure ("blockiness") and aliasing vs. sharpness. Less attention has been given to the presampling aperture, although it is known that too narrow an aperture gives rise to moiré patterns in the case of pictures with high contrast periodic detail.

When filtering is implemented with the usual type of Gaussian aperture, it is found experimentally that the reconstruction aperture should be large enough so as to fill in the space between lines or points (in analog systems the aperture should be as small as possible in the direction of scanning, since aliasing does not occur) while the scanning aperture should be about half as large. Further insight into these trade-offs can be gained by examining the spectrum of the sampled signal in more detail.

4.3 The Two-Dimensional Spectrum

The theorem can readily be extended to two dimensions. If a signal is limited in bandwidth to B_x in the x direction and B_y in the y direction, then it can be shown to be equal to

$$v(x,y) = \sum_{m=-\infty}^{\infty} \sum_{n=-\infty}^{\infty} v\left(\frac{m}{2B_x}, \frac{n}{2B_y}\right) \text{sinc}(2B_x x - m)\text{sinc}(2B_y y - n) .$$

As in the one-dimensional case, the spectrum of the sampled image is found by convolving the spectrum of the original image with that of the sampling waveform. Since the latter is periodic in x and y, it has an impulsive spectrum periodic in ω_x and ω_y. As shown in Fig. 4.5, the convolution causes a replica of the image spectrum to be centered on each spatial harmonic of the sampling signal. Aliasing is avoided, permitting "exact" reconstruction, if the image spectrum is band-limited so that

$$B_x \le 1/(2\Delta_x), \quad \text{and} \quad B_y \le 1/(2\Delta_y) .$$

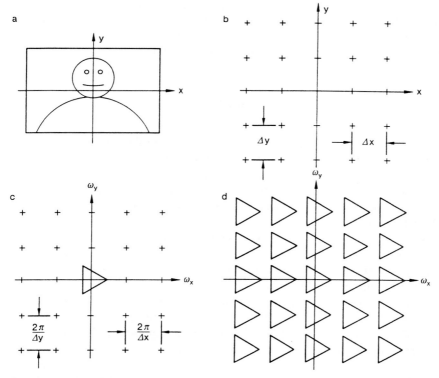

Fig. 4.5a–d. Spectrum of a sampled image: (**a**) an image; (**b**) a sampling signal; (**c**) the spectra of each; (**d**) the spectrum of the sampled image

Improvements in operation are possible by varying the presampling and interpolation filters.

The desirable features of a filtering/sampling system are not hard to describe. For a given sampling density, the highest sharpness is desired, along with minimum visibility of the sampling structure. At the same time, image artifacts including stair-step rendition of diagonal edges ("jaggies"), ringing, and moiré patterns, should be minimized.

Since each of these properties is related to a number of different physical parameters of the system, it is not easy to make the best trade-offs. Sharpness requires high response within the allowed spatial frequency band, while the elimination of moire patterns due to aliasing requires small response outside the band. The implied sharp cutoff leads to overshoots and ringing. Smooth rendition of edges implies an isotropic or circularly symmetrical filter, but this is inconsistent with the usual Cartesian sampling matrix. It is also possible to use a hexagonal sampling matrix, as discussed below.

It is clear that it is impossible to have maximum sharpness with neither aliasing nor ringing, all at the same time. Since the latter is least acceptable, some aliasing and loss of sharpness must be accepted by using a more gradual roll-off in the filters. The resulting distortion cannot be minimized by using mathematical concepts alone. Since image quality is subjective, the effects can only be judged using visual criteria. In the following sections, we shall reconsider the shape of the sampling and interpolation apertures, using both spatial domain and frequency domain analyses to improve our insight.

4.4 Interpolation of Digital Images

It is easier to see the application of visual concepts to the total problem if we break it into two parts and attack the second part first. Assume a set of discrete samples from which it is desired to interpolate either to a continuous image or to a second discrete image with so high a sampling density that interpolation is noncritical. This is a valid problem in its own right in the case of enlargement of digital pictures.

Let us arbitrarily limit our technique to linear filtering of the samples. The usual interpolation apertures are unity at the samples, are zero at the neighboring samples, and have mirror symmetry about the 50% point so as to sum to a constant for a uniform field of samples.

Square (sample and hold), linear (or bilinear if two dimensions), raised cosine, and cubic spline [4.4] interpolation functions are the favorites of digital processors, while the Gaussian is often found in optical equipment [4.5]. In Fig. 4.6 these are compared in the space domain, frequency domain, and by sample pictures, with the "ideal" low-pass filter (ILPF). In order to avoid the ringing of the ILPF, all the others have slower roll-offs in frequency. They all have lower sharpness and varying degrees of aliasing and visibility of sampling structure.

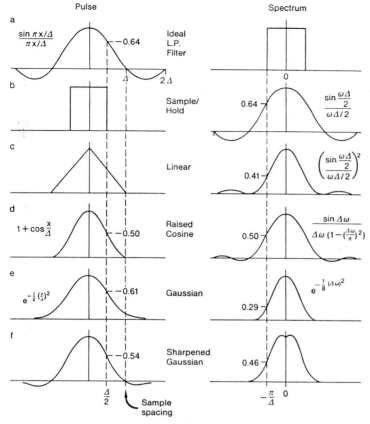

Fig. 4.6a–f. A comparison of interpolation functions: (**a**) ideal low-pass filter; (**b**) sample-and-hold; (**c**) linear; (**d**) raised cosine; (**e**) Gaussian; (**f**) sharpened Gaussian

It should be noted that much subject matter does not have strong periodic components at high spatial frequencies, and hence is not troubled very much by aliasing. However, sharp edges, especially if nearly horizontal or vertical, do exhibit aliasing as "staircase" distortion. The complete elimination of this effect without much loss of sharpness requires such a sharp-cutting filter as to result in ringing, which is almost always undesirable.

One index of sharpness is intersymbol interference, i.e., the "spillover" of information from one sample to the other. Clearly it should be reduced if possible, but it is only one of the elements that must be traded off to achieve the best result. It is also desirable to have minimum visibility of sampling structure, which requires low response at the sampling frequency. In the space domain this means at least that the pulse should sum approximately to a constant in a uniform field. Circular symmetry tends to give equal rendition of edges at all angles. High sharpness demands a low level of intersymbol interference and good response in the passband. Separability helps the computation problem.

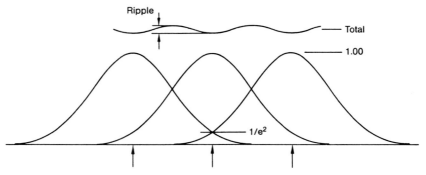

Fig. 4.7. The Gaussian interpolation function

From the visual viewpoint, it is quite permissible to have some waveform distortion, so long as it is compatible with normal visual processes, especially if this permits better performance in other factors. Although the ringing of a low-pass filter is destructive of image quality, single properly shaped and proportioned overshoots at edges can actually improve the quality. This result is facilitated by filtering in the lightness domain so that overshoots are equally perceptible at all luminance levels.

An interpolation function of this nature is the sharpened Gaussian [4.6]. A specific example illustrates its operation. As shown in Fig. 4.7, if the width of the basic Gaussian interpolation pulse is such that its amplitude is e^{-2} at the neighboring samples, the intersymbol interference is 13.5%, the sampling frequency ripple only 4.3% peak-to-peak[3] and the aliasing less than that of the bilinear filter. Thus this function, which is both circularly symmetrical and separable, produces smooth images, although of rather low sharpness.

The intersymbol interference in one dimension can be reduced to zero by prefiltering with a sharpening filter of the form

$$\delta(t) - e^{-2}[\delta(t - \Delta) + \delta(t + \Delta)]$$

whose frequency response is

$$\frac{1 - 2e^{-2}\cos\omega\Delta}{1 - 2e^{-2}} \ .$$

This is equivalent to replacing the single Gaussian with the sum of three spaced Gaussian pulses. The impulse response and step response are shown in Fig. 4.8, and sample pictures in Fig. 4.9.

---

[3] For any reasonable filter, the sampling structure can be removed completely in blank image areas by small perturbations of the impulse response of the interpolation filter, designed to cause the interpolated image in a field of uniform samples to be a constant. This process is called "normalization" [4.7].

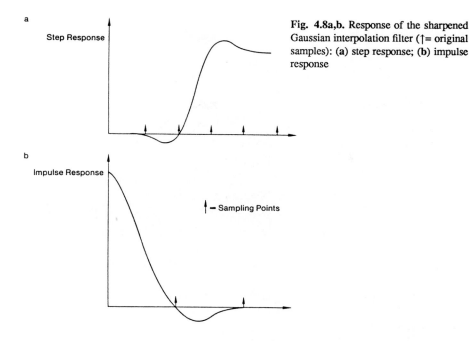

a Step Response

b Impulse Response

↑ – Sampling Points

Fig. 4.8a,b. Response of the sharpened Gaussian interpolation filter (↑= original samples): (a) step response; (b) impulse response

In two dimensions, the sharpened Gaussian is not quite circularly symmetrical, since it has four "dimples" at the neighboring samples. Nevertheless, it clearly is superior to any of the others.

4.5 The Presampling Filter

Neglecting sampling for the moment, if we consider the cascade of a low-pass filter, a sharpening filter, and a low-pass interpolation filter, they may be commuted since the system is linear. The new spectrum is the product of the image spectrum with that of the cascade of the three filters. Thus there is no reason on this basis for the input filter to be different from the output. Since we have already found that the Gaussian makes a good output filter, it follows that the input filter should be the same. The convolution of two Gaussians with each other is Gaussian with $\sqrt{2}$ times the width. Removing the intersymbol interference with such a wide Gaussian results in higher overshoots. It is probably desirable, therefore, to reduce the aperture widths a little, producing some sampling frequency ripple, to accept some intersymbol interference, and to filter in the lightness domain, in order to maintain the overshoot at a tolerable level.

The description above adequately describes the effect of the filters on the baseband response, and suggests which parameters can be traded off for best results. If we now add the effect of sampling, we see that additional spectral components are produced that overlap the baseband.

93

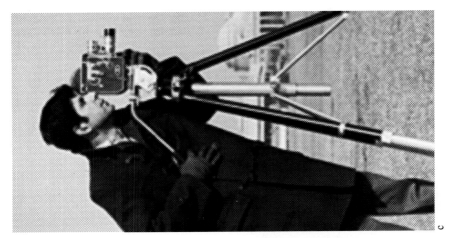

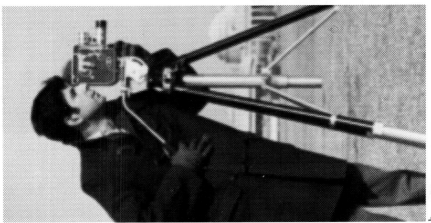

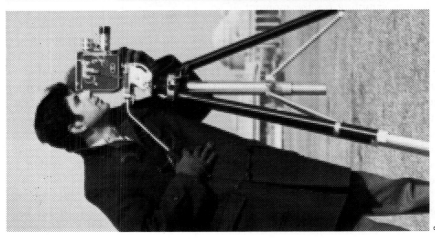

Fig. 4.9a–c. For caption see opposite page

It is the function of the interpolation filter alone to remove the visible sampling structure. Thus it should have a very low response at $\omega = (2\pi/\Delta)$, but should not unduly disturb the baseband response. The sharpened Gaussian has a very low response in the region of the sampling frequency, which is perhaps why it works so well.

Spectral overlap in the region of half the sampling frequency is responsible for aliasing. Since the power density at these frequencies depends on the original image spectrum as well as the filters, some subjects, especially those containing high spatial frequency repetitive components of high contrast, show strong moiré patterns, while most continuous tone images do not. Experience shows that over the range of subject matter found in news pictures, and without the sharpening technique mentioned above, the presampling aperture should be about half the diameter (twice the bandwidth) of the interpolation aperture. Some success has been achieved by *Roetling* with removing the occasional moiré patterns by adaptive techniques [4.8].

4.6 Hexagonal Sampling

In the search for isotropy and circular symmetry in sampling and interpolation filters, it is worthwhile noting that the retinal receptors of most animals are arranged in a hexagonal matrix. This may very well be a result of geometrical rather than visual imperatives, since the hexagonal array gives the highest packing density of circular objects. Hexagonal sampling [4.9] may be achieved by interleaving sampling points on successive lines with the appropriate relative spacing. Referring to Fig. 4.10, when $d_y = d_x$ we obtain the 45° arrangement used in the monochrome halftone process. This is preferred both because the 45° dot pattern is less visible and because better rendition of nearly vertical and horizontal lines, so common in man-made objects, is obtained. When $d_x = \tan 30° d_y$ we obtain the hexagonal grid, as shown in Fig. 4.11. For the same number of samples per unit area as the square grid, $d = \Delta/\cos 30° = 1.075\Delta$. The minimum separation of spatial harmonics is $2\pi/d \cos 30° = 1.075(2\pi/\Delta)$. Thus for a given degree of intersymbol interference, a circular presampling aperture can be 7.5% wider. Such an aperture has a 7.5% narrower spectrum, but the closest harmonics are 7.5% farther away, which gives substantially less aliasing. For example, for 13.5% intersymbol interference with a Gaussian pulse, the amplitude of the spectrum at one half the sampling frequency is 20%, rather than 29% as in the Cartesian grid. A sharpened Gaussian in the hexagonal matrix would

---

Fig. 4.9a–c. Comparison of several interpolation filters. The original image 512 × 968 pels, was prefiltered, subsampled by a factor of four, and then interpolated up to the original size. A given presampling filter was used in each case: (a) sample-and-hold, (b) bilinear, (c) Gaussian. In (c), a degree of sharpening was applied to the subsampled image

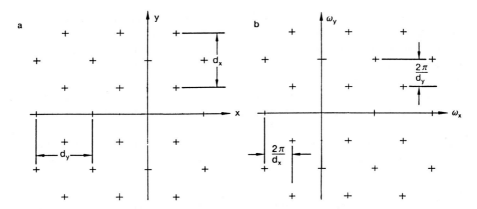

Fig. 4.10a,b. Interleaved sampling: (a) the sampling pattern, (b) its spectrum

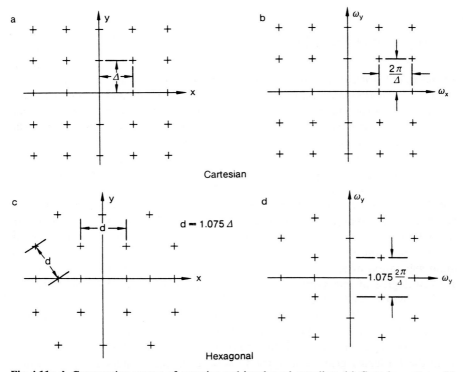

Fig. 4.11a-d. Comparative spectra of cartesian and interleaved sampling: (a) Cartesian pattern; (b) Cartesian spectrum; (c) hexagonal pattern; (d) hexagonal spectrum

96

have six equally spaced "dimples" rather than four, and so be significantly more isotropic.

In spite of this mathematical prediction of the advantage of hexagonal over Cartesian sampling, the experiments that have been carried out [4.3] indicate no marked improvement. In the case of sampled typography, the Cartesian raster seems to be superior. For a given interpolation function, the most significant index of quality yet found is simply the number of samples per unit area.

4.7 Quantization

The result of sampling a filtered image is a train of narrow pulses. The amplitudes of these pulses constitute a sequence of numbers that is the discrete representation of the image. The total amount of data in this representation is the number of samples per image (or per unit area) times the number of binary digits per sample. As pointed out previously, dictates of cost require using the least number of bits per sample. This gives rise to quantization noise having the appearance of visible false contours if fewer than about eight bits per sample are used.[4] The required number falls to between six and seven if the luminance signal is converted to the lightness scale by a suitable nonlinear transformation.

Any method that reduces the visibility of quantizing noise and therefore permits the use of fewer bits per sample is a legitimate way to reduce the total required channel capacity for a picture. Two methods in addition to the optimum distribution in the tone range have been shown to work.

4.7.1 Amplitude of the Quantizing Noise

The mean square value of quantizing noise is easily calculated, by reference to Fig. 4.12, if the assumption is made that the amplitude probability distribution is constant across each quantization step. In that case all errors between $d/2$ and $-d/2$ are equally probable.

Since

$$\int_{-\infty}^{\infty} P(e)de = 1 \quad \text{and}$$

$$P(e) = \frac{1}{d}, \quad \frac{-d}{2} < e < \frac{d}{2}$$

$$P(e) = 0, \quad \text{otherwise, then}$$

$$\overline{e^2} = \int_{-d/2}^{d/2} \frac{e^2}{d} de = \frac{d^2}{12} .$$

---

[4] This figure may be as low as 4 or as high as 11, depending on subject matter, signal-to-noise ratio, and sampling density.

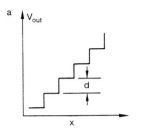

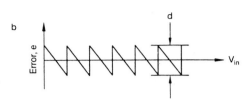

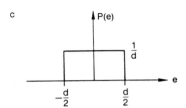

Fig. 4.12a-c. Calculation of quantization noise: (a) quantizer characteristic; (b) quantization error; (c) probability distribution of quantization error

This can also be calculated on a time average basis by considering a ramp input signal, $e = kt$, as representative of signals with uniform probability distributions. The error within the period centered at the origin is then $e = -td/T$. Thus

$$\langle e^2 \rangle = \overline{e^2} = \frac{1}{T} \int_{-T/2}^{T/2} \frac{t^2 d^2}{T^2}\, dt = \frac{d^2}{12} \ .$$

Since the maximum signal is $2^n d$ in the case of n bits per sample, the ratio of peak signal to rms noise is $2^n / \sqrt{12}$. The following table shows the SNR for the usual range of n. [SNR, dB $= 20 \log_{10}(\text{SNR})$]

n	SNR	SNR, dB
2	13.9	22.8
3	27.7	28.9
4	55.4	34.9
5	110.9	40.9
6	221.7	46.9
7	443.4	52.9
8	886.8	59.0

Thus, nearly 60 dB SNR is required to make quantizing noise invisible on the luminance scale and about 45 to 50 dB on the lightness scale. Substantially higher levels are invisible in the case of purely random noise.

4.7.2 Randomization of the Quantization Noise

A method for converting quantizing noise to purely random noise of essentially the same rms value was discovered by *Roberts* [4.10]. This method works so

Fig. 4.13. A pseudorandom noise generator

well, is so easy to implement, and has so few disadvantages, that it is hard to see why it is not universally used. All that is required is to add to the signal, before quantizing, a random noise of uniform amplitude probability distribution and of peak-to-peak amplitude equal to one quantization step. Both the original signal and the noise can be analog, or the entire operation can be carried out digitally. For example, if a signal quantized to ten bits is to be represented by only three, the noise can be seven bits. Since it is inconvenient to transmit the noise, apparently random (but actually deterministic) sequences that are long enough so as not to repeat in a noticeably short distance, are usually used. These sequences can be generated by a shift register with feedback [4.11], as shown in Fig. 4.13.

The result of this kind of noise processing is precisely the same as if a pseudorandom noise of uniform distribution and of peak-to-peak amplitude equal to one quantum step were added to the unquantized signal. Thus the false contours are not merely masked. They are eliminated both in theory and in fact, and replaced by an equal amount of purely random noise. This can be shown heuristically by considering Fig. 4.14, where a greatly enlarged diagram of a portion of a quantizer is shown. An input of value V would normally produce the output V_{k+1}. The addition of the noise causes the input to the quantizer to vary between $V + d/2$ and $V - d/2$. The output of the quantizer is then sometimes V_k and sometimes V_{k+1}. When the noise is subtracted, the output can have any value between $V + d/2$ and $V - d/2$, with equal probability and with an average value of V. Since the same result is achieved for any value of V within the range, the effect is to produce an output signal of the same average value as the input, with a superimposed noise equal in rms value to the quantizing noise. False contours are completely eliminated and in any uniform area of the original, the average output is an unbiased estimate of the input.

Input plus noise between c and s is quantized to V_{k+1} and rendered between j and g after subtraction of noise. Values between b and c are also quantized to V_k but are rendered between g and h. Values between a and b are quantized to V_k and rendered between e and j.

To avoid clipping of the first and last step, the entire signal must be reduced in amplitude by one part in 2^n, resulting in some overall decrease of SNR, depending on n.

If only the input noise is used, the false contours are spatially shifted in a random manner, but the output still consists of only 2^n levels. If only the output noise is used, the false contours are somewhat masked, and the overall noise level is increased. If the input and output noises are not identical, the noise level is increased and the contours are obscured but not eliminated. If the input and

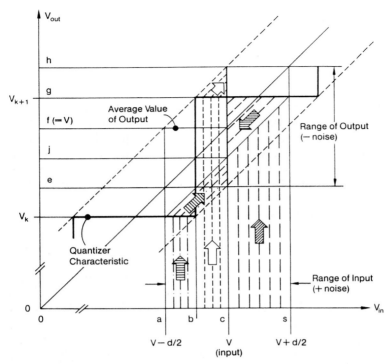

Fig. 4.14. A heuristic explanation of Roberts' method

output noise are identical but too small, the contours are partially removed. If the noises are identical but too high, the contours are totally removed and the noise level is raised. If the original picture is noisy, the contours are less visible in any case.

An example of Roberts' noise processing is shown in Fig. 4.15. It can readily be seen that at two or three bits per sample, quantizing noise is ordinarily very visible but it can be eliminated completely and replaced by random noise by this method. In Fig. 4.16, another important effect is shown. The visibility of both quantizing noise and purely random noise depends markedly on the subject matter. With the printing method used in this book, 6 bits per sample suffice to eliminate quantizing noise for any picture. At 3 bits per sample, the computer-generated image of a sphere shows prominent quantizing effects that can be eliminated by randomization. The crowd scene, however, shows no quantizing at all at 3 bits per sample.

Fig. 4.15a-d. An example of Roberts' pseudorandom noise processing: **(a)** 2 bits, no noise, **(b)** 3 bits, no noise, **(c)** 2 bits, PRN, **(d)** 3 bits, PRN

4.7.3 Filtering the Noise Spectrum

If noise, quantizing or otherwise, is added to a signal in transmission or processing, its visibility may usually be decreased by pre- and postfiltering, in addition to that required by the sampling procedure. There are two reasons for this. One is the spatial frequency response of the observer and the other is the characteristic falloff of the video spectrum at higher spatial frequencies. The filters achieve their effect by shifting some of the noise from the middle spatial frequencies, where it is most noticeable, to higher and lower spatial frequencies, where it is less visible [4.12]. A decrease of perceptible noise corresponding to an increase of 1.5 to 2 bits per sample results. A limitation of this technique results from the tendency of the peak-to-peak amplitude of signals to increase when the high frequencies are emphasized.

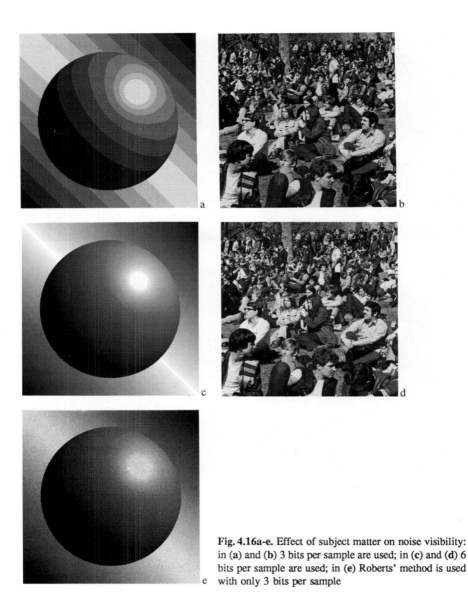

Fig. 4.16a-e. Effect of subject matter on noise visibility: in (a) and (b) 3 bits per sample are used; in (c) and (d) 6 bits per sample are used; in (e) Roberts' method is used with only 3 bits per sample

4.7.4 Combination of Filtering and Randomization

Roberts' original work included a companding step in addition to PRN. *Wacks* [4.13] studied the possibility of using filtering in addition. A complication arises in that the preemphasis of spatial high frequencies not only increases the dynamic range, as mentioned above, but also changes the probability distribution of the input to the quantizer. This would increase the quantization noise and decrease the efficiency with which the Roberts technique transforms it into random noise. This difficulty can be partially corrected by using additional nonlinear transformations before and after the quantizer. While this does increase the complexity of the process, the improvement is substantial. The combination of all three techniques results in a quality improvement roughly equivalent to two bits per sample.

4.8 Lessons for the System Designer

The conversion between the continuous domain of natural objects and images and the digital domain of modern signal processing has many pitfalls for the unwary. These conversions cannot generally be accomplished simplistically, i.e., by straightforward analog $\to$ digital and digital $\to$ analog transformation, without serious image degradation. The final image quality, for a given number of bits per frame (or per second or per unit angle subtended at the eye), can be increased substantially in most practical cases by careful consideration of both mathematical and perceptual principles. For example, the use of ideal low-pass filters is not a satisfactory solution because of ringing.[5]

In applications where maximum picture quality is desired, the presampling and interpolation filters must be chosen with care. Our experience has been that a purely mathematical approach does not give good results. A perceptual approach, in which a conscious tradeoff is made among the several elements of image quality, including sharpness, aliasing, and the visibility of the sampling structure, is more fruitful. The sharpened Gaussian filter seems to give the best performance. Interestingly enough, its response is much like that of the human visual system.

Although not discussed in this chapter, adaptive interpolation techniques appear to be quite promising. Pictures are not homogeneous. Much of the spectral energy is concentrated at low spatial frequencies. Sharp edges are important, but occur only in relatively small portions of most images. The best interpolation filter at a sharp high contrast edge is surely not the best in a blank image area. In all likelihood, it will prove possible to develop methods for choosing an optimum shape of interpolation function at each point, depending on some local measure of detail content. Such adaptive filters should be able to make a better tradeoff than the nonadaptive filters that are, after all, but a special case.

With respect to quantization, the failure to use randomization in nearly all cases is very hard to justify. Randomization by Roberts' method clearly improves picture quality without increasing channel capacity, and is very inexpensive. Unless the quantized image is to be further coded by some statistical method, or more bits per sample are used than are needed, the omission of randomization should be considered an error in design.

---

[5] Images that do not ring when processed by such filters must not contain significant energy at the upper end of the Nyquist baseband. In that case, consideration should be given to using a lower sampling density.

5. Compression

In this chapter we deal with data compression using methods beyond the optimized sampling and quantization techniques discussed in Chap. 4. After a brief introduction to statistical methods orginally due to Shannon, the special but important case of graphical (two-level) images is presented in some detail, culminating with a treatment of the international standard CCITT system. Continuous-tone coding methods, with emphasis on still monochrome images, are then presented. Emphasis is placed on the importance of using perceptual ideas in the design and analysis of these methods.

5.1 Introduction

In previous chapters we discussed the general problem of achieving maximum picture quality with minimum data [bits per second, per image, per unit area, or per picture element (pel or pixel)]. The methods used included companding to make quantization noise uniformly visible throughout the tone scale, conversion of quantization noise to uncorrelated random noise by Roberts' method, and the use of appropriate presampling and interpolation filters.We now take up the question of whether it is possible to do even better. The motivation for this study is that even with the methods described, picture data is often so voluminous as to require excessive storage, transmission, and computation facilities.

Frequently the best way to reduce the volume of picture data is simply to reduce the spatial, temporal, and/or tone-scale resolution of the image to that just essential to the task. If this is done adaptively, according to the image being processed, it may be possible to achieve large compression with little loss of essential data. For example, if high spatial- and low tone-scale resolution is used only for complicated, stationary images, high tone-scale and low spatial resolution for simple pictures, and adaptive frame replenishment or motion compensation for slowly changing images, significant reduction can be achieved. On scene changes, where adaptive schemes would be expected to fail, advantage can be taken of the fact that the new image may be built up slowly without serious degradation.

When these methods do not suffice, or when their implementation becomes impractical, it is necessary to look in other directions for economies. There are two possibilities. One is to make use of the statistical correlation that exists in image data because of the nature of images, and the other is to find methods to describe images numerically that are more efficient, in visual terms, than simply measuring the intensity at each pel. Thus data compression methods are often divided into "information preserving" or "lossless," i.e., statistical methods, and "lossy," i.e. techniques that make use of psychophysical phenomena. The former has as its goal the "exact" reproduction of images, while the latter exploits properties of the human observer to produce equivalent (but not identical) images with less data or subjectively better images with equal data.

This division of coding methods into two classes is somewhat deceptive. Analog signals cannot be reproduced exactly by any real system, and any digital representation of an analog image is merely an approximation. Furthermore, there is no universally agreed-upon "natural" way to digitize images. The better methods, in fact, do rely very much on visual properties. There is nothing sacred about a particular digitization of an image, and therefore there is no guarantee that a change in a digital representation must necessarily degrade an image. In many cases, changes can improve the quality. In spite of this, it is still meaningful to ask if it is possible to reproduce a particular digital image representation exactly.[1] In fact, in most cases some compression is possible. For example, in the case of high resolution two-level graphics such as typography and line drawings, economically important reductions are possible and some such coding systems are in commercial use.

For continuous-tone images, purely statistical coding methods have so far not proved very effective. There are a number of good reasons why they are not likely to become so in the future. One of the most important is that signficant texture is often no larger in amplitude than random noise. Noise sets an upper limit to the efficiency of all lossless systems, since channel capacity must be used to transmit the noise faithfully. Lossy systems, on the other hand, have been moderately successful, and are likely to become more successful in the future, as we sharpen our knowledge of those aspects of visual response most pertinent to image coding. This is particularly true in the coding of moving images. As a practical matter, the cost of picture processing hardware is falling rapidly, while its speed is increasing, making possible kinds of processing it would not have been fruitful to consider just a few years ago.

Topics not covered in this brief review include transform coding [5.1], color coding [5.2], and coding of motion pictures and television [5.3]. There is likewise no treatment of rate-distortion theory, which, in my opinion, is of more theoretical than practical value. In any event, no significant work on this subject has been done in our laboratory. Color coding [5.4] is dealt with in Sect. 7.8. With respect to transform coding, some of the earliest work in this field was done at MIT by Prof. T.S. Huang and his students. However, until the development of the discrete cosine transform, it was my opinion that the

---

[1] No quotation marks are needed here – digital signals, unlike analog signals, can be reproduced exactly.

image quality obtained by transform coding was unacceptable for commercial use. Not until the recent work of *Chen* and *Pratt* [5.5] has the combination of good quality and significant compression been obtained. Even there, a good part of the saving comes from the substantial elimination of detail in blocks of low luminance and/or very low contrast.

Coding of moving images has only recently become a topic of study at MIT. It promises to be a very fruitful field with important commercial implications. This subject is dealt with in Chap. 8.

5.2 Information-Preserving Coding

5.2.1 The Basis for Statistical Coding

In a conventional digital transmission system, each possible value of each sample is assigned a constant-length binary code. For example, a 128-character "alphabet" can be represented by a seven-digit code. This ignores the statistical relationships that may exist in the information to be transmitted. In the Morse code, on the other hand, short codes are used for common characters such as "i" and "e" and long ones for uncommon characters like "x" and "q". In this way a net saving in transmission time is achieved for English or other messages having English letter frequency. Should the Morse code be used for equal letter frequency messages, or worse still, for messages with contrary statistics, the "saving" might become a loss with respect to what would be achieved with a constant-length code. A complication of variable-length codes is that characters are not sent through the channel at a constant rate, so that buffering is needed.

In picture transmission, as in alphanumeric transmission by Morse code, if the "message" is made up of "characters" that occur with unequal frequency, and if the frequency or probability distribution of characters is known, then a code can be designed that produces a saving in average transmission time, or average number of channel bits per picture. The comments concerning buffering apply to this case, as do a number of other complications that will be mentioned below.

From the operational point of view, an important distinction between coding systems depends on whether they use a fixed number of bits per unit area or per unit time. Certain adaptive systems, which, for example, may exchange spatial and tone-scale resolution, are in the latter category. Although they can, in a certain restricted sense, be considered statistical, they require little or no buffering. Other systems, often called "elastic," transmit at a constant channel data rate, using a buffer to average over the simple and complex portions of the image. The compression ratio thus depends on the statistical properties of the image source. Inelastic systems are much easier to protect against loss of synchronization due to channel noise and on the whole are less difficult to implement. It is not to be expected that nonelastic systems can achieve the same compression as elastic systems, since they cannot average the simple areas with the complex. In comparing systems, therefore, it is essential to keep in mind whether elasticity is used. Most transform coding systems are elastic; most nonelastic systems can have a higher compression ratio if made elastic.

In Table 5.1, several examples are given of the savings achievable using the Shannon-Fano statistical code [5.6]. In all the cases shown, the signal has eight possible values of unequal probability. The code is derived by first arranging the values in decreasing order of probability, dividing them into two groups with as nearly equal total probability as possible, and assigning 0 and 1 as the initial digit for the two groups. The process is continued until exhaustion. The uncoded rate is three bits per sample in all cases, but the coded rates are 1.984, 2.5, and 1.3 bit/sample in the first three cases. It is a general rule that the more unequal the distribution, the higher the coding efficiency. A theoretical upper limit is set by the entropy or "information content" of the distribution

$$H = -\sum_{i=1}^{m} p_i \log_2 p_i \ .$$

The reasonableness of this measure of information is evident from the following argument. We first define the information content as the number of binary digits required to specify a sample from a distribution. For uniform distributions having M equiprobable levels or characters and using natural binary code, $H = -\log_2 p$, where $p = 1/M$. Thus we can think of such a distribution as generating or transmitting information at that rate. For nonuniform distributions, the average amount of information per sample is obtained by averaging this information over the distribution according to the probability of occurrence of each character.

The real meaning of the measure of information does not stem from such a heuristic argument, but from the fact that codes can be found which, in the limit

Table 5.1. Some examples of the application of the Shannon-Fano code

Case 1

I_k	p_k	Natural binary code	Shannon-Fano code	Number of occurrences in 128	Total number of bits	$-p_k \log_2 p_k$
0	1/2	000	1	64	64	0.5
1	1/4	001	01	32	64	0.5
2	1/8	010	001	16	48	0.375
3	1/16	011	0001	8	32	0.25
4	1/32	100	00001	4	20	0.156
5	1/64	101	000001	2	12	0.094
6	1/128	110	0000001	1	7	0.109
7	1/128	111	0000000	1	7	0.109
Total					254 bits	1.984 bits/sample.

Uncoded rate = 3 bits/sample.
Coded rate = 254/128 = 1.984 bits/sample.
Entropy = $-\sum_k p_k \log_2 p_k = 1.984$ bits/sample.

Case 2

I_k	p_k	Natural binary code	Shannon-Fano code	Number of occurrences in 14	Total number of bits	$-p_k \log_2 p_k$
0	1/2	000	1	7	7	0.5
1	1/14	001	0000	1	4	$1/14 \log_2(1/14)$
2	1/14	010	0001	1	4	$1/14 \log_2(1/14)$
3	1/14	011	0010	1	4	$1/14 \log_2(1/14)$
4	1/14	100	0011	1	4	$1/14 \log_2(1/14)$
5	1/14	101	0100	1	4	$1/14 \log_2(1/14)$
6	1/14	110	0101	1	4	$1/14 \log_2(1/14)$
7	1/14	111	0110	1	4	$1/14 \log_2(1/14)$
Total					35 bits	2.404 bits/sample.

Uncoded rate = 3 bits/sample.
Coded rate = 35/14 = 2.5 bits/sample.
Entropy = 2.404 bits/sample.

Case 3
Like Case 2, but p(0) = 0.9, p(1), ... p(7) = 0.1/7.
Uncoded rate = 3 bits/pel.
Coded rate (same code as Case 2) = 1.3 bits/sample.
Entropy = 0.750 bits/pel.

Case 4
Like Case 3 but coded in pairs.

I_k	p_k	Shannon-Fano code
00	$(0.9)^2$	1(1 bit)
0X (7 cases)	$(0.9 \times 0.1)/7$	00XXXX (6 bits)
X0 (7 cases)	$(0.9 \times 0.1)/7$	00XXXX (6 bits)
XX (49 cases)	$(0.1/7)^2$	01XXXXXX (8 bits)

Coded rate = 0.98 bits/pel.

Case 5
Like Case 3 but coded in blocks of 3.
Coded rate = 0.913 bits/pel.

Case 6
Like Case 3 but coded in blocks of 6.
Coded rate = 0.82 bits/pel.

of infinitely large blocks, require no more than this amount of data for exact transmission, and that no codes can be found that can transmit at a lower rate.

In Table 5.1, the entropy is also calculated. In Case 1, the S-F code achieves the theoretical rate, but in Cases 2 and 3 it does not. The reason is that in the first case, the code words are just $\log_2 p_i$ in length, which is possible only if the distribution can be divided in such a way that the total probability in each group is $1/2^n$ where n is an integer. This is not possible in Case 2 and is even less possible in Case 3. In such cases, the S-F code is not necessarily the best. A *Huffman* [5.7] code always is best, as it can be proved that any rearrangement of the correspondence between the symbols and the code groups decreases its efficiency.

To achieve coding efficiencies close to the theoretical rate, block coding can be used. In Case 4, the distribution of Case 3 is used, but characters are transmitted in pairs. The distribution now has 64 character pairs, of which the highest probability is 0.81. This reduces the coded data rate from 1.3 to 0.98 bits/sample. Coding in blocks of three reduces the rate to 0.913 bits/sample, while blocks of 6 reach 0.82 bits/sample, quite close to the entropy limit of 0.75 bits/sample. As a rule of thumb, the blocks should be long enough so the highest probability is no greater than 0.5.[2]

It is obvious that block coding is necessary in most cases to achieve high efficiency with the S-F code. Since the minimum code length is one bit, the maximum compression factor is $\log_2 M$ no matter how low the entropy. Therefore M must be large, which is generally possible only by block coding. Two-level facsimile, where the uncoded data rate is just one bit/pel, is a case in point.

5.2.2 Coding Based on Conditional Probability Distributions

If the message to be coded is the intensity of individual samples of an image, little coding efficiency is to be expected. While for any particular picture the luminances are bound to be somewhat nonuniform in probability, some pictures are dark, some light, and overall a nearly equal distribution of all values is to be expected. Changing the code for each picture is awkward, and in any event, except for very unusual pictures, the entropy would not be very low. In pictures, the main statistical correlations are found in the relative intensity of neighboring points, rather than in the distribution at one point. If $p_i(j)$ is the probability of getting the jth intensity at one point when the previous point has the value i, then $-\sum p_i(j)\log_2 p_i(j)$ is the information generated by a sample when the previous sample has the value i. Averaging this over all values of i we have

$$H_x(y) = -\sum_{i,j} p(i)p_i(j)\log_2 p_i(j)$$
$$= H(x,y) - H(x) \ .$$

Here, $H_x(y)$, the entropy of a sample knowing the value of the previous sample, is equal to $H(x,y)$, the entropy of the two taken as a pair, less the entropy of the first element in the pair, $H(x)$.

Actually, statistical correlation in images extends over more than two adjacent elements [5.8]. The ultimate entropy that sets the limit to coding efficiency, no matter how complicated the code, depends on the conditional probability

---

[2] Note that this type of block coding does not exploit any character-to-character correlation, if it exists. Such correlation can be exploited by measuring the probability distribution of the blocks.

distribution of sample values taking into account the total statistical dependence of that sample value on all others. As the order of the governing distribution becomes higher, it becomes more and more difficult to measure, the message sample must get longer to be representative, and the code becomes harder to implement. When using a lower-order distribution to make up a code, there is no guarantee that some other distribution of the same order might not give better results. For example, if the present sample value is predicted by a linear combination of previous values, the probability distribution of the difference between the actual and predicted value will be highly nonuniform and the entropy rather low. Coding the difference signal may give better results in some cases than codes based on the second order distribution $p_i(j)$. It can be implemented by entropy coding the transmitted values in a DPCM system, as will be described later.

5.2.3 Noise as a Limit to the Efficiency of Entropy Coding

A fundamental limitation to the effectiveness of any compression system that exactly reproduces the digital representation of an image is that noise, which is always present to some extent, has some entropy. The entropy of the sum of an image plus a superimposed uncorrelated noise clearly cannot be less than that of the noise alone. For example, if the rms value of additive Gaussian noise is equal to one quantum step, and if the image is completely blank, then 27% of the pels will have unit error in the PCM representation, 5% will have two-step error, etc. Thus the entropy of the PCM signal

$$H = -\sum p \log_2 p$$
$$= -[0.68 \log_2 0.68 + 0.27 \log_2 0.27 + 0.05 \log_2 0.05 + \ldots]$$
$$= 1.12 \, \text{bits/pel} \, .$$

Other values are plotted in Fig. 5.1. In cases where some preprocessing is performed on the signal before coding, such as differencing or linear prediction, the limit is even higher because the noise in general will be increased by the preprocessing. For practical values of SNR these limits are highly significant. Thus we can confidently predict that lossless statistical coding is very unlikely ever to achieve high efficiency.

5.2.4 The Variation of Entropy with Sampling Density

The channel capacity required for uncoded transmission of an image is proportional to the square of the linear sampling density. Thus an $8.5'' \times 11''$ graphics page requires 0.935 megabits at 100 samples/inch but 8.42 megabits at 300 samples/inch, and so on. Yet if the word "information" has any meaning in this context, it is clear that there is some resolution beyond which no more real in-

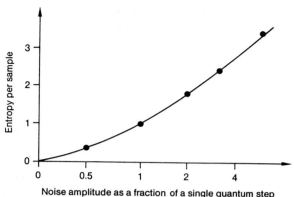

Fig. 5.1. Entropy of a blank signal as a function of additive noise

Entropy per sample

Noise amplitude as a fraction of a single quantum step

formation is added to the signal. There is, in fact, just so much information or entropy in a picture, and can be no more in its digital representation.[3]

The physical processes by which images are formed, as well as the size of the objects depicted, dictate that the intensities of nearby points, at typical sampling densities, have strong statistical correlation.

The degree of correlation between adjacent samples depends on the relationship between the sample spacing and the size of typical depicted objects. When the sample spacing is larger than the object size, samples are uncorrelated and the entropy per pel is nearly $\log_2 M$. Therefore, as the sampling density rises from a very low value, the entropy increases in proportion to the area, since each new sample adds new information. However, as the sample spacing rises into the region where the spacing becomes comparable to or smaller than the object size, the correlation between adjacent samples becomes higher. Finally the entropy per unit area becomes constant, although at different levels depending on image complexity, as shown in Fig. 5.2a. The entropy per pel is the area entropy density divided by the pels/unit area, giving the corresponding curves on a per pel basis, in Fig. 5.2b. The maximum possible coding efficiency, shown in Fig. 5.2c, is the uncoded data rate (bits/unit area), which is proportional to area scanning density divided by the entropy per pel. Thus maximum coding efficiency is relatively constant at low resolution but rises in proportion to area scanning density at high resolution.

The significance of these considerations is that the efficiency of optimum coding systems depends very strongly on sampling density. *The only condition under which the efficiency is independent of resolution is where the latter is low.*

A measure of the goodness of a coding system is the extent to which the variation of its compression ratio with resolution follows the theoretical relationships just described. For example, suppose there existed a perfect one-dimensional code that permitted the transmission of a digitized image, line by line, with no more

---

[3] We are disregarding the effects of noise in the analysis, and assuming that all signal variations represent wanted detail.

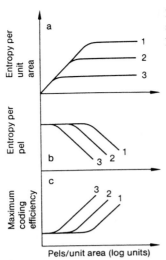

Fig. 5.2a-c. Variation of coding efficiency with sampling density (log units both axes): (*1*) complicated picture, (*2*) average picture, (*3*) simple picture

data than the entropy as measured with a suitable high order (one-dimensional) probability distribution. If the resolution were doubled in both directions, and the coded data for each line remained constant, the total data for a page would double. The uncoded rate would quadruple. Thus no one-dimensional system increases its efficiency with resolution according to the theoretical model, and therefore cannot be near optimum for very high resolution systems.

Another way to look at this phenomenon is to note that as the resolution increases, picture quality and entropy increase. As the point of diminishing returns is reached, entropy stops increasing, the uncoded data rate still goes up, and the possibilities for coding improve. With the current interest in computer manipulation, storage, and transmission of high definition TV and graphic arts quality imagery, both continuous-tone and graphics, we may expect to see the development of many systems that would have been impractical for lower resolution applications such as standard TV and office facsimile.

The considerations discussed here have rarely been mentioned in the literature on image data compression until quite recently. *Ericsson* and *Dubois* [5.9] have dealt with the case of transform coding of high definition television signals, while *Bodson* et al. [5.10] have measured the compression ratio of binary facsimile systems as a function of sampling density. Coding efficiency previously was most often expressed in terms of bits/pel and usually compared with 8-bit PCM without regard to image resolution. If the viewing conditions are standardized, e.g., "NTSC television viewed at four times picture height", this measure is suitable. In general, however, it is not, on account of the influence of sampling density on picture statistics as well as on the visibility of image degradation in lossy coding.

5.3 Graphics Coding

We often think of typography and line drawings as a class of images quite distinct from continuous-tone pictures since they apparently are just two-toned. On close examination such images are seen, in fact, to take on all values of reflectance between black and white. Nevertheless, since we would be perfectly satisfied with just two levels if the resolution were high enough, the classification is valid. Both types of copy share the problem that major degradation can occur in the sampling and quantization process before any data compression is attempted. Quality requirements for graphics are easier to assess, since there is a substantial body of experience with a variety of commercial systems.

The useful range of graphics resolution is about 70–1500 pels/inch, as indicated in Table 5.2. Below the former level, serious illegibility occurs with nearly any copy, while above the latter, output equal to the finest products of the printing industry are obtained. At the lower resolutions, gray scale contributes significantly to quality [5.11]. Older office facsimile apparatus operated at about 100 pels/inch with analog gray scale and often with continuous (unsampled) transmission in the scan direction. The quality was far superior to that obtainable with completely digital transmission at one bit per pel, 100 pels/inch in each direction. Gravure printing is effectively sampled at about 150 pels/inch in each (diagonal) direction, with a continuous grey scale.It gives highly acceptable copy, though not equal to book work. So-called "letter quality" is achieved by a number of 1 bit/pel laser printers such as the Xerox 9700 (300 pels/inch)and the Canon LBP-10 (240×240 or 240×480). In this case, the quality is enhanced since the type is computer-generated and therefore aligned with the sampling raster, but there is also some quality loss due to the electrostatic output process. Phototypesetters for the newspaper industry operate in the 500 to 800 pels/inch range, with magazine quality achieved at the upper end of the scale. The best book work requires in excess of 1000 pels/inch, although in this case the output device as well as the sampling standards probably influence the final result.

An early method of transmitting graphics was by means of the parametric equations of motion of a stylus tracing out the copy. The Telautograph, invented by Elisha Gray in 1890 and now almost unheard of, was once a widely used method of sending handwritten messages, especially over short distances. Vector drawing CRT displays, now mostly displaced by raster scan systems, exploit this

Table 5.2. Relationship between resolution and quality

Legible	Message fax	96–100 lpi analog
Legible	Modern fax	100 × 200 dpi digital
Newsphotos	Wire services	150 lpi analog
Letter quality	Office of the future	200–400 dpi digital
OCR	Typesetting	250 dpi digital
Typography	Newspapers	600 dpi digital
Halftones	Newspapers	720 dpi digital
Ultimate quality	Books, magazines	1000–2500 dpi

technique. Especially suitable for sparse copy, a number of effective compression methods were developed to specify all the lines and curves needed to describe a complete drawing.

For the special case of typography, a list of the designations (names) and locations of the characters on a page is by far the most compact description. Body type in a newspaper page, for example, can easily be represented in ASCII (a far from efficient code) at about 200 bytes per square inch, while about 60,000 bytes are required for uncoded facsimile representation. Needless to say, no coding system has ever been discovered that approaches this 300 to 1 ratio. The efficiency of the character representation has led to some interesting proposals to incorporate optical character recognition in facsimile coding [5.12]. For nontypographical copy, such as halftones, some attempts have been made to develop an "alphabet" from which graphical images could be built up [5.13].

We now concentrate our attention on copy that has been sampled on a rectangular grid and quantized to one bit per sample, and particularly on the problem of coding such digital images so that they can be stored or transmitted with fewer bits per page. In some copy, the fraction of black pels is very small, in which case we can send the location of just the black pels. This simple system can readily be improved by sending the addresses, not at random, but in order. An end-of-line code (EOL) enables the skipping of blank lines and the transmission of only the horizontal component of the address. A next step is to send not the absolute, but the relative location of black points. The probability distribution of relative addresses is highly peaked and therefore amenable to statistical coding. Finally, it will usually be found better to transmit only the number of pels (run lengths) between edge points, i.e., black-white and white-black transitions, rather than the run lengths between black points. The black runs have a different probability distribution from the white runs, but this can easily be handled. The use of a Huffman or Shannon-Fano code or a truncated version of one of these is the basis for a large number of different coding systems.

5.3.1 One-Dimensional Run-Length Coding

The technique of transmitting data relative to the distance between significant pels has come to be known as run-length coding and has developed an extensive literature [5.14]. Because of the difficulty of implementing variable-length codes and in part because of their fragility in the presence of channel noise, many systems using constant-length codes have been developed. Some of these amount to truncations or quantizations of Huffman codes, but some attempt to regain the efficiency lost by using a constant length by adapting it to the copy being scanned. A reasonably efficient method is as follows:

Prescan each line, recording the location of the first and last black pel and the average run length in between. If the line is blank, send an EOL. Using a repertory of perhaps eight codes (six different code word lengths, one "long run indicator" and one uncoded mode) transmit a word after the EOL to signify the code to be used and the location of the first black pel. For each short run, transmit one code group. For runs too long to be transmitted in that manner, transmit the long-run code followed by the run length in natural binary code. When the last black point is reached, transmit an EOL. For typical copy this results in a compression ratio of five to ten at 200 pels/inch.

Many variants of this scheme have been used with similar results. Since the efficiency of all of them increases with scanning density, and since such one-dimensional methods do not take advantage of vertical statistical correlation, they are generally designed to have higher horizontal than vertical resolution. For a given number of samples per unit area, it is best to have equal resolution in both directions. However, there is only small loss in having twice the resolution in one direction than in the other [5.15].

5.3.2 Two-Dimensional Coding

a) Edge-Point Tracking. Exploitation of the vertical correlation in graphical images has been accomplished in a number of ways.One that works well with line drawings consists of locating edge points on each line relative to corresponding edge points on the previous line [5.16]. This can be done crudely by simply subtracting edge points on one line from those on the line before. In the case of upright rectangular outlines, only the corners would then be transmitted. In the more general case of slanted or curved outlines, it is more effective to transmit the line-to-line shift in position of corresponding edge points on successive lines. Special codes are required at the top and bottom of outlines where new edge points appear or old ones vanish.

These systems, especially when used on typography, become more efficient above several hundred pels/inch, since at lower resolutions the overhead of starting and stopping outlines may be higher than the savings due to edge point tracking.

b) Block Coding. Another method of utilizing vertical correlation is to divide the page into two-dimensional blocks, using a single code per block [5.17]. If the probability of all possible patterns of black and white in the block is measured, a $S/F/H$ code can be used. In the limit as the block becomes equal to the page size, this method achieves ultimate compression.[4] The system can be made practical by using smaller blocks, by using algorithms rather than a code book for coding and decoding, and by adapting the block size and code to the local copy complexity. All of the latter methods benefit by transmitting edge points rather than black points, although the "ultimate" scheme mentioned does not.

The block size can be adapted by measuring the number of edgepoints per unit area, using large blocks for small counts and vice versa. A useful algorithm, sometimes called a quadtree, is to start with a full-page block, transmitting a zero if the block is blank and a one if it is not. If not, the block is subdivided and

---

[4] Except, of course, for page-to-page correlation. This technique is now often called vector coding or vector quantization [5.18].

the process continued to completion. At about 120 pels/inch, the best of these methods achieves a compression ratio of four to twelve. As in other cases, the poorest results always are obtained with dense typewritten copy.

c) Contour Coding. Most of the previous methods are based on proceeding through the page, more or less in raster fashion, with perhaps some detours to pick up nearby detail, as in block coding. Contour coding [5.19] first identifies and locates sets of connected edge points, and then codes and transmits each contour in its entirety before proceeding to the next. For reasons of economy of storage, some systems have been built in which contours are tracked, coded, and decoded in raster format, or "scanline order," as it is sometimes called. Although page-size contours can be handled, the method is particularly applicable to typography, where, at high resolution, it is superior to other techniques [5.20].

On a rectangular raster, there are eight possible directions to the adjoining contour point, so that 3 bits per point are needed. However, since all directions are not equiprobable, a $S/F/H$ code improves the efficiency [5.21]. Higher efficiency is achieved by fitting the longest possible straight lines, subject to an error criterion[5] [5.22]. Allowing circles as well as straight lines is a further improvement, while the best results are achieved by the fitting of splines, i.e., second or higher order polynomials with continuous derivatives [5.20,23]. In such cases the starting point and parameters of each spline are transmitted or stored. A property of spline fitting is that the contours may be decoded on rasters of arbitrary resolution. The method is therefore the most compact way to specify very high quality copy, such as book style typography, typically rendered at 1000 or more pels/inch.

5.3.3 Some Practical Coding Systems
and Their Performance as a Function of Resolution

Uncoded transmission of two-level graphics requires one bit per pel. As long as the image is to be transmitted on a pel-by-pel basis, no saving is possible, even though the first order entropy per pel may be considerably less. For example, if only 1% of the pels are black, $H = 0.08$ bits/pel. Block coding in Shannon's sense, i.e., representing all of the elements of a block by a single code, but assuming the elements are uncorrelated, permits transmission rates closer to the first order entropy. In order to use all of the statistical relationships within each block, it is necessary to find the probability of each unique arrangement of black and white pels within the block, as in vector quantization. Even this method does not use any of the statistical relations that might exist among blocks.

Whenever other than separate pels are to be transmitted – for example, run lengths – there is an appropriate governing probability distribution, in this case the run-length distribution, and an associated entropy. A code is then required

---

[5] Contour coding lends itself to lossy coding, which is sometimes very effective, while most of the other graphics coding methods do not, as errors would propagate through the page.

that represents the data to be transmitted, in this case the run length, by words composed of binary digits. The closer the code used conforms to the distribution, the closer the transmission rate is to the entropy.

The total coding process is thus seen to consist of two steps – the transformation of the pel-by-pel page description into one more appropriate to the image structure, such as run lengths, and the coding of the transformed data in accordance with its probability distribution. Since there is no analytical method for predicting optimum transformations, most coding schemes are constructed intuitively and then checked by measuring the governing probability distribution. Since it is not practical to measure the probability of each different possible page and certainly not the relationship of successive pages, there is always the possibility that some new and more efficient system will be invented.

A useful measure of goodness of coding systems is the variation of coding efficiency (uncoded data rate/coded data rate) with scanning density. This will now be calculated for several representative coding systems.

a) **Transmit the Positions of all the Black Points.** For a page $N \times N$ pels, $2 \log_2 N$ bits are required per point. If the probability of black is p, then $2pN^2\log_2 N$ total bits are required and the compression ratio is $1/(2p \log_2 N)$. The break-even point is $p = 1/(2 \log_2 N)$. Thus the compression ratio of this system actually *falls* as the resolution increases.

b) **Transmit an EOL and the Horizontal Positions Only.** For an $n \times (n-1)$ page, $N \log_2 N$ bits are required for the EOL's using one of the 2^N codes, and $pN^2\log_2 N$ for the data. Total data is $(N + pN^2)\log_2 N$ and compression ratio is

$$\frac{N^2}{(N + pN^2)\log_2 N} = \frac{1}{(p + 1/N)\log_2 N} .$$

This is about twice as efficient as (a) but also does not increase properly with N. The basic difficulty with both methods is that as N increases, so do the number of points to be transmitted. A better method is to transmit only transition or edge points, since except for extremely low resolution, the number of edge points per line is independent of N.

c) **Transmit an EOL and Edge Points Only.** As in (b), $N\log_2 N$ bits are needed for EOL's and $K\log_2 N$ bits per line for K edge points/line. The total data required is $(N + NK)\log_2 N$, the compression ratio is $N/[(1 + K)\log_2 N]$ and the break-even point is at $K = (N/\log_2 N) - 1$. The efficiency now increases with N, but slowly. Note that for typewritten English text, the number of edge points on an 8" typed line is between 20 and 160.

d) **Transmit an EOL and Transmit Run Lengths Rather Than Edge Positions.** This is similar to (c) but more efficient since the distances between edge points are smaller than edge positions and can be represented by shorter codes. Since the longest representable run is now shorter than the longest possible run, some runs do not terminate at edge points and require special codes. This can be done in many ways, such as adding one video level bit, or by using one code group to mean "long run to follow" and then giving the length in $\log_2 N$ bits.

e) Entropy Coding of Run Lengths. If the probability distribution of run lengths is measured and an efficient code devised, the total number of bits/line is proportional to $\log_2 N$. This is because the number of runs/line is constant, but the number of possible run lengths is proportional to N and therefore the average code length is proportional to $\log_2 N$. The compression ratio is proportional to $N/\log_2 N$.

f) Contour Coding. Except for very low resolution, the number of contours, C, on a given page is a constant. The total number of contour points is proportional to N. If we transmit the starting point of each contour (total data generally negligible) and three bits per contour point, the total amount of data is proportional to CN. The compression is proportional to (N/C), which is a better than the position coding methods. If we fit curves to the contours, and if the error tolerance is independent of N, then the total number of bits required is proportional to $\log_2 N$. In this case the total number of parameters per curve is unchanged by N, but the required accuracy increases, so that the number of bits per parameter increases as $\log_2 N$. Thus the compression ratio is proportional to $N^2/\log_2 N$, which is very near to the theoretical relationship. The best results to date using this method have been obtained by *Coueignoux* [5.20], who has coded book-quality typography using 200–800 bits/character.

The fundamental reason why (f) is better than (e) is that the latter ignores statistical relationships in the vertical direction, which generally are as significant as those in the horizontal direction. Numerous other methods of exploiting vertical correlations have been devised. It would appear that none could have better efficiency than contour coding, although some, such as edge point tracking, might well be easier to implement.

5.3.4 CCITT Standard Codes

In order to permit the introduction of commercial digital facsimile compression systems, a study committee was formed by the International Consultative Committee for Telegraphy and Telephony (CCITT) to receive submissions from interested parties and to make recommendations. Extensive tests were carried out using a set of eight standard documents. Factors in the decision included coding efficiency, error performance, complexity of implementation, and availability of a system free of patent restriction. The goal was a system that permitted transmission in an average time of about one minute at 4800 bits/sec for an A4 document using 1728 samples per line (about 8/mm or 200/inch) and 1188 lines per page (about 100 lines/inch). The asymmetry came about because in the 1-*D* case it takes little more channel capacity to use the higher resolution in the scan (coded) direction while the resolution is substantially improved. This would also be true in the other direction in the 2-*D* case if the code were an efficient one. However, it had been decided to make the 2-*D* code an extension of the 1-*D* code. It therefore needs very nearly twice the capacity to use 200 lpi in the vertical direction. This was made an option, along with a number of other features [5.24].

a) The One-Dimensional Modified Huffman Code. Runs of length 0 to 63 each use one code word, 2 to 12 bits long, based on Huffman coding of the separate black and white run-length distributions. Longer runs require two code words, one word for the highest submultiple of 64 samples within the run, again Huffman coded separately for black and white, and one word for the remainder as for the shorter runs. An EOL code, ten zeros followed by a one, permits resynchronization even if an error has been made since the largest number of consecutive zeros in the other codes is six. This code achieves an average compression factor of about 7 for the eight documents and about 4.5 for the most dense, a French-language single-spaced typewritten page.

b) The Two-Dimensional READ Code. This scheme, the "relative element address designate" code, developed in Japan, achieves an improvement of about 35% over the 1-*D* code at low resolution and about 49% at high resolution, using the least conservative choice of parameters.

The principle of the code is that if corresponding edge points (i.e., black-white transitions) on successive lines are less than 4 pels apart horizontally, "vertical mode" coding is used in which the edge location on one line is designated by its separation from the corresponding edge on the previous line. If the separation is 4 or more pels, then "horizontal mode" is used in which the edge is located relative to the previous edge on the same line. "Pass mode" is used where, by a simple logical test, it is determined that there is no corresponding edge point on the previous line. An uncompressed mode is optional for very complex lines. Every Kth line can optionally be 1-*D* coded as mentioned above, with $K = 2$ being recommended for low resolution and $K = 4$ for high resolution. This limits error propagation but of course reduces the compression factor.

Although the READ code is not very efficient, it does make possible one minute transmission, on the average, with 200 lpi in both directions, using rather simple equipment with good error performance. This probably settles the matter with respect to Group Three systems. However, with the development of Group Four systems, not as yet defined but probably about 300 lpi and with more powerful processing available, there may still be scope for further developments.

5.4 Coding of Continuous-Tone Images

5.4.1 Lossless Coding

In purely statistical coding, an image is digitized in a conventional manner and the resulting data stream compressed by exploiting the statistical relationships. This technique is on the whole rather unattractive, not only because "real" digitized continuous-tone picture signals do not have enough redundancy to be compressed more than two or three times even by complicated codes, but because the implementation of such schemes, even in these days of MSI, LSI, and cheap memory, is awkward.[6] Pictures tend to have nonuniform information content, so that the coded representation must be buffered in order to be transmitted at a constant rate. Provision must be made for graceful failure when the buffer overflows or underflows. Special precautions must be taken to avoid loss of synchronization due to channel errors. These problems are especially serious in TV applications, but less so in facsimile where the exact time taken to transmit any one document is not significant. In this case, the stack of documents to be transmitted can be thought of as part of the buffer.

a) **Run-Length Coding.** In spite of these problems, several systems have been proposed and implemented or at least simulated. In one method, the picture is divided into contiguous areas of constant luminance. Each area is then transmitted as a whole. When carried out in one dimension, this amounts to run-length coding. In two dimensions, it is "blob" coding, probably best attacked by transmitting data from which the blob contours can be constructed. Clearly this is most efficient for coarsely quantized, high resolution images. At least the quantization steps must be substantially larger than the noise level [5.25].

b) **Bit-Plane Coding.** A second method, usually called "bit-plane" coding, [5.26] resolves the picture into a small set of binary images, each of which is then transmitted as if it were a separate graphics page. Contour coding works best. The number of contours is decreased by a significant factor if reflected binary (Gray) code is used rather than natural binary code. Even so, only the first three or four most significant digits have good statistics. The higher order bit planes seem to be nearly pure noise, although they contribute a great deal to the quality. The basic reason why bit-plane coding is inefficient is that many of the contours in the separate images do not correspond to actual outlines of objects, but rather are artifacts of the quantization process. Other methods, discussed below, attempt to isolate object outlines. If successful, much higher efficiency can be expected.

c) **Predictive Coding.** High correlation among luminance values at nearby sample points means that such values can usually be predicted accurately [5.27]. For

---

[6] VLSI will eventually change this.

example, in a conventionally scanned image, the nearby preceding points can be used for the prediction. The difference between the actual and predicted value is governed by a marginal distribution of the overall high order probability distribution. It is highly peaked near zero and has low entropy. Therefore the error signal can be transmitted efficiently and the picture reconstructed without error by entropy coding.The accuracy of the prediction and thus the efficiency depends both on the picture statistics and the noise level. Low-noise images of collections of large, untextured objects are best. Typical predicted television images would be expected to have entropies of no less than two bits/pel.

Although pure statistical coding, as defined here, has not been very successful, it does show promise when combined with some lossy preprocessing.In this case, one seeks an image digitization that not only is more efficient than simple PCM, but creates a data stream of low entropy which is amenable to subsequent lossless coding. For example, if some process could ignore noise and then pick out only the visually important features, one would expect that the output of the process would be an excellent candidate for coding.

5.4.2 Approximation in the Space Domain (Waveform Coding)

a) Interpolation. Straightforward sampling and digitization can be thought of as an approximation technique using a sample-by-sample fidelity criterion. Rather than transmitting pel values, one can use any of a wide variety of other techniques to perform the approximation, working either in one or two dimensions. A smaller set of pel values obtained by subsampling can be transmitted, and the intermediate points interpolated [5.28]. Additional information can be sent as necessary to remove excessive interpolation errors. Rather than using a regular array of subsamples, which is bound not to take best advantage of the image structure, it is possible to start in one corner, fitting, for example, lines or surfaces described by polynomials (splines) to the data points according to some error criterion. Parameters of the splines can then be transmitted.

These methods work best if the entire picture is done in one piece, which requires a large buffer. Dividing the picture into blocks is less efficient since the blocks may cut across image features. In addition, block boundary artifacts may occur at error levels otherwise invisible.

All of these techniques require buffering for worthwhile efficiency and are therefore in the class of lossy statistical systems.

b) Differential PCM. A form of predictive coding is shown in Fig. 5.3. What is transmitted is the coarsely quantized difference between the current pel value and the prediction of that value as produced at the receiver from the transmitted differences. This can be a lossless system if the difference, or error, is quantized to the same fineness as the original signal. In that case, no saving is made (in fact a penalty is incurred since the error has a dynamic range twice that of the input signal) unless entropy coding is used.

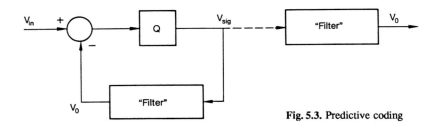

Fig. 5.3. Predictive coding

An effective use of this idea that does not require buffering is differential pulse code modulation (DPCM) [5.29]. It is based on the observation that errors are not equally noticeable everywhere. They stand out in blank areas and are more tolerable in busy areas or near sharp transitions. The difference signal is subjected to a tapered quantization process having few steps, as shown in Fig. 5.4. Those near zero are comparable to the quantization interval of the original signal. The predictor is simply an integrator, which means that each point is predicted to have the same value as its predecessor. The savings come exclusively from coarse quantization of the error signal – typically three to five bits per sample. Large errors follow sharp transitions, but the slowly changing areas are rendered accurately.

DPCM is a generalization of delta modulation [5.30], a system that was developed for voice transmission. The advantages of delta modulation, which is the one-bit version of DPCM, lie in its extreme simplicity and its ability to produce audio quality equivalent to PCM at a lower data rate. The available design parameters are the sampling rate and the step size. If the latter is small, good rendition is achieved in slowly varying areas but slope overload occurs at sharp transitions, which fortunately do not generally occur in voice signals. Delta modulation is ineffective for video transmission since the sampling rate must be raised so much to get acceptable step response that the savings disappear.

A potential problem in all difference transmission systems is accumulation of errors in the integrator. No matter what the cause, such errors produce un-

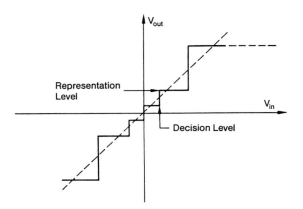

Fig. 5.4. A tapered quantizer: The quantizer may have a zero decision level or a zero representation level. A common implementation is a cascade of nonlinear amplifier, linear quantizer, and a compensating nonlinear amplifier

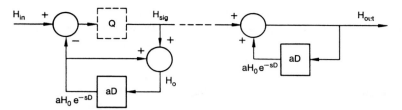

Fig. 5.5. DPCM with previous-value prediction and "leaky" integrator

acceptable picture distortion in the form of streaks together with drift of the dc level. Noise performance can be improved by using a "leaky" integrator, i.e., one in which the loop gain is less than unity, as shown in Fig. 5.5. This improves performance in the presence of channel errors, since such errors eventually die out as they circulate around the loop. However, as shown in Fig. 5.6, this worsens the quantizing noise in relatively blank areas ("granular" noise) by lowering its spatial frequency. It would appear to be a better practice to use channel noise protection and a true (unity gain) integrator.

Assuming that we sample at the Nyquist rate, the optimization of DPCM consists of choosing the loop gain and designing the quantizer [5.31]. The minimum step must be as small as possible to reduce granular noise in blank areas and the maximum must be made as large as possible to minimize slope overload. The ratio of steps cannot be too large or else oscillations occur after edges. A great deal of effort has gone into quantizer design, but it turns out that those designed by intelligent trial and error are about as good as those using the most sophisticated mathematical methods.

In the simplest DPCM systems using a fixed number of bits per sample, an even number of steps is naturally available. Usually no zero step is used, as a result of which, in blank areas, the output oscillates between two values. It may be thought that a zero level would be an easy way to eliminate this granular noise, even at the expense of giving up one of the maximum levels. At three or four bits per pel, it is usually found that inclusion of a zero level requires the next higher levels to be so large that fine detail and low contrast texture are

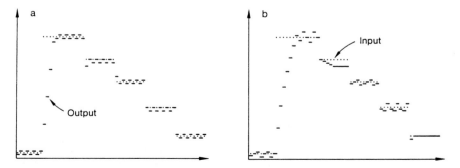

Fig. 5.6a,b. Effect of a "leaky" integrator: (a) unit gain integrator, (b) integrator with a gain of 0.8

123

wiped out. The use of two symmetrically disposed levels renders such detail, albeit noisily. When the signal is to be entropy coded, the number of levels used does not as significantly affect the channel capacity, in which case enough are used to get good quality even with a zero step.

A final type of distortion in this system is "edge business," which is primarily a phenomenon of moving pictures, but that can sometimes be observed in still pictures in which edges are rendered differently on successive scan lines. There is some difference of opinion as to the cause (and the cure) of this problem [5.32]. There may be more than one cause. Quantizing noise in this case is picture-dependent, as a result of which it may change radically line-to-line or frame-to-frame. It may also be accentuated by aliasing that manifests itself by a lack of shift-invariance. Edges that are slightly displaced on successive lines or frames may be reproduced quite differently. Finally, it may be caused by too large a ratio between successive quantizer steps.

To achieve quality equivalent to eight-bit PCM in the simplest type of DPCM system described requires about 27 properly placed quantizer levels, in the case of Picturephone[7] images intended for video conferencing applications [5.33]. For standard TV signals 24–32 levels have been reported required [5.34]. Systematic comparison of basic DPCM with PCM for other applications, such as high resolution facsimile, has not been made, as far as we know.

c) Predictive DPCM. In Sect. 5.4 c, the use of predictive coding, i.e., the transmission of the difference between the current sample value and a prediction based on the values of a number of nearby points, was shown to reduce the entropy. While the reduction of entropy does not directly improve the efficiency of inelastic[8] DPCM (the savings come from coarse quantization of the error signal based on a visual criterion), reduction in the amplitude of the error signal by sophisticated prediction permits even coarser quantization. Whether the quantizing noise so introduced is more or less objectionable than in the case when the prediction is based simply on the previous pel, must of course be settled by psychophysical tests.

The commonest predictor is a linear combination of sample values, either previous element or previous line or both. It is found that no more than three are very effective [5.35].[9] Some nonlinear predictors have been tried and do not seem to be substantially better.

d) Adaptive DPCM. In the system of Fig. 5.3, a fixed quantizer is used and the decision as to which level is transmitted depends only on the present and

---

[7] Copyright Bell Laboratories

[8] Inelastic systems can be defined as those using a constant number of bits per unit area or per second in the channel. They are to be distinguished from elastic systems, using buffers, in which the information rate can be averaged between simple and complex areas.

[9] Recent results in TV coding at the BBC [5.36] indicate that in the special case of coding composite color TV signals, many more points can be used effectively.

predicted sample values. In one form of adaptation, different quantizers are used depending on some measure of local picture complexity, or "activity" [5.37]. It has been found that a net reduction in the visibility of quantizing noise can be accomplished by tailoring the quantizer. For example, in relatively blank areas, small steps are better, while in complicated areas large steps are preferred. In some schemes, some housekeeping information must be sent to inform the receiver of the code, while in others, the code is derived from the transmitted information itself. For example, two maximum values in a row could be made to trigger a change [5.38].

Adaptation of the quantizer can be accomplished by selecting from a set of quantizers according to some local measurement on the image. While this is most efficient, an easier method is simply to change the gain before and after a fixed quantizer. This method guarantees the absence of artifacts as long as the gain changes continuously from pel to pel. Various criteria can be used to select the gain. A rather successful example of this technique is the work of *Zarembowitch* [5.39] in which the picture is divided into blocks and an adaptation factor, which is sent as side information, selected for each block. The factor used for each pel is found from the set of block factors by linear interpolation, eliminating all block artifacts. Excellent quality 512×512 images were obtained at slightly more than 2 bits/pel. This was reduced to about 1.5 bits/pel when averaged over the frame by elastic coding.

Another form of adaptation, called delayed encoding, [5.40] bases the quantization decision on an error criterion operating over several successive pels. Since there are no code changes, no extra information need be sent. This technique has the effect of partially randomizing the quantization noise, always a good technique to minimize its appearance. As we shall see below, easier and more effective methods are available to do this.

In the more general predictive form of DPCM, the predictor can be selected adaptively, for example by changing the set of samples on which the prediction is based. In one case, the prediction was switched from vertical to horizontal depending on local detail [5.27 (*Graham*)]. In lossless coders this may have some advantage, but in lossy coders artifacts are likely at the points where the prediction is changed. A continuously variable prediction or quantization is much less likely to produce artifacts.

e) Entropy-Coded DPCM. Small prediction errors are more likely than large ones. Hence additional saving can be accomplished by using a $S/F/H$ code, although this makes the system elastic and requires a buffer. Further savings can be accomplished by changing the code according to a measure of local activity at the expense of substantial complication [5.41]. As much as 30% reduction in entropy is reported by this method. Note that in such systems, the image quality depends on the basic DPCM process, since it is assumed that the statistical coder itself is lossless [5.42]. This gives a direct relationship between image quality and channel capacity, since the DPCM quantizer can always be chosen for high fidelity reproduction. One would expect, in fact, that as the number of steps in the quantizer is made larger and larger, the entropy of the error signal would eventually become constant, as no new "real" information would be added by the extra levels.

f) Randomized DPCM. The visibility of quantizing noise in DPCM is accentuated because it is correlated with the image data. This was recognized at an early date and a number of attempts have been made to reduce the effect. One method is to preemphasize the high frequencies before coding so that the corresponding deemphasis filter after coding blurs the contours that tend to echo sharp transitions. Too much emphasis aggravates slope overload. Another method is the use of "noise feedback" predictors, which show small but significant gains [5.43].

A direct method of randomizing DPCM quantizing noise is to use Roberts' method of introducing pseudorandom noise (PRN) before the quantizer and subtracting it afterward [5.44]. Computer simulation of this method [5.45] shows a marked reduction in the visibility of granular noise, which is now decorrelated. This effect is sufficiently effective that a less tapered quantizer can be used than in normal DPCM, so that slope overload can be decreased.

g) The Effect of Camera Characteristics on DPCM. The design of the DPCM quantizer involves the balancing of slope overload and granular noise. A linear quantizer ameliorates the former defect and a highly tapered quantizer ameliorates the latter. The permissible degree of tapering depends on the stability of the feedback loop, the visibility of the edge noise due to the larger steps, and the maximum pel-to-pel signal change to be expected. Typical good quantizers have a maximum representation level of only 25%, which means that four pels are required for a black-to-white transition. This is permissible if the signal is from a normal TV camera such as a vidicon. However, when the camera or scanner has a resolution fully appropriate to the scanning standards, much faster transitions are found and DPCM system performance is much poorer. We have found in our own computer simulation studies, where we use a very high resolution laser scanner as a signal source, that we must reduce the MTF at the Nyquist frequency to no more than 50% to obtain useful compression with any DPCM system. Even at that, the response is better than most TV cameras.

To my knowledge, this phenomenon is never mentioned in the DPCM literature. As improved cameras are developed in connection with high-definition TV and are applied to standard TV, as is likely, the use of DPCM, at least when the signal is sampled at no higher than the Nyquist rate, will be less rewarding.

h) The Relationship Between DPCM and PCM. Consider a linear differential transmission system as shown in Fig. 5.5, but in which the quantizer in dotted lines is omitted. Note that the integrators at receiver and transmitter are identical. At the adders

$$H_0 = H_{\text{sig}} + aH_0\,e^{-sD}\,,$$

while at the subtractor

$$H_{\text{sig}} = H_{\text{in}} - aH_0\,e^{-sD}\,.$$

Thus,

$$H_0 = H_{\text{in}}\,,$$

and therefore

$$H_{\text{sig}} = H_{\text{in}}(1 - a\,e^{-sD}) = (1 - a)H_{\text{in}} + aH_{\text{in}}(1 - e^{-sD})\,.$$

When a is nearly unity, the transmitted signal is essentially a pel-to-pel difference (derivative) signal. As the "leak" is increased, the proportion of baseband signal transmitted along with the derivative is increased until, for low loop gain, the system is equivalent to ordinary PCM. Considering the elements inside the dotted boxes of Fig. 5.7a as linear filters, then for all values of a, DPCM is identical to PCM with filters. We need not limit ourselves to the simple recursive filters of Fig. 5.5, but can consider more generalized filters as in Fig. 5.7b.

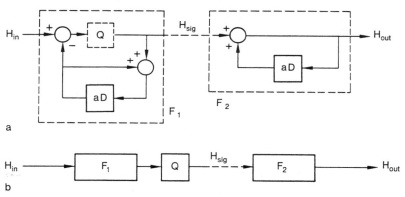

a

b

Fig. 5.7a,b. A comparison of DPCM and PCM: If the quantizer of a DPCM coder (a) is assumed linear, then the entire encoder is simply a linear filter, F_1. The topology is thus equivalent to a generalized PCM system (b)

The only essential difference between DPCM and PCM is the placement of the quantizer. In DPCM it must be inside the loop for two related reasons. One is to prevent the accumulation of quantization noise. The other is to permit the accurate recovery of the dc and low-frequency components of the input signal. The penalty for this location of the quantizer is heavy: F_2 must be causal, as a result of which all of the quantizing errors are located after any luminance transition. In addition, the filter is recursive, which restricts the design in many ways, especially when attempting to operate in two dimensions.

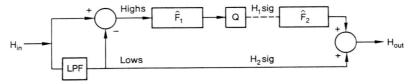

Fig. 5.8. A two-channel system

Removal of the quantizer from the loop is possible if the integrator is required to recover only the high frequency signal components. This can be done if a separate channel is provided for the dc and low frequency components as shown in Fig. 5.8.

At very low cost in channel capacity and equipment complexity, the lows channel can provide a high SNR dc component completely free of digital artifacts. The filters and quantizer in the highs channel can be designed with fewer constraints, as a result of which the quantization errors can be confined to local areas and can be distributed isotropically.

i) Two-Channel Systems. Early work on two-channel systems was motivated by the differing SNR requirements of the low- and high-frequency components of video signals, either one- or two-dimensional. *Kretzmer* [5.46] developed a system for real-time TV that used a lows band of 500 KHz, sampled at 1 MHz, 7 bits/sample, plus a complementary highs band sampled at 7 MHz, 3 bits/sample. *Schreiber* et al. [5.47] used a similar lows band, but synthesized the highs from a coarsely quantized, entropy-coded, pel-to-pel difference signal. *Graham* [5.21] and *Pan* [5.22], by computer simulation, demonstrated similar systems using coarsely sampled but finely quantized two-dimensional lows plus synthetic highs reconstructed from coarsely quantized spatial gradients. Contour coding was used for gradient transmission.

All of these two-channel systems obtained their savings by coarse quantization and, in the two-dimensional case, contour coding of the highs. While substantially free of slope overload, the two-dimensional versions had poor texture reproduction when the contour threshold was set high enough to result in low data rates. Somewhat "lumpy" rendition of sharp boundaries was sometimes encountered. The one-dimensional versions showed some horizontal streaking like DPCM.

In our most recent work, we have specifically attempted to use the two-channel technique to overcome the main limitations of DPCM while at the same time preserving its principal advantages. We have also addressed the problems of the previous attempts to use the method [5.48].

The basic difference from DPCM is the provision of a separate channel for the two-dimensional lows signal, as shown in Fig. 5.8. Note that the signal that is quantized for transmission in the highs channel is the difference between the baseband (original) and lows signals. Thus, in a sense, the lows signal takes the role of the predictor in DPCM. (It is probably not as effective a predictor.) An important result of this arrangement is that the filtering and quantization of the highs signal are totally unconstrained by the causality and stability considerations so important in DPCM. The use of a finite impulse response filter with circular symmetry distributes the quantization noise isotropically. In addition, the response can be chosen in such a way as to produce a pleasing edge transition when edge enhancement is used. DPCM permits little enhancement of any kind, since accentuating the highs before coding increases slope overload, while accentuation after decoding decreases the SNR.

A substantial improvement over previous work has resulted from the application of Roberts' pseudorandom noise processing to a properly tapered quantizer in the highs channel. This method eliminates the problem of texture distortion of the earlier two-channel systems that used coarse quantization, since the quantizer is stepless, albeit noisy. Quantization noise is completely dispersed as random noise uncorrelated with the image. As such, it is less visible than the typical granular noise of DPCM. The tapering, moreover, distributes the noise spatially so that it is smaller in blank areas, where it is more visible, and larger in complicated areas, where it is less visible, as shown in Fig. 5.9.

When appropriate companding is used before the coder, so that the noise is also equally visible throughout the tone scale, useful amounts of compression can be achieved with no nonlinearities (corresponding to slope overload in DPCM) at all. For 512×512 pictures, fair quality is obtained at 2 bits/pel for the highs and good quality at 3 bits/pel. Lows require about 0.3 bits/pel. A notable point about this system is that the only difference between the input and output images is additive random noise, which is equally visible over the entire image. As the compression ratio is reduced, the errors tend toward zero.

The two-channel systems lends itself to adaptive quantization, as does DPCM [5.49]. A useful technique is to measure the largest value of the highs signal in blocks, typically 5×5, to multiply the signal into the quantizer by a factor f, always larger than one and that just does not cause overload, and then to divide the quantizer output by the same factor, thus reducing the noise by the factor f. In blank areas, the value of f is typically 16–32, so that the noise reduction is significant. In fact, the process reduces the blank area noise to such a large extent that reverse companding can be used. The result is that better edge rendition occurs because the outer levels of the quantizer are finer. A complete system is shown in Fig. 5.10, where the components in the dotted box are required for adaptation.

The two-channel system can also be entropy coded, since the probability distribution of the highs signal is highly nonuniform and its entropy is low. *Sharpe* [5.50] used 11 to 17 levels (as compared with the 24–32 required in DPCM) to obtain "perfect" pictures without noise processing. He then performed hybrid coding on the quantized highs signal, using run-length coding for the zero level and modified Huffman coding for the others. For three typical 512×512 subjects, the coded data rate was a little more than one bit/pel.

Although more research is required to quantify the performance of this system, it appears to have an advantage over DPCM systems of comparable com-

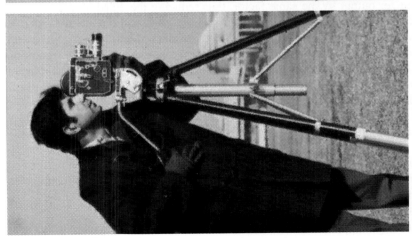

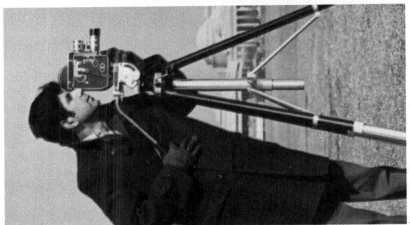

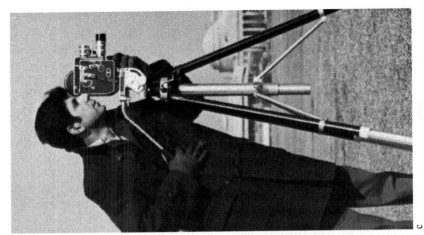

Fig. 5.9a–c. For caption see opposite page

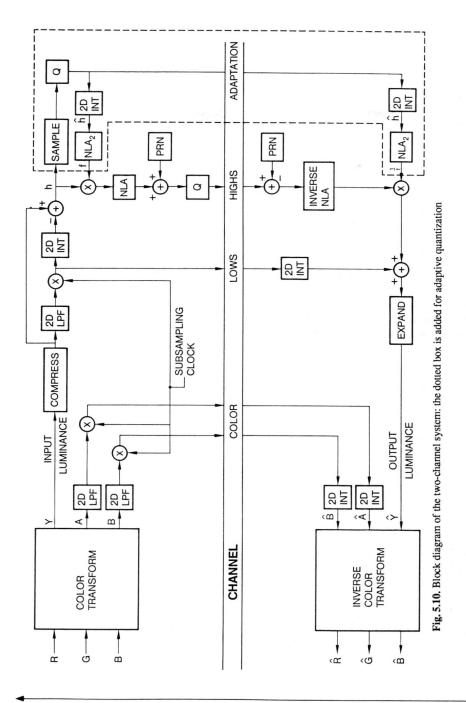

Fig. 5.10. Block diagram of the two-channel system: the dotted box is added for adaptive quantization

---

Fig. 5.9a–c. Comparison of several coding systems: (**a**) 2-channel system, adaptively companded with 2 bits/pel for the highs, (**b**) DPCM, 2 bits/pels, (**c**) randomized DPCM, 2 bits/pel

plexity of at least one bit/pel, and much better performance in the face of channel errors.

The systems of [5.48,49] do not use contour coding for the highs transmission, which had proven to be the source of the distortion in the earlier work [5.21,22,47]. Recently, efforts have been made to retain the very high compression ratio obtainable by contour coding of edge information that is probably more efficient than straightforward entropy coding of the highs data. Texture distortion is avoided by using a third channel to transmit the latter information explicitly [5.51]. While this work has not been carried far enough to make definitive judgments at this time, there is at least a possibility that it will prove the most efficient of all intraframe coding methods.

Another promising approach is the multichannel "Laplacian pyramid" of *Burt* and *Adelson* [5.52]. In this method, the two-dimensional spectrum is divided into a series of band-pass channels, much in the way that some psychophysicists believe the human visual system is organized. In some ways an extension of the systems mentioned previously, it features a fast computation algorithm as well as tailoring of the sampling and quantization of each band in accordance with the data. Although it can be thought of as a kind of crude transform coding, the components are perceptually meaningful. Like the contour-texture method, it has not yet been fully developed.

5.5 Lessons for the System Designer

The first step in designing any image transmission system is the appropriate choice of basic system parameters: the spatiotemporal sampling densities, pre- and postfilters, bits/sample, and, of course, the method of randomization of coding noise. Should further compression be required, some of the methods of this chapter can be used.

Choosing the best compression system for a particular application on the basis of the literature is a frustrating task. Although most authors now agree that compression ratios must be linked to image quality, in all but a few papers, the quality of reproduction is too low for evaluation. In addition, the great variety of subject matter and equipment used make comparison difficult. Many authors ignore the effect of sampling density and viewing conditions (quoting the data rate as bits per pel without qualification) and very few seriously deal with the influence of cameras and display devices.

The principal exception to these deficiencies is in broadcast television, where careful subjective testing with standardized equipment is quite common [5.53]. It is interesting to note that in this field, entropy coding of differential signals had been the clear winner, at least until quite recently. Transform coding, in spite of nearly twenty years of effort, had not been able to match differential coding, where very high quality has been produced at 6 Mb/s, and, with somewhat reduced motion, at 1.5 Mb/s. This has changed with the development of systems that use two-dimensional adaptive discrete-cosine transform coding applied to the motion-compensated frowe-to-frowe prediction error. See Chap. 8. In any event, adaptive methods are certainly the methods of choice, regardless of the type of coding used.

It is the author's belief that multichannel coding system show great promise, as they allow the transmission characteristics to be tailored most precisely to the known characteristics of the human visual apparatus. (No doubt we need to know more about the HVS to optimize these systems.) Adaptive quantization, and perhaps adaptive filtering, can be used to fit the system to the characteristics of the picture information. Entropy coding of the resulting image representation should minimize the channel capacity without affecting the quality. Explicit channel error correction is preferable to compromising the image quality for better noise perfomance.

6. Image Processing in the Graphic Arts

One of the most productive applications of electrical engineering and computer science in the last twenty years has been in the preparation of information, i.e., text, pictures, and line drawings, destined for reproduction on paper. In this chapter, after a brief review of typography and printing technology, we present a discussion of traditional and modern methods of producing pictures by printing. Not surprisingly, we find that many of the principles discussed in earlier chapters are directly applicable to this problem. In particular, we find that perceptual phenomena, especially as they relate to the halftone process, must be given careful consideration to get good results.

6.1 Introduction

Without question, the great majority of permanent images are produced on printing presses. While printing technology has always attracted the attention of inventors, mechanics, and artists, in recent years the economics of printing has motivated a substantial application of the most sophisticated modern technology. The resulting revolution in the production of the printed page merits serious study by the image processing community. In few fields is there comparable opportunity to combine art and science in the service of education and culture.

The proper role for modern technology in printing cannot, I believe, be understood without at least some knowledge of its history, which is actually quite brief, since there was practically none in the West before the fifteenth century. This is rather curious, since both the Greeks and Romans, not to mention the Chinese[1], had ink and paper, together with ample mechanical skill, and many even older civilizations used stamps to make repetitive patterns in clay. Nevertheless, books were laboriously copied by hand and usually devoid of illustrations. The latter is readily understood, since pictures can withstand few generations of manual copying without serious degradation, while alphabetical characters, which may be thought of as very coarsely quantized pictures, are much less prone to error and distortion. This leads to such curiosities as a Greek botany text without pictures.

---

[1] Movable type was in use in China during the Soong Dynasty (ca. 1041) but whether it spread throughout the Orient is not clear. In any event, Marco Polo did not bring it back in 1275 along with gunpowder and pasta.

Two methods came into use, apparently at about the same time, for making reasonably accurate repeated copies of images. Intaglio is the generic name given to processes in which scratches (depressions) in surfaces, most often metal, are filled with ink, which is then transferred to paper by pressing the latter against the surface with a soft pad. Usually the entire surface is first inked and then wiped clean, leaving ink in the depressions. This process undoubtedly was derived from the art of decorating metal objects, such as vases and shields, with engraved patterns. For centuries, there was no better way to make printed pictures, so that the word engraving became synonymous with good picture making. Many newspaper camera and plate-making departments are still called the "engraving room" although no engraving, i.e., the cutting of lines in the plate with a sharp tool, is or ever was practiced.

Wood-block printing seems to have appeared at about the same time as intaglio. In a sense the opposite of engraving, material is removed from the block where no ink is to be printed instead of vice versa. The relief surface is inked and the image then transferred in much the same manner. The composition of text was immeasurably simplified with the invention of movable types by Gutenberg, but images were still hand carved for the next 400 years.

Many improvements in relief printing (now called letterpress) and intaglio (now called gravure or rotogravure) were made in the period before the first application of photographic techniques around 1850. These included stronger, faster, and more accurate presses, better paper, improved inking methods, as well as advances in plate making itself. Woodcuts and metal cuts became finer and finer, but the gravure plates were inherently capable of better detail, since an inked line corresponds to a depression rather than a ridge. In particular, the ability to make very thin lines made it possible to simulate shadings of tone by regulating line spacing. Since the arrangement of such lines is always visible, engravings have a particular character, depending on the style of shading, which came to be prized for itself.

A significant step in both types of plate making was the early discovery that material could be removed by acid etching.[2] In this method the metal surface is covered with wax or other acid-resistant material, which is then scratched away with a stylus to form the image. Then acid is applied, which removes some unprotected metal, producing black lines in gravure but white lines in letterpress, corresponding to the paths of the stylus.

Lithography, an entirely different type of printing using flat (planographic) printing surfaces, is attributed to Senefelder in 1797. Literally "writing on stone," it was originally carried out (and still is by some artists) by drawing with oily or waxy material on flat stone having a specially prepared ground surface. The stone is then dampened with water and covered with oily ink. The latter adheres to the

---

[2] Technically, the word etching is reserved for chemical material removal and the word engraving for mechanical removal, i.e., cutting with a tool. However, both words are often misused, even by graphic arts workers.

drawing but is repelled by the wet portion of the surface. The ink is transferred as with letterpress or gravure.

The development of photography contributed in many ways to the improvements in printing. Of course it provided original images resulting from the mechanical operation of a camera rather than the skill of an artist, a change that has not invariably been an improvement. More significant is the role of photographic methods in transferring images of any kind to printing plates.[3] The discovery of photoresists such as albumen (still sometimes used), which harden after exposure to light, make it possible to make relief plates photomechanically. In this process the metal is covered with a photoresist and exposed to a negative line image, for example, of a drawing. The resist becomes hardened on exposure, so that the unexposed material can be washed away, leaving the metal uncovered. Acid etching removes the metal, after which the remaining resist is removed. A positive image is used to make a gravure plate.

The final step in the elimination of handwork from plate preparation was the development of the halftone process for the simulation of shadings, so that continuous-tone images might be reproduced purely mechanically. This was a mechanical version of aquatints and mezzotints, in which a very fine semirandom pattern of interlaced black and white marks was engraved or etched by working a plate, either bare or covered with a chalky acid resist, with a fine tool that might have a single point or many points to create a stipple effect. As early as 1850, Talbot, one of the pioneers of photography, conceived of the use of a crosshatch screen, sandwiched with a negative, to expose a photoresist coating on a metal plate. After washing, the plate could be etched to produce either an intaglio or relief plate. The first newspaper picture of this type was printed in 1880. Ives invented the cross-ruled glass screen in 1891, and by early in the twentieth century, high quality halftones were regularly being made using such screens.

The spread of education from a wealthy elite to common men could not have occurred without the development of movable type and wood-block illustrations. Cheap books on science, technology, travel, and the practical arts were among the first products of that combination even before the end of the fifteenth century. The development was essentially completed by about 1910, when good quality halftones could be reproduced from original photographs without depending entirely on the skill of the artist. Accurate inexpensive color reproduction was the only missing element at that time.

---

[3] The earliest application of photography to printing was probably made by Nièpce, who utilized the photosensitivity of bitumen in 1826, and by *Talbot*, the inventor of the negative-positive process, who by 1850 was making good quality steel intaglio plates [6.1].

6.2 Type Composition

Modern printing methods have been strongly influenced by the manner in which the characters making up the text are assembled. This is especially true in newspapers, where a large quantity of material must be set in a short period of time. In the four centuries following Gutenberg, there was no other way than hand setting of raised type for letterpress. Lithography and gravure never were practical for text before the advent of photographic, i.e., photomechanical methods of plate preparation, although a small amount of type could be incorporated by treating it as an image.

6.2.1 Type Casting Machines[4]

Although skilled hand compositors work at a speed astonishing to the layman, it is very slow compared to that achieved by a modern typist. The high error rate is a handicap, and the necessity of returning the used type to its case results in an added cost. When the idea of the keyboard appeared in the nineteenth century in connection with the early typewriters, many inventors attacked the problem, leading eventually to the development of type casting machines[5] of which there are, or were, two main kinds – those that cast one character at a time, and those that cast an entire line at once. A key element in both is that the machines themselves use molds, or matrices, rather than type, and that the matrices recirculate in the machines. The cast type is melted down after use, rather than returned to the case. Finally, the matrices to be used are selected in response to keystrokes.

The Monotype machine has over two hundred characters arranged in a rectangular array, in the form of molds cut into the surface of a hard metal block. The block is pneumatically positioned so that the selected character receives molten type metal through a tube. The cast character hardens almost immediately, after which it is removed and put in a channel. When the channel is filled nearly to the desired line width, spacers push the words apart to even the right hand edge (justification). The lineful of type is transferred to the chase, where the entire page is assembled. Type fonts are changed by using different matrix blocks.

The compressed air that drives the matrix block is controlled by thirty-one rows of holes in a paper tape. The tape is prepared on a separate keyboard unit with one key for each different character, or "sort". The large number of sorts is due to the requirement of more than one alphabet in most kinds of printing. For example, italics, boldface, small capitals, etc., in addition to a large number of punctuation marks, may be used.

---

[4] This discussion is limited to Roman alphabets and those others (such as Arabic and the scripts used in the Indian subcontinent) having no more than several hundred characters. Chinese and similar scripts used in Japan and Korea, which require thousands of characters, are just now being automated by photocomposition.

[5] Type casting, formerly the main printing method for books and newspapers, has become virtually obsolete in developed countries.

The line-casting machines, of which there were once many kinds, have individual character molds arranged in long channels in magazines, each of which contains perhaps 90 different character sorts. Two to four magazines may be on a machine at once, with provision for quick change of magazines and for the use of separate "pie" characters for rarely used sorts. In response to key strokes, the molds slide down the channels into a receptacle that holds a lineful. When the line is nearly full, tapered wedges are inserted between the words for justification and type metal flows down to cast the entire "line-of-type" in one piece.

Although the name of Ottmar Mergenthaler is usually associated with the Linotype machine, he was but one of many inventors who made successful machines. He was, however, by far the most successful entrepreneur, buying up all his competitors except Intertype. Corporate descendants of both companies still exist, although line casting machines are nearly obsolete, and the companies have turned to newer techniques.

The Linotype machine can operate a good deal faster than its keyboard, so that eventually separate five-level paper tape perforators came into use for preparing copy. Since there are many controls on the line caster other than the keyboard, many of the tape characters are typesetting commands rather than characters to be printed. The particular commands needed for the Linotype machine, such as "upper rail" to select a magazine, "quad left" to push the type to one side, etc., have (unfortunately) remained in the typesetter's command dictionary long after the Linotype itself went out of service.

6.2.2 Photocomposition

A development of even more significance to printing than the tape- or keyboard-operated type casting machines is photocomposition. First proposed by *Higonnet* and *Moyroud* [6.2], it translates a keystroke sequence into a photographic image of the copy to be printed by means of projecting onto the output film or paper, in sequence, images of master characters stored on a rapidly rotating disk. Characters are selected by flashing an intense light at the appropriate moment during rotation, while the placement of characters on the line and page is accomplished by moving the optical system and the output material, respectively. Photocomposition is often called "cold type" to distinguish it from the older "hot metal" composition.

The development of the "first generation" photocomposition machine was carried out in a rather unusual manner. A nonprofit organization called the Graphic Arts Research Foundation, headed by Dr. Vannevar Bush of MIT, was set up by the US printing industry. After successful prototypes were developed about 1950, Photon Corporation was formed (since absorbed) and production commenced. Eventually many other companies entered the field, and this method of composition became nearly universal in the developed world.

The first machines were followed by a second generation that use cathode ray tubes for display and a TV camera to scan the master characters. Then came digital storage of the master characters to eliminate the TV camera, and recently laser scanners for display have begun to come into use. The preparation of digital type fonts of sufficiently high quality to satisfy the exacting requirements of printers proved to be more difficult than at first expected. As a result, it became a specialty of a few individuals and companies, along with the coding of the fonts so as to minimize storage space [6.3].

Of equal significance to the mechanical improvements has been the increase in "intelligence" through the use of computers and software, which has enabled typesetters to do hyphenation and justification ("H&J"), to accept typesetting commands for formatting and font selection, and even to organize and position all of the elements of a complete page, except the graphics. This process, properly called page composition, is sometimes confusingly called pagination, a term used in much of the printing industry to mean assigning page numbers.

At the present time photocomposition systems are available from less than $5,000 to nearly $200,000. The more expensive machines set thousands of lines of copy per minute and have dozens of fonts and sizes on line at once. Many of these machines produce typographical quality fully the equal of the best metal type. The enormous reduction in time and cost of composition, together with a much lower error rate, have made obsolete not only the venerable Linotype, but also the highly skilled compositors needed to key the complicated tapes with all their special commands. This has created a serious human problem in the printing industry that has only recently been finally resolved.

The technological impact of photocomposition on printing is closely related to accompanying developments in other aspects of printing technology, and in particular the methods of plate making, to be discussed below.

6.2.3 Text Editing Systems

As mentioned above, since even the earliest photocomposition machines required some internal logic to take the place of the mechanical operations of the linecasters and the decisions of the operator relative to word spacing, etc., it was not a big step to start making this logic more powerful. Even the simplest systems now require little more knowledge to prepare tapes for straight text than to type. As more elaborate composition tasks were added, the "front end" of these machines became so large as to warrant division into a separate system, leaving to the output device only the task of printing the specified characters in the specified locations.

An important concept in computer systems that feed typesetters has been to "capture" the original keystrokes as copy is created. This means that where possible, text is created at a computer terminal. In some cases, optical character recognition (OCR) machines may be used for the input of copy typed on standard typewriters. Copy originating at a distance may be sent directly to computer storage. In any event, editing – i.e., additions, deletions, rearrangements, and even rewriting, is done at video display terminals (VDTs). Captions and subheadings may be added, space indicated for pictures, preformatted layouts used for repetitive copy such as classified ads, type fonts and sizes selected, copy rearranged to fill a given space, etc. In short, virtually every function that once was done by hand can now be done interactively with computer assistance or entirely algorithmically. In addition, many ancillary functions not directly involved with printing, such as billing, preparation of indices, statistical studies, etc., can be carried out in separate computer runs or simultaneously. Like photocomposition itself, these developments have also displaced many skilled people while at the same time reducing costs for printers.

6.2.4 Page Composition

As photocomposition has developed practically to perfection, and as "front end" systems have become more ambitious and successful, numerous proposals and some attempts have been made to automate the entire composition process, from text and graphics input to full page output, either on a negative or perhaps an actual printing plate. There are two obstacles that have so far prevented such systems from becoming cost effective. One is the vastly larger volume of data required for images as compared with type, and the other is the inability, thus far, of computer systems to provide the same easy overview of the entire composition task as do conventional methods.

With respect to the volume of data, text is normally kept in ASCII form right up to the typesetter. In this form, no more than about 200 bytes/in^2 are required to specify typical body type for books or newspapers. Even this is not negligible – perhaps 3 megabytes for a thick newspaper. Uncoded graphics, on the other hand, at a resolution suitable for newspapers or magazines, may require 100 times that for a rather modest proportion of the page area. Rapid manipulation of this data so that the editor can easily see the effect of moving blocks of copy from page to page has not proved simple.

It may be that the difficulty with economical page composition stems from the viewpoint of the systems designers, who are primarily text-oriented, and who therefore tend to look on the pictures as special forms of text files. In any event, the pressure for solutions to this problem is great and we may expect to see a great deal of progress in the coming years.

6.3 Modern Plate Making

6.3.1 Letterpress

a) Hot Metal. In this most traditional form of newspaper printing, still used by some of the large metropolitan dailies, the text is prepared in the form of column-wide slugs of wrong-reading raised type on a Linotype machine, together with headlines, etc., prepared on a Linotype machine or Monotype, or even hand set. The pictures and display advertising are usually in the form of metal blocks, in which the halftone dots have been chemically etched into the metal. All the material for the page is locked into a chase, taking care to get all the printing surfaces at the same height ("type high"). The whole assembly may then be inked by hand or roller, and a proof taken by pressing a sheet of paper onto the inked surface. For small, non-newspaper work, the case is put into a sheet-fed press and used directly. For printing on a web-fed rotary press, which uses a continuous roll of paper (a "web") and in which the printing plates are wrapped around a cylindrical roller, a curved copy of the printing surface must be made. This is done by the stereotyping process, in which a papier-mache mold is made by pressing a damp cardboard mat against the type under heavy pressure. The mat is then removed and dried while bent around a drum. Finally, a cylindrical plate called a stereotype is cast in type metal using the mat as a mold. Often several copies of each mold are made and run at the same time ("two up" or "four up") so that more impressions (copies) can be made than is possible with a single press plate.

b) Cold Type. When using photocomposition (cold type) the output of the type-setter is a photographic image of the printed column. To make a relief plate, the photocomposition copy is first assembled into pages by pasting it with rubber cement or wax onto a page-sized "mechanical" or "flat" together with screened positive pictures ("Veloxes") and positive images of the advertising. The mechanical is photographed onto a page-sized negative using high contrast "lith" film. The negative is then placed in contact with a flat plate, usually magnesium, which has been covered with a photoresist. The plate is exposed through the negative to a very bright light in a "plate burner". The photoresist is developed in such a way that a protective layer is left on the area exposed to light. The plate is finally chemically etched away where not protected, leaving a relief image. The etched plate is the equivalent of the assemblage of metal type in the chase, and further steps are carried out in the same way.

If the etched plate is thin enough, it may run on a rotary press directly. Alternatively it may be used as a pattern plate for stereotyping, in the same manner as the hot metal locked in the chase.

Objection to the traditional letterpress-hot metal system is based both on its inability to use photocomposition easily and on the cost and messiness of the metal stereotype process. Deep etching required to make relief plates from film images is also objectionable on ecological grounds. Many newspapers, for these reasons, have shifted to offset. Other letterpress newspapers have turned to photopolymer (plastic) plates. These are metal-backed plastic sheets covered with a photoresist. After exposure in a plate burner, just as magnesium plates, they are processed and then etched with hot water, rather than acid. They are usually run directly and not used as pattern plates. When the press is to be run "two up," two plates are often made rather than making stereotypes. In general, plastic plates give poorer quality than magnesium, but the pressure to abandon metal etching is very strong and development of the plastic processes is still proceeding.

6.3.2 Lithography

Although actual stones are still used by artists to create original lithographs, the modern version of this process uses thin aluminum plates wrapped around the cylinder(s) of a rotary press, which may be sheet- or web-fed. On the printing press, the plate is first dampened with a water solution and then inked with greasy ink. The resulting ink image is offset[6] onto an intermediate roller or "blanket" and then transferred to the final page. A recently developed form of direct lithography, called "Di-Litho," which is mainly used when letterpresses are converted to lithography, transfers the ink image directly to the final page. Offset uses a right-reading plate while Di-Litho, like direct letterpress, uses a wrong-reading plate.

a) Offset/Cold Type. This is the most popular contemporary printing method, especially in newly converted newspapers and the newer suburban papers. The paste-up and full-page negative are prepared in much the same manner as described above. A photosensitive offset plate is then exposed through the negative[7] in the plate burner and the plate then processed in such a manner as to produce an essentially flat printing surface with the ink-attracting-repelling properties previously mentioned. For printing multiples, several plates must be exposed to each page negative. In the case of pictures, it should be noted that it is the practice of some printers, both offset and letterpress, to make separate, individual halftone negatives of each picture, and to strip these into the full page negative of the text. "Stripping in" is quite expensive in skilled labor and is used in the belief that it is required in order to get good image quality. It is true that the making of the page negative takes more care if pictures are included on the mechanical in addition to "line art" (line art or line work refers to copy that is just black and white with no intermediate gray tones to be rendered). However, it should be noted that correct exposure and development of the page negative are also required if the full typographical quality of the line work is to be retained.

b) Offset/Hot Metal. This is a rarely used combination in which a high quality proof is taken from the locked-up chase, and this proof is photographed to get the full page negative. The negative is then used to prepare an offset plate.

c) Di-Litho. When a conventional hot-metal letterpress newspaper is faced with converting to cold type and a cleaner plate process, most have either gone to plastic plates, which are quite expensive and not yet of very good quality, or have replaced their entire printing plant and shifted to offset. An alternative to which some newspapers have turned is direct lithography. This has the advantage

---

[6] Hence the common name. In fact, offset may be used with letterpress or gravure where it is advantageous to keep the paper away from the plate.

[7] Especially in Europe, positive-working offset plates are sometimes used in which a positive film image is employed in the plate burner.

of permitting conversion, rather than replacement, of the printing presses, and uses much cheaper plates (about \$0.70/page versus \$2–\$4 for plastic plates). As mentioned previously, these systems use wrong-reading lithographic plates otherwise identical to offset plates.

This process is too new to say definitely how well it will work in practice. Its cost advantages are obvious, but it also has all the disadvantages of offset, in particular water on the press and difficult ink control, without one of the main advantages – namely the use of the offset cylinder that keeps the paper away from the plate. Some very successful installations have been made, nevertheless.

6.3.3 Gravure

While the traditional forms of intaglio are still used for art prints, the process has also been mechanized. The modern form of the process has become of great commercial importance. By some accounts it is the fastest growing form of printing. Often called photogravure or rotogravure, it is carried out on high-speed, web-fed rotary presses. The printing surface is in the form of a cylinder, generally copper and often very large, having an array of small etched or engraved cells, typically 150–200 per inch.

The cylinder is rotated in a bath of ink and is wiped clean by a "doctor blade" as the surface emerges. Paper is then fed against the cylindrical surface, picking up the ink from the cells. Since the paper does not come in contact with the inside walls of the cells, they do not wear out. The surface of the cylinder, which does wear down, may be repeatedly replated with chromium. Thus gravure cylinders can make millions of impressions with very accurate metering of ink. This accurate ink transfer, which is virtually independent of speed, makes the process very suitable for color printing. It is used for mail-order catalogs and Sunday newspaper supplements, quality magazines, and even low-quality publications where the runs are very long. There is very little wastage of paper or ink. Only the long time and high cost involved in preparing the cylinders prevents this process from becoming the most common.

a) **Etched Cylinders.** Unlike handmade intaglio plates, the ink-holding depressions in rotogravure cylinders must be small cells. It is thought that the doctor blade would not work reliably if it encountered lines in the surface, especially parallel to the blade.

Several methods are used for etching, but they all have in common the coating of the copper with a photoresist that is exposed to the image either before or after being placed on the cylinder, wash-off of the unexposed resist, and chemical etching to remove material where no light has struck. To ensure that the walls between the cells are intact, separate exposures are usually made to a positive film transparency and to a transparent line screen (opaque square dots). Some experiments have been made with prescreened images in an effort to reduce the cost of cylinder preparation.

Since the cell pattern is imposed on the entire surface, line work as well as continuous tone material is "screened." This reduces the sharpness of print as compared with letterpress and lithography. The effect is not as bad as one might expect, for two reasons. One is that the shapes of the etched cells tend to follow the edges of characters. In addition, the ink flows on the paper to form solid strokes. With an appropriate choice of font and size, highly acceptable results are obtained, although admittedly inferior in resolution to the best of other forms of printing.

b) Engraved Cylinders. A remarkable machine, the Helio-Klischograph, made by Hell, engraves the cells in gravure cylinders by means of a battery of diamond styli. These operate at 3600–4000 cells per second. For a typical cylinder eight feet long and four feet around, capable of printing thirty-two magazine pages, eight styli are placed along the cylinder. Each moves in and out, cutting four pages as the cylinder rotates, engraving the entire cylinder in about an hour.

Because of the geometry of the Helio, only diamond-shaped cells can be cut. As a result, in color printing, it is not possible to use conventionally angled screens for the different separations. This causes some unwanted color shifts when printing on coated paper. It is also not possible for the cell shape to conform to type stroke boundaries as well as in etched cylinders. These deficiencies are compensated for by the cleanliness, speed, and consistency of the process. In addition, since the signals that drive the cutting heads are of necessity in electrical form (the video signals come from a companion scanning drum), the Helio lends itself to both analog and digital signal processing for improved sharpness and more accurate tone scale control [6.4].

c) Electronic Engraving. In an effort to reduce the cost of cylinder preparation, a number of attempts have been made to cut gravure cells with electron or laser beams. The potential attractiveness of such processes is that they can combine the speed and cleanliness of the Helio with even more accurate control of cell size while at the same time permitting arbitrary cell shapes. Especially in the case of computerized prepress systems, free-form cells should permit typographical quality much closer to that achieved in letterpress.

These attempts so far have not been successful, in part because copper removal takes a large amount of power and lasers have very low power efficiency. A system under development by *Crosfield* [6.5] avoids this problem by using plastic cylinder surfaces and cutting a continuous, width/depth modulated furrow in a helical path around the cylinder. It remains to be seen whether this will prove practical.

6.4 The Halftone Process

6.4.1 Introduction

Image reproduction processes using opaque ink or that are incapable of modulating the amount of transparent ink that is deposited at each point cannot reproduce a continuous gray scale. The effect of intermediate gray tones is achieved in such cases by a fine pattern of dots or lines. When viewed from a suitable distance, the fine marks become invisible or at least inconspicuous and the observer perceives the average reflectance, which depends on the proportion of paper covered by ink. This method was used before the application of photography to printing. In the case of engravings and etchings, it was carried to a very high level of perfection, although generally the shading lines were quite visible. Even so, the desired effect was achieved. The type of shading marks used was simply considered an aspect of the style of the artist.

Shading marks fine enough to be virtually invisible at normal viewing distances, and thus capable of giving a nearly perfect illusion of a continuous tone scale, were made possible by the cross-ruled glass screen. By the beginning of the twentieth century, fine quality monochrome pictures were commonplace. The principal improvements that have been made since include the contact screen, which requires less skill on the part of the camera operator, and the development of electrical screening methods. The latter hold promise of lowered cost, improved quality, and perhaps most important, compatibility with computerized typesetting.

6.4.2 The Photographic Halftone Process

Conventionally, halftones are produced in a process camera in which a sharply focussed image of the original is projected onto a sheet of very high contrast lithographic film ("lith film") through an out-of-focus screen. This screen originally was Ives' glass ruling set somewhat in front of the focal plane, but is now generally a contact screen placed in the focal plane directly in front of the film. The contact screen simulates the out-of-focus shadow of the glass screen. In either case, the intensity of the light exposing the film is modulated by the repetitive transmittance of the screen. Since the film has a sharp threshold, dots are produced whose size and shape depend on the image, the screen and the exposure.

It is customary to refer to the "dot percentage," which is the percentage of area covered by ink. Thus the tone scale of the process can be expressed as dot percentage versus reflectance of the original. In the case of glass screens, the tone scale can be manipulated almost at will, although at considerable cost in time and trouble, by using a number of exposures, with and without the screen, without and without the image, and with various degrees of screen defocus. In the contact screen, overall dynamic range is adjusted by the control of the "main" (screen plus image) and the "flash" exposure (screen without image). Highlight

145

contrast is sometimes adjusted by a no-screen exposure ("bump") as well. Certain dyed contact screens permit contrast adjustment by separate exposures through different color filters.

Inspection of screened images or some simple modeling of the process quickly shows that it is only in uniform image areas that the dots are of a uniform shape characteristic of the screen – for example, square, round, or elliptical. In areas having fine detail, a wide variety of dot shapes as well as sizes is produced. These shape variations are such as to follow image contours. It is even possible to produce two or more marks within one dot area if the image has sufficiently small detail of very high contrast. If the detail contrast has been artificially enhanced by unsharp masking or an equivalent process, it is possible to obtain essentially unscreened reproduction in those image areas consisting just of high contrast graphics. This phenomenon greatly improves picture sharpness and detail rendition over what would be obtained if each dot were replaced, say, by a dot of equal area.

The influence of detailed dot shape on picture quality is well known and completely accepted by printers. Halftone scanners made by the Hell Corporation maintain a resolution of 2–3 times the dot density for this reason. The PDI *Compudot*[8] system changes the shape and position of the dot in accordance with a measure of local brightness gradient. ECRM's *Autokon** electronic process camera mimics the operation of a conventional camera, in which many black/white decisions are made within the area of each dot.

6.4.3 Fundamental Considerations in Halftone Reproduction

a) Choice of Screen Ruling. Setting aside for the moment the fine-detail considerations mentioned above, we consider the problem of how to design or select a screen for a particular printing process. Since we depend on the eye to average the array of dots to produce an effect similar to viewing the original, the finer the screen, the more closely the image can be examined without the dot structure becoming obtrusive. Obviously, if it were possible, we would choose a very fine screen in all cases since that would give imagery most like the original. A limitation on the achievable fineness relates to the dynamic range of the process. Naturally, the image cannot be whiter than the paper or blacker than the ink. But within this range, the dynamic range is further limited by the smallest black and white dots that can be printed reliably. This in turn depends on the ink, press, paper, and skill of the operators.

Using the best coated paper with all other factors optimized, printing press results can be about as good as photographs on glossy paper: 100 to 1 (2 log units) or slightly better. Newspapers, however, barely achieve 20 to 1 (1.3 log units). In letterpress printing, an additional problem occurs with isolated highlight dots, which are bound to occur if dots are allowed to drop out in any light area. Such isolated dots tend to accumulate globules of ink and print very

---

[8] Compudot and Autokon are trademarks.

large[9]. For this reason, letterpress printers usually try to keep a minimum dot everywhere. Lithographic printers, both offset and direct, may or may not keep dots everywhere, but in both cases, there is a smallest reliable dot. To achieve a given dynamic range with a given minimum dot requires a minimum dot spacing. Aiming for an unrealistic dynamic range (comparable to the paper/ink ratio, for example) results in too coarse a screen. Likewise, choosing an unreasonably fine screen reduces the dynamic range. It is this trade-off that has resulted in the widespread use of 65 lpi for letterpress and 85 or 100 lpi for offset newspapers. Both letterpress and offset can easily go to 150 lpi or greater using coated paper and better presses. Commercial printers most often use 133 or 150 lpi. In any event, there generally is a screen pitch that produces the best overall results for a given printing situation.

c) Tone Reproduction with Conventional Screens. As in most other image reproduction processes, gray-scale reproduction is of great importance in the acceptability of a halftone. In addition to what may be called the geometrical factors, discussed below at some length, the effect of the printing paper is so strong that it cannot be overlooked. Even taking into account the actual reflectivity of ink and paper rather than assuming zero and 100%, the measured average reflectivity over a uniform field of dots is lower, and sometimes much lower, than indicated by the "dot percentage" or area coverage of the ink. [6.6] The reason for this is that light is never reflected just from the paper surface. Some light penetrates into the body of the paper, is scattered, and emerges through the surface dispersed over an area that can be as large as or larger than a halftone dot. The limit of this effect is reached with fine screens and relatively translucent paper, in which the ink has two equal chances to absorb light, first when it enters and second when it leaves, doubling the absorptance. The effect can easily be seen by first examining a positive halftone transparency by transmitted light, and then viewing it by reflection when it is in firm contact with a sheet of white paper.

Except for extremely coarse screens, this effect is always important in halftone printing. The resulting overwhelming tendency to darken midtones must be counteracted at some other point in the process. Since the degree of correction required depends on both the paper and the screen ruling, it is hard to treat the problem except empirically. Therefore, in the following discussion, it is assumed that such correction will be included in the process in addition to any manipulations used for other reasons.

The tone scale achieved with a particular screen using a single exposure is determined in part by the screen transmittance profile. Consider for the moment a one-dimensional screen as shown in Fig. 6.1 and assume a positive-working photographic process of very high contrast. (Actually, two negative-working stages would normally be used.) The output is therefore white wherever the product of image intensity and screen transmittance exceeds the film threshold. Expressed

---

[9] With plastic plates, small dots are not physically strong enough to avoid distortion on the press.

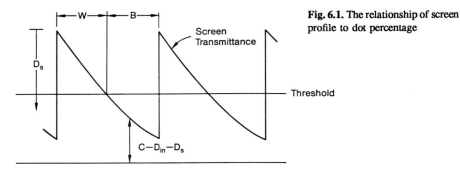

Fig. 6.1. The relationship of screen profile to dot percentage

mathematically the white condition is

$$E - D_{in} - D_s \geq K,$$

where D_{in} is the reflection density of the original, D_s is the screen density, E is a constant depending on the exposure, and K is the film threshold in log units. The fraction of the output image covered with ink is $B/(B + W)$. Clearly the relationship between the dot percentage, $100B/(B + W)$, and the input density is found by interchanging the axes of the graph of the dot profile.

Note that the effective density range of the screen is identical to the density range of an original which, with proper exposure, produces an output that just encompasses the 0–100% dot range. Thus we speak of "short range" and "long range" screens. Between the screen limits, the shape of the tone rendition curve can be adjusted by using screens of appropriate profile. Sections of steep density profile result in low incremental contrast in the output and vice versa. In the case of glass screens, the effective profile can be changed by varying the defocus of the screen shadow by changing the screen-to-film distance and/or the lens stop. In contact screens this property is built in.

Questions of resolution aside, the tone scale of a halftone is not altered if the screen is cut into very small sections and arbitrarily rearranged. For a given original density and exposure it is obviously the fraction of such small elements exceeding a given density level that determines the fraction of corresponding points in the halftone which are black. Thus the integral of the amplitude probability distribution of the screen governs its tone scale, a fact useful in the design of electronic screens.

The variation possible in tone reproduction by the use of "main", "flash", and "bump" exposures is best discussed in connection with a specific example. In order to get a nearly linear reproduction, in log units, the screen transmittance must be inversely proportional to 100-(dot percentage), as shown in Fig. 6.2. This screen has a range of one density unit, so that with the main exposure only, the full 0–100% range of dot sizes is produced from a unit range of original density. If we further assume that the halftone is to be printed on paper of 100% reflectance with totally black ink, then the visual halftone density is simply the negative logarithm, base 10, of the fraction of area that is white.

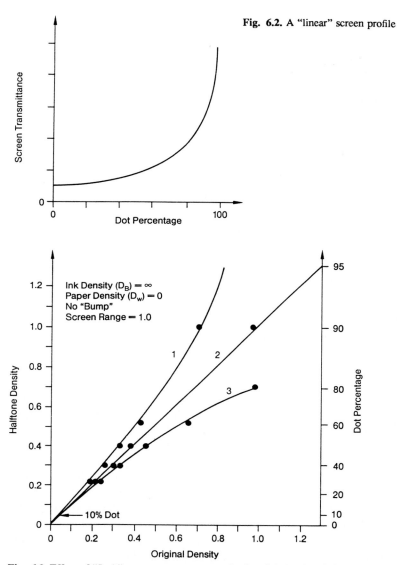

Fig. 6.2. A "linear" screen profile

Ink Density (D_B) = ∞
Paper Density (D_w) = 0
No "Bump"
Screen Range = 1.0

Fig. 6.3. Effect of "flash" exposure on tone reproduction. Ink density = inf., paper density = 0, screen range = 1.0, no "bump". (*1*) Main = 10, flash = 0; (*2*) main = 8.9, flash = 1; (*3*) main = 7.8, flash = 2

Figure 6.3 shows the halftone density calculated on this basis, for three different combinations of main (image plus screen) and flash (screen only) exposures so proportioned that an original density of 0.04 is always reproduced as a 10% dot. The effect of the flash is to increase the effective screen range, so that higher shadow densities can still be reproduced, although at lower incremental contrast. Printers think of the flash as putting a dot in the shadows, which would otherwise go completely black, or "block up".

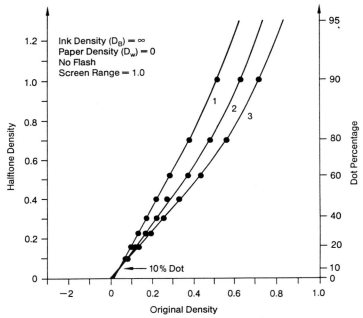

Fig. 6.4. Effect of "bump" exposure on tone reproduction. Ink density = inf., paper density = 0, screen range = 1.0, no "flash". (*1*) Main = 5.45, bump = 0.5; (*2*) main = 7.73, bump = 0.25; (*3*) main = 10, bump = 0

Figure 6.4 shows the effect of a "bump" (image only) exposure, again with the two exposures proportioned so as to maintain the highlight dot. Here the effect is to change the contrast in the highlight to middle tone area, while leaving the shadow contrast unchanged. Bumping is not usually done with contact screens because of the difficulty of removing the screen without disturbing the film.

In Fig. 6.5 the combined effect of bump and flash is shown. In these three cases, varying amounts of bump exposure are used with the main and flash being proportioned to reproduce 0.04 density as a 10% dot and 1.0 density as a 90% dot. The use of three exposures permits a change in the overall curve shape while preserving the end-points. Since the bump exposure can only darken the midtones, the basic screen curve must itself be sufficiently light.

In order to facilitate the curve-shape control made possible by the use of this method, commercial equipment is available that automatically calculates and controls the three exposures based on three density readings of the original.

The assumption of ideal paper and ink is, of course, unrealistic. If we assume perfect dot formation and ignore the physical properties of the paper on which the halftone is printed, then the visual density for given ink and paper as a function of dot percentage can be calculated from

$$D = -\log_{10} R = -\log_{10}[RR_w + (1 - R)R_b] \; ,$$

150

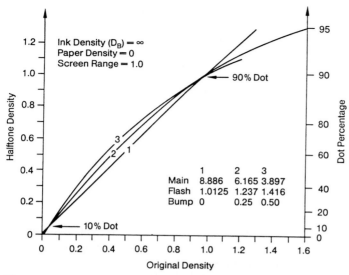

Fig. 6.5. Effect of combined "bump" and "flash" on tone reproduction

where R_w and R_b are the paper and ink density, R is the fraction of paper that is white (ideal reflectance), and D the visual density.

Figure 6.6 shows the effect of imperfect ink, assuming 100% paper reflectance. The incremental shadow contrast is reduced because as the dot becomes smaller and smaller, the amount of light reflected by the "black" ink may

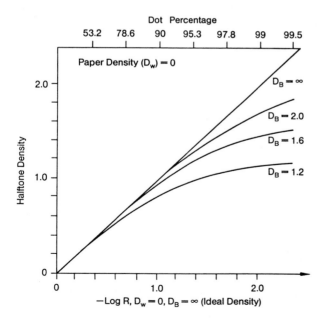

Fig. 6.6. Effect of ink density (D_b) on halftone density

151

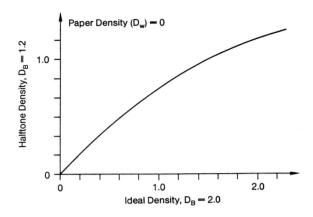

Fig. 6.7. Effect of ink density on halftone reproduction

reach or exceed the amount reflected by the white dot. Better shadow contrast can be obtained at the expense of highlight and midtone contrast by predistorting the dot percentage as a function of original density.

Because of this effect, it is very easy to be misled about the ultimate quality of printed halftones when a proof is examined that is printed with a much higher D_{max} (black density) than that of the final ink. An example is shown in Fig. 6.7 in which the density of a halftone with $D_b = 1.2$ is compared to another with $D_b = 2.0$. The higher ink density extends the useful tone scale further into the shadows, but this effect is lost when the halftone is copied onto paper with ink of lower density.

Figure 6.8 shows the effect of paper density. Darker paper darkens the highlights and midtones, while leaving the contrast in these areas unchanged. If the

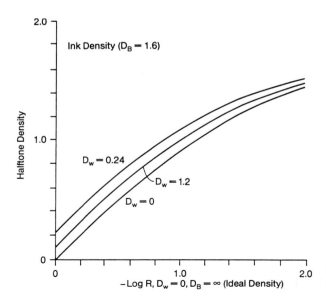

Fig. 6.8. Effect of paper density on halftone density

picture is an intermediate, such as a photo to be reproduced, it is usually better to allow this darkening, since it can readily be compensated for in the next stage by increased exposure. However, the visual quality will often be unacceptable. For this reason, in the case of final prints, it is regular practice to predistort the tone scale so as to render the midtones at about the correct density, even though this flattens out the highlights, reducing tone separation and sometimes giving a chalky appearance.

It is very unfortunate that printers and photo editors tend to ignore the fact that many of the pictures they look at are intermediates, to be further reproduced. The attempt to get best "eyeball" quality at every stage of a picture reproduction system invariably reduces the quality of the final result.

It should be clear from the foregoing that while a good deal of flexibility is available in photographic screening methods, this is accessible only to highly skilled operators. Much experience and attention to detail, a thorough understanding of all the processes involved, and stable chemical processing are all required to achieve the desired result. In addition, attractive output pictures in printing processes of limited dynamic range can be obtained only by adjusting the tone scale for each picture. This requires so much trial and error that it is more an art than a science as usually practiced.

c) Dot Size Change in Successive Stages. Each image passes through a number of stages in going from original to press print, as shown in Fig. 6.9. Many of these stages can be characterized as analog image transfer followed by a sharp threshold process.

The result, as can be seen from the waveforms of Fig. 6.10, is for the contrast to rise, thus losing tone separation in highlights and shadows. This occurs because all of these analog processes have limited spatial frequency response. If the

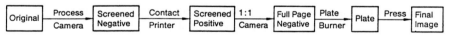

Fig. 6.9. A typical multistage printing process

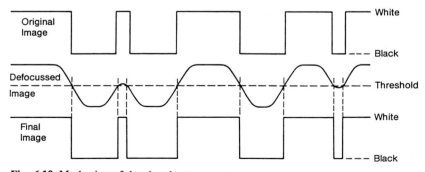

Fig. 6.10. Mechanism of dot size change

threshold is properly set, large dots, whose density reaches its full value at dot centers, do not change in size. Small dots, however, both white and black, do not reach full density and are made smaller, leading to an unwanted rise in contrast.

This effect, which is usually called "dot gain," must be allowed for in preparing the original halftone that is, therefore, often made of quite low contrast. Unfortunately, since virtually every printing situation is different, halftones must be individually tailored in each case. If this is not done, loss of quality – usually too low or too high a final contrast – results. In cases where individual tailoring is not possible, as in the preparation of national advertising distributed to newspapers and magazines already screened in camera-ready form, coarser screens are often used in order to allow a margin of error in printing.

6.4.4 Electronic Screens

a) History and Fundamentals. In spite of the fact that quality halftones can readily be made with the conventional photographic process by craftsmen with adequate skill, there have been many attempts to replace the manual process by various electronic techniques. The motivation has usually been higher speed, lower cost, and a more consistent product. There are even some claims that the more sophisticated techniques of modern image processing might provide better quality.

Early electronic screening systems were hampered by the misconception that halftones are simply sampled images, and that, in order to avoid aliasing, the spatial bandwidth must therefore be limited to one-half the sampling frequency before screening. In these systems, the original image was usually averaged over the repeating area of the screen pattern and a dot printed of appropriate size so that the average reflectance of the halftone was equal (or related in the desired way) to the average reflectance of the original. For example, a repertory of dots of different sizes could be stored and the appropriate one fetched in accordance with the local average reflectance. Since the effect of dot shape was ignored, any shape was assumed to be as good as any other shape.

While these methods do make halftones of a sort, the sharpness is so inferior to conventional screened images in the screen rulings normally used that such processes were never commercially successful.

Nearly all successful electronic screening methods are based on an analogy to the process camera [6.7]. Recall that the halftone is white wherever the product of original image intensity and screen transmittance exceeds a threshold that depends on exposure and film sensitivity. Taking the logarithm of this relationship gives us the fundamental equation

$$E - D_{\text{in}} - D_{\text{s}} \geq K \,,$$

where D_{in} is the reflection density of the original and D_{s} the screen density. This relationship can be implemented physically by the simple comparator shown in Fig. 6.11, in which the constants are absorbed into the screen signal. Actually, this "model" greatly simplifies tone scale considerations as compared to the

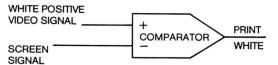

WHITE POSITIVE
VIDEO SIGNAL

SCREEN
SIGNAL

+
COMPARATOR
−

PRINT
WHITE

Fig. 6.11. Comparison Method of Electronic Screening

photographic case if the picture signal is proportional to the image reflectance rather than density. Then all that is needed for a linear tone scale is that the screen signal have a uniform amplitude probability distribution between those values of the picture signal corresponding to minimum and maximum reflectance.

A straightforward method of implementing this model is with a three-drum scanner. Two drums are used to obtain the video and screen signals respectively, by scanning the original and a contact screen (each either reflective or transparent) while the third drum is used to expose the output material using a modulated light source. The light source is turned on and off as the video is greater or less than the screen signal. It is not a big step to derive the screen signal from some kind of more compact, less easily damaged memory than from a physical screen.

As described, the scanner is much like an analog television system, discrete along one axis and continuous along the other. Thus the resolution in one direction is determined by the line spacing and in the other by the bandwidth of the various circuits. The optical spot size affects both directions. The system can readily be made discrete in both directions by sampling the video and screen signals, preferably by a clock locked to the rotational speed. Such discretization is mandatory if digital memory is to be used for the screen signal.

A question immediately arises as to the required resolution of such a scanner. Obviously the recorder section must have a fine enough resolution to delineate the dots in sufficiently fine increments and suitable perfection of shape. The scanning resolution can be different from the recording resolution and in some cases an advantage can be obtained this way. For example, if each dot covers a maximum of eight scan lines, eight lines can be recorded simultaneously from one line of video information. Likewise, in the direction of scanning, all the comparisons required across a dot can be made, if desired, from a single video sample. At the other extreme, the comparison for each picture element (pel) of the output image can be made independently using video and screen signals unique to that pel.

More generally, we can think of the process as the overlay of two separate rasters – the discrete video image and the discrete screen signal. The latter must have enough resolution to delineate dots with sufficient accuracy. The former may have the same or much lower resolution. A very important question in the design of such systems is the appropriate relative resolution of these two images.

The question is most often put as, "How many video samples are required per halftone dot?" However, there are theoretical reasons as well as some experimental evidence for believing that this is the wrong question. It appears that the required video resolution is actually independent of the screen pitch, relating instead primarily to the visual acuity of the observer.

155

In the case of continuous tone (unscreened) images at normal viewing distance, picture quality improves up to 200 or perhaps 250 samples per inch, depending of course on the subject matter. The spectrum of a coarse halftone version of such a picture includes the entire baseband as well as harmonics of the screen frequency and many alias components. It is perfectly clear that the presence of the higher baseband components, even though corrupted by heavy aliasing, greatly improves the subjective quality. We may therefore speculate that the resolution requirement of a picture to be screened may simply be that of the original itself. Thus, 65-line screens should have 4 or 5 samples (one dimensional) per dot, 100-line screens about 3, and 150-line screens about 2 for the best possible resolution [6.8].

The rendition of spatial frequencies equal to or higher than the screen frequency is highly nonlinear. Fine details of sufficiently high contrast to break up the dots are reproduced rather well, while low contrast details are lost. For this reason, surprisingly large degrees of accentuation of image components in this frequency range produce very well worthwhile improvements in picture sharpness, without at the same time showing undesirable artifacts.

b) Digital Screens. We shall confine our attention in this section to digital screens that are the functional equivalent of photographic contact screens. The screen signal is a periodic function of two space variables, is designed to be used in the comparison technique, and does not depend on the video signal. More complicated screens have been proposed in which, effectively, a large number of dots of different shape and size are stored, the correct one being fetched by some complicated function of all the video values within the dot area or even in the neighborhood. The essential point about the screens to be discussed is that each dot encompasses all the black pels of the next smaller dot. Screens that do not have this characteristic and therefore cannot be implemented by the comparison methods are likely to result in a discontinuous tone scale when the halftone is reproduced. This may be the reason why such electronic screens have not been used commercially.

Repeating Area. The screen information is normally stored in a digital memory that is accessed by the discrete horizontal and vertical addresses on the page, modulo the dimension of the repeating rectangle. In general, the smallest repeating area is twice that of a 50% dot, i.e., the sum of a 50% white dot plus a 50% black dot. In the case of dots that are symmetrical about the 50% point, the memory can be as small as a single 50% dot, the second half of the signal being obtained by reflection. In both cases, the 2×1 rectangle must be offset on alternate rows. In Figs. 6.12–13, examples are given of both kinds of dots, together with logic necessary to obtain the correct screen signal.

In some cases, it may be advantageous to use a larger repeating area. Note that the theoretical maximum number of gray levels is one plus the number of pels in the repeating area.[10] At the expense of introducing some unwanted lower spatial frequency component due to differences between adjacent dots, the number of steps may be increased by increasing the repeating area to encompass

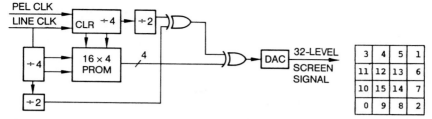

Fig. 6.12. Screen signal generation from the smallest possible memory comprising a single 50% dot: the second half of the dot is obtained by bit inversion (one's complement)

3	4	5	1
11	12	13	6
10	15	14	7
0	9	8	2

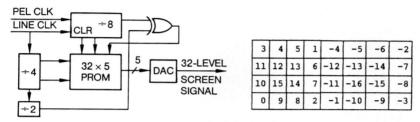

Fig. 6.13. Screen signal generation from a larger memory comprising two 50% dots; this method permits asymmetrical (black/white) dots and somewhat simpler logic

3	4	5	1	-4	-5	-6	-2
11	12	13	6	-12	-13	-14	-7
10	15	14	7	-11	-16	-15	-8
0	9	8	2	-1	-10	-9	-3

four or more 50% dots. Experiments indicate that at least 50 and preferably 100 or more pels in the repeating area are needed for a good gray scale. The use of four, and perhaps as many as sixteen, 50% dots in the repeating area generally does not introduce a significant low frequency component. An example is shown in Fig. 6.14.

Dot Shape. The succession of dot shapes to be produced is achieved by assigning ordinal numbers to the pels in the repeating area. Figs. 6.15-16 show "square dot" and "round dot" screens.

If the memory contents are identical to the ordinal numbers, and if the video signal is at least as finely quantized, and if the pels are reproduced black or white of the ideal shape, then the screen will produce a linear tone scale[11] and all possible dot shapes will occur. If the video is more coarsely quantized, some dots will not occur, as is also the case if the memory contents are a coarsely quantized version of the ordinal numbers. This is done deliberately if the effect known as "posterization" is desired, as shown in Fig. 6.17.

---

[10] Since these gray levels are, in principle, equally spaced in reflectance, amplitude quantization will first show in the shadows. To avoid this, a larger number of levels is required than would be the case if the quantization occurred in the lightness scale. Actually, the darkening of the tone scale due to scattering in the paper causes the darker levels to be more closely spaced in reflectance, reducing the tendency of quantization noise to become visible first in the shadows. In many practical cases, quantization noise is nearly equally visible everywhere, even with ostensibly uniform spacing of levels in the screen.

[11] Except for the darkening phenomenon previously mentioned.

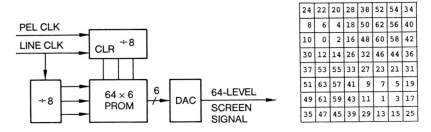

24	22	20	28	38	52	54	34
8	6	4	18	50	62	56	40
10	0	2	16	48	60	58	42
30	12	14	26	32	46	44	36
37	53	55	33	27	23	21	31
51	63	57	41	9	7	5	19
49	61	59	43	11	1	3	17
35	47	45	39	29	13	15	25

PEL CLK

LINE CLK → CLR ÷8

÷8 → 64 × 6 PROM → 6 / → DAC → 64-LEVEL SCREEN SIGNAL

Fig. 6.14. Screen signal generation from a larger, square memory comprising four 50% dots. This permits a larger number of gray levels and the simplest reconstruction logic

81										79
	80	56	54	52	50	51	53	55	64	
	78	49	47	45	43	44	46	48	61	
	76	42	24	21	20	23	25	30	59	
	74	40	19	6	7	8	10	27	57	
	72	38	17	5	0	1	9	26	58	
	73	37	16	4	3	2	11	28	60	
	75	39	18	15	13	12	14	29	62	
	77	41	36	34	31	32	33	35	63	
	79	71	69	67	65	66	68	70	81	
64										80

Fig. 6.15. Half of a 163-level "square dot" screen for a 162-location (9×18) memory: 15.5% and 50% dots outlined

98	110	106	93	86	77	82	89	102	111	100	72	64	52	48	56	68	76
109	97	74	66	54	46	50	62	70	99	112	104	91	84	79	88	95	108
101	69	57	42	33	21	30	38	58	75	107	115	122	130	138	126	118	114
92	61	37	25	17	9	13	26	43	67	94	117	133	141	150	146	134	123
81	49	29	16	8	4	5	18	34	55	87	125	145	153	158	154	142	131
80	45	24	12	3	0	1	10	22	47	78	137	149	157	161	159	151	139
85	53	36	20	7	2	6	14	31	51	83	129	144	156	160	155	147	127
96	65	41	28	15	11	19	27	39	63	90	121	136	148	152	143	135	119
105	73	60	40	32	23	35	44	59	71	103	113	120	128	140	132	124	116
111	100	72	64	52	48	56	68	76	98	110	106	93	86	77	82	89	102
99	112	104	91	84	79	88	95	108	109	97	74	66	54	46	50	62	70

Fig. 6.16. A 163-level "round dot" screen for a 162-location (9×18) memory: 13%, 50%, and 87% dots outlined

Fig. 6.17. A five-level posterized picture

If the range of screen values exceeds that of the video values, then dots will be present even in the extreme highlights and shadows, as is usually preferred in letterpress printing. If the video range exceeds the screen range, then highlights and shadows will become solid white and black.

If it is desired to maintain absolutely a minimum highlight and shadow dot this can be achieved by setting the memory contents to zero and to some maximum value in the corresponding areas, detecting these values, and forcing the output to be black or white regardless of the video values, as shown in Fig. 6.18.

Another way to maintain a reasonably uniform extreme dot, but still allow it to drop out under certain circumstances, is to use the scheme shown in Fig. 6.19,

159

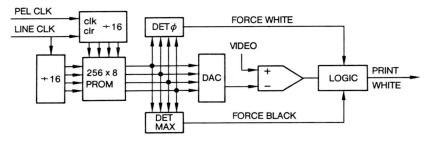

Fig. 6.18. A 16×16, 256-level screen with preset minimum highlight and shadow dots

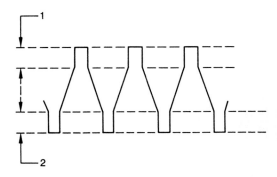

Fig. 6.19. Screen profile that produces uniform highlight and shadow dots in spite of some variation in video level: (*1*) range of highlight video, (2) range of shadow video

where the same value is used at all locations corresponding to the minimum dots, this value being substantially different from the other values, but to omit the detection, force-white, and force-black logic.

Tone Scale. Particular tone reproduction may be achieved by nonlinear amplifiers (NLAs) in the video and screen signal leads as shown in Fig. 6.20. Some obvious conclusions may be drawn about these transformations. The effect of a given NLA in the video path may be exactly compensated by a transpose (x and y axes interchanged) in the video path. Thus the effect of an NLA in one path is identical to that of the transpose NLA in the other path. A desired transformation can be achieved equally well by either method, although quantization noise may be different.

An NLA in the screen path is equivalent to replacing the stored values by the transformed values. Some care may be needed in doing this because of the

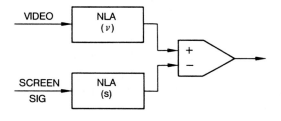

Fig. 6.20. Control of tone scale by nonlinear processing of video and/or screen signals

160

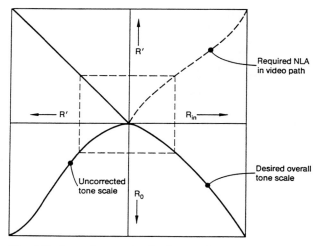

R'

Required NLA
in video path

R'

R_in

Desired overall
tone scale

Uncorrected
tone scale

R_0

Fig. 6.21. Jones diagram showing how to derive the nonlinear characteristic required to achieve a desired overall tone scale

quantization effects. For example, if there are more pels in the repeating area than there are levels in the quantized video, some choice can be exercised so that only certain preferred dot shapes occur.It may, for example, be desirable to avoid dots with single pels protruding on one side.

A common problem is designing a screen with built-in compensation for some unwanted distortion occurring elsewhere, for example in the printing plate. A Jones diagram [6.9] such as in Fig. 6.21 is helpful in providing a graphical solution.

This diagram, or an equivalent computer process, permits the easy derivation of a nonlinear characteristic which, when combined with one or two given non-linearities, permits the achievement of the desired overall characteristic shown in Quadrant IV. To use this diagram, one first makes an overall tone reproduction test using no NLAs and a screen memory whose stored values are equal to or-dinal numbers. This is plotted in Quadrant III as R_0 vs. R'. (Note that $R_{in} = R'$ with no NLAs.) Quadrant II simply has a 45° line. A compensating NLA for the video path is now constructed in Quadrant I. If it is desired to incorporate this characteristic in the stored values, the transpose NLA is used to derive these numbers from the ordinal numbers. The curves given are for an actual case of a pseudorandom screen used to produced simulated continuous tone pictures in a binary printing process.

Joining Corners. Because of the finite spatial frequency response of the various stages in the halftone process, there tends to be a discontinuity in tone repro-duction at points in the tone scale where the corners of dots join. This can be alleviated to some extent in photographic halftones by causing the four corners of the dots to join in sequence instead of all at once. Some care is required in doing this, however, since the resulting asymmetry implies some energy con-

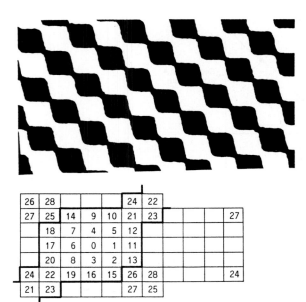

Fig. 6.22. An "elliptical" screen

26	28				24	22				
27	25	14	9	10	21	23				27
	18	7	4	5	12					
	17	6	0	1	11					
	20	8	3	2	13					
24	22	19	16	15	26	28				24
21	23				27	25				

tent at lower spatial frequencies and because the misshapen dots are less able to follow detail in the original. Elliptical screens in which the corners join two by two, and in which 45° chains are formed at the 50% point, are sometimes used. The corresponding operation is possible in digital screens of the type shown in Fig. 6.22.

c) Multiple-Resolution Screens. The flexibility of electronic screens permits easy experimentation with new types. While these types can, in principle, be generated with contact screens, the difficulty of achieving exactly what is wanted by photographic means makes the latter procedure impractical in many cases.

Multiple-resolution screens attempt to circumvent the trade-off between minimum dot size, dynamic range, and screen gauge. Given a smallest reliable highlight and shadow dot, the dynamic range depends on the screen gauge. The fewer dots/in, the higher the possible ratio of highlight/shadow reflectance. Choosing a finer gauge reduces the dynamic range, while choosing a higher dynamic range coarsens the dot structure. Many suggestions have been made to get around this limitation, in part, by spacing the dots farther apart where they are small, i.e., in the highlight and shadow areas, and closer together, rather than larger, in the midtones. The two-dimensional nature of the page prevents a simple continuous change in dot spatial frequency. What can be done is to keep a fairly fine screen in the midtones, dropping out alternate dots in those density ranges where they would otherwise become too small. The "Respi" screen, Fig. 6.23, the photographic version of this idea, has not seen much use, presumably because of the pronounced cane-like texture that occurs in areas where two dot sizes alternate.

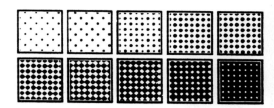

Fig. 6.23. Respi Dual Dots, with conventional square dots [6.10]

An electronic screen using this principle is shown in Fig. 6.24, along with a sample picture in which all the dots have been enlarged by a factor of two to show the effects more clearly. Here the cane effect is eliminated since entire dots drop out as units, rather than alternating in size. An undesirable effect is the amplitude quantization that accompanies sudden dot dropout. An additional problem is that the 50% dots contain many fewer pels than would normally be used in such a fine screen. As a result, the corner joining effect is more pronounced. The effect is alleviated by using a repeating area of sixteen 50% dots, paying particular attention to the order in which the corners are joined. This is done in such a manner that, at any density level, the corner joins that have been made are as evenly distributed throughout the repeating area as possible. It is necessary to check any such proposed screen for unexpected coarse patterns at any density level.

This screen, which can be implemented in a 256×8 memory, is, as normally used in the Autokon camera, 130 dots/in in the midtones and 65 dots/in in the highlights and shadows. The minimum dots are 2×2 pels (0.003″ square). The midtone dots are 4×4 pels. Just above and below the 50% dot, the corners are joined in a pseudorandom sequence selected to minimize the appearance of "chains" or other highly visible, low spatial frequency patterns. Although a screen with a 16×16 repeating area can theoretically have 257 gray levels, in this case many fewer are used so as to eliminate dot patterns having asymmetries that would produce visible texture.

d) Random Screens. A popular method of hand shading in early gravure work was accomplished by using an engraving tool with a random arrangement of multiple cutting points. Such "mezzotints" are simulated with contact screens having random or pseudorandom transmissions. The printability of such screens in any particular printing process depends on the size of the minimum dot.

In order to produce practical random screens by electronic methods, the minimum dot must be controllable. Although this can be accomplished by several techniques, the method shown in Fig. 6.25 has proved to be particularly effective.

Both PRN shift registers have sixteen stages, but the first produces a 16-bit sequence to serve as the starting state of the second, while the second output produces a sequence of 8-bit numbers that are decoded at the output. In order that successive outputs are independent, both registers are multiply shifted for each clock pulse input.

Fig. 6.24. Multiple-resolution screen

118	50	34	110	142	182	166	158	126	50	34	102	134	184	168	150
74	1	1	42	206	229	229	174	74	10	10	42	208	238	238	176
90	1	1	58	222	229	229	190	90	10	10	58	224	238	238	192
126	82	66	98	130	214	198	146	114	82	66	106	138	216	200	162
162	188	172	130	98	54	38	114	146	186	170	138	106	54	38	126
212	256	256	180	78	19	19	46	210	247	247	178	78	28	28	46
228	256	256	196	94	19	19	62	226	247	247	194	94	28	28	62
146	220	204	138	106	86	70	122	154	218	202	130	98	86	70	114
114	50	34	106	138	184	168	154	122	50	34	98	130	182	166	146
74	10	10	42	208	238	238	176	74	1	1	42	206	229	229	174
90	10	10	58	224	238	238	192	90	1	1	58	222	229	229	190
122	82	66	102	134	216	200	150	118	82	66	110	142	214	198	154
154	186	170	134	102	54	38	118	150	188	172	142	110	54	38	122
210	217	217	178	78	28	28	46	212	256	256	180	78	19	19	46
226	217	217	194	94	28	28	62	228	256	256	196	94	19	19	62
150	218	202	142	110	86	70	126	158	220	204	134	102	86	70	118

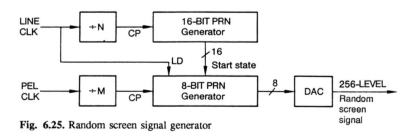

Fig. 6.25. Random screen signal generator

Fig. 6.26. Picture produced by the algorithm of Fig. 6.25

Two different PRN generators are provided. The generating polynomial is selected by a theoretically guided experimental procedure so that the sequences are free of noticeable structure in their length of 2^{16} pels, an area of 12×90 inches at 722 pels/in. The first binary generator, which is clocked at the line-scanning rate divided by N, produces the starting state for the second generator. The latter, which is clocked at the pel rate divided by M, produces a 256-level PRN, identical for N scan lines. Thus the page is divided into $N \times M$ pel squares in each of which the PRN is constant. Selecting N and M greater than one permits adjusting the minimum dot size to suit the printing process. In the sample picture shown in Fig. 6.26, the minimum dot has been enlarged. It should be viewed at about twice normal distance to visualize the effect that would be produced by a high quality printing process.

e) Continuous-tone Screens.

Although the conventional variable-sized, constant spacing dot screen is in most cases a satisfactory method of simulating a continuous gray scale in photomechanical image reproduction, there are some two-level displays, such as plasma tubes, and some paper printing methods, in which very small dots are feasible. In such cases other algorithms can produce images nearly free of visible screen structure. A pseudorandom screen with a very small repeating area, independently discovered by a number of investigators but first properly described by *Bayer* [6.11], works very well. It is designed so as to minimize the low spatial frequency content of the patterns that appear in blank image areas. The horizontal and vertical dimensions of the repeating area are powers of two. Several such screens are shown in Fig. 6.27, together with a sample picture.

2 X 2

0	2
3	1

4 X 4

0	8	2	10
12	4	14	6
3	11	1	9
15	7	13	5

8 × 8 CT SCREEN
(First few pels shown)

0	32	8		2		10	
48	16				18		
12		4	14		6		
3		11	1		9		
	19			17			
15		7	13		5		

Fig. 6.27. Continuous-tone screen and sample picture

There are many other algorithms available for rendering continuous tone images in a binary process [6.12]. The relative quality depends very much on the printing resolution. For example, the error diffusion process [6.13] is probably better for low resolution, while the Bayer algorithm is probably better for high resolution. A defect of the former, which in general has lower noise and higher detail resolution than the latter, is the appearance of periodic irregular contours in blank areas. These may be made much less visible by randomization of the diffusion process [6.14].

6.5 Lessons for the System Designer

In this short chapter, only a brief discussion of this interesting subject could be presented. What should be evident, even from this introduction, is that the graphic arts provide a wide variety of challenging problems for the electrical engineer and computer scientist. A message that I have tried to get across is that, in order for the scientist to be successful in this field, he must understand what the artists and craftsmen have accomplished with existing techniques. It has not been proven, and it may not be true, that high technology, and in particular computer technology, invariably gives a better result than traditional methods. For example, computers have not yet been able to compose pages that meet ordinary aesthetic standards. Computer systems cannot, even today, permit the same kind of rapid access to, and last minute alterations of, the entire news content of a newspaper as is possible with a manual paste-up system. On the other hand, computer-based systems often do a better job of color printing than otherwise possible, and they permit some image-assembly or image-distortion tasks that are impossible manually. Word processing systems are a nearly unqualified success, although most are not as easy to learn or to use as desirable. A great deal more work must be done to make the more powerful systems easy to use, especially by nontechnical employees.

Many of the failures of computer-based systems in this field have been due to trying to do more than reasonable in one step. Especially if the final system is to be large and complex, an incremental approach is desirable. The targets for technological solutions should be chosen carefully, and the changes made small enough to be digested by the user in easy steps. The quality of the final product and the reliability of the equipment should be at least as high as achievable by ordinary methods, and there must be some definite advantage of cost or convenience. Users are rarely attracted simply by novelty of approach. They normally demand all of the capabilities they now have, plus more if possible, and do not want to change their working methods very much. They want the capital costs to be recoverable very quickly. When these stringent requirements are met, they become very enthusiastic boosters of high technology, but not before.

7. Color

The widespread use of color in photography, television, printing, motion pictures, art, and everyday life testifies to its practical and aesthetic importance. The efficient and successful incorporation of color into imaging systems requires detailed knowledge of the technology and methods of color reproduction, as well as some acquaintance with color perception. It is important to realize that it is entirely possible to design and build excellent color reproduction systems without accepting (or even understanding) any of the several theories of color perception. It is, however, absolutely necessary to learn some experimental facts about human response to colored stimuli.

The sensation of color is associated with the spectral distribution of radiant energy reaching the eye. Thus, a colored light may be specified completely, in physical terms, by its spectral distribution, $E(\lambda)$. However, it has been known at least since the time of Newton that two colored lights might look alike although they were spectrally quite different. This is known as a "metameric match."[1] Without the phenomenon of metamerism, color reproduction would be exceedingly difficult, since it would then be necessary to duplicate the spectrum in order to achieve the same human response.

For our purposes, there are two basic questions. How do we *match* a given input color with the limited range of spectral distributions of intensity available in a reproduction system, and, given the spectral distribution (i.e. the physical description) of a light, what does it look like? The second question is much harder than the first and is by no means a closed issue, even today, so we shall begin by dealing with the question of color matching. The question of appearance is dealt with extensively by *Evans* [7.1]. To an important degree, appearance is secondary for reproduction systems, since if the output matches the input in the sense used here, the reproduction will almost always be deemed highly satisfactory. Because many systems have a smaller output dynamic range and color gamut than that of the input material, exact matching is often not physically possible. In that case, some colors must be changed and the appearance becomes important, as it also does when the output is viewed under conditions very much different from the input.

---

[1] In this chapter, double quotes ("") are used for definitions, single quotes for special meanings of ordinary words, and italics for emphasis.

7.1 Color Matching

Experiment shows that almost every color of light can be matched by a mixture of three "primary" lights. The rules of color matching have been known about 150 years, although accurate data have only been available in this century. Different observers may obtain slightly different matches, but the results are independent of brightness and the state of adaptation over a wide range of photopic illuminations. The necessary amounts of primaries are called the "tristimulus values" of the unknown color in terms of the specific set of primaries. The matching is generally done in a device called a "colorimeter." A split field is viewed with the unknown color on one side and the controllable primary mixture on the other. The more brilliant and pure the primaries, the wider the gamut of matchable colors, but with almost *any* set,[2] a match can be achieved to *all* colors, by adding one or two of the primaries to the unknown for those that cannot be matched otherwise. In that case, the tristimulus value is regarded as negative.

The results of such matching experiments lead to some very simple relations, called Grassman's Laws, which express the linearity of color matches – if two colors match a third, they match each other, and the sums and differences of matching colors also match.

7.2 Color Mixtures

Before considering the numerical specification of color, it is well to mention the fact that there are several different ways to mix colors. Note that when light of a single wavelength is transmitted or reflected, the wavelength is unchanged; only the intensity is affected. Thus, when light from several sources falls on a screen, the spectral intensity of reflected light is the sum of that emanating from each source. Such mixtures are "additive" and obey Grassman's Laws. On the other hand, when light from one source passes through a succession of transparent colored filters, the spectral intensity of the transmitted light is the product of the spectral intensity of the incident light and the spectral transmittances of all the filters. These mixtures are called "subtractive." They are not amenable to any analysis simpler than actually calculating the precise physical characteristics of the transmitted light.

Examples of additive mixtures include that of a color TV picture tube where the three primary phosphors are side by side, as well as the temporal mixtures of flicker photometers and spinning discs. The physics and psychophysics of such additive mixing systems are usually such that additivity is perfect. Subtractive

---

[2] The three primaries must be chosen in such a way than none can be matched by a mixture of the other two. If all three are "pure," i.e., composed of single spectral colors, and distributed through the spectrum, this is, fortunately, never possible. The gamut achievable with positive amounts of the primaries can, of course, be extended by using four or more, rather than three. Three good primaries give a sufficiently large gamut for most purposes.

mixtures, in addition to tandem filters, include multilayer color film and mixtures of dyes in solution. The physics of many subtractive processes is complex.

There are some mixtures more complicated than either additive or substractive processes. For example, in paint mixtures, some light penetrates through all the pigments before reflection from the substrate, while some is scattered or reflected from single pigment particles. Thus the mixture partakes of both additive and subtractive qualities. Some pigments mix mostly subtractively. Mixtures of printing inks in the halftone process, wherein the image is made up of a large number of variable-sized spots, are very complicated since when the dots happen to fall side by side, the mixture is additive, while when they happen to superimpose, they are mostly subtractive. As a result of these complications, the color-mixture laws of inks are not easy to describe accurately. In our discussion, we shall deal first with additive mixtures, since they exhibit the fundamental principles of colorimetry most clearly.

7.3 Numerical Colorimetry

A history of the development of the science of colorimetry is given by *Wintringham* [7.2]. The objective of this science is to permit the calculation of the additive mixture of primaries necessary to match a given color, as specified by its spectrophotometric curve. The basic principles are the following:

Every real colored light is the physical sum of a number of essentially pure spectral components. Thus it can be uniquely described by a function $E(\lambda)$ – the power per unit wavelength, as a function of wavelength. If we know how much of each of three primaries is needed to match unit power for every infinitesimal spectral band, we can match $E(\lambda)$ simply by adding up the matches of each spectral component.

The manner in which this is done, partly empirically and partly analytically, is as follows:

In the colorimeter, match an arbitrary intensity of "white" light (equal energy per unit wavelength) with variable amounts of the three primaries. These amounts are designated as unit amounts. Equal-energy white (Illuminant E) is very near to the quality of outdoor illumination, which is composed of yellowish sunlight plus bluish skylight. (For convenience, the light is sometimes replaced by "Illuminant C," a slightly different color, which is an incandescent lamp operated at a certain temperature plus a certain filter, intended to simulate daylight.) The purpose of this normalization is to cancel out small differences between observers, and to bring the numerical value of the tristimulus values into a reasonable range.

The normalized primaries are then used to match very narrow bands of unit energy from 400 to 700 nm. We thus obtain three functions of $\lambda - R(\lambda)$, $G(\lambda)$, and $B(\lambda)^3$, shown in Fig. 7.1, which are the tristimulus values of unit amounts of pure spectral colors. These color mixture curves (CMC's) are usually obtained with the aid of an instrument called a monochromator, which uses a prism to

---

³ Footnote see opposite page

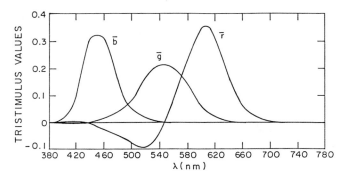

Fig. 7.1. Tristimulus values of the pure spectral colors in terms of the CIE narrow-band primaries at 435.8, 546.1, and 700 nm. White reference is Illuminant E

disperse white light and a slit to pick out a narrow spectral region. A commonly used set of primaries consists of narrow bands of light at 435.8, 546.1, and 700 nm.

We can now *calculate* the tristimulus values for any light $E(\lambda)$ as follows:

$$R_E = \int E(\lambda)R(\lambda)d(\lambda) ,$$

$$G_E = \int E(\lambda)G(\lambda)d(\lambda) ,$$

$$B_E = \int E(\lambda)B(\lambda)d(\lambda) .$$

The meaning of the numbers we calculate in this way is that they tell us how much of each of these three primaries is required to match $E(\lambda)$.

7.3.1 Graphical Representations

Having thus been able to specify any real color by three numbers, it is possible to plot each color as a point in a three-dimensional space. It is clear that all points on a line through the origin have the same ratio of primaries and hence differ in brightness, or luminance only.[4]

The color aspect or "chromaticity" of such lights can be described independently of luminance by the points where such lines intersect the plane

---

[3] Called $\bar{r}$, $\bar{g}$, and $\bar{b}$ by Wintringham, and frequently called 'color mixture curves' by other authors. Each distinct set of primaries has a corresponding set of CMCs, which are objective entities, fully specified by their spectral curves. If these curves are nonnegative, they also can be thought of as representing physically realizable colors. Note that these 'colors' corresponding to the CMCs are *always* different from the colors of the primaries. All six cannot be physically realizable.

[4] It is not true in general, however, that two lights that differ only in intensity will appear to be of the same color. For example, yellow and brown are light and dark colors made up of the same band of wavelengths. In this connection, light and dark are relative, and not absolute levels. A yellow patch surrounded by a very bright white border appears brown.

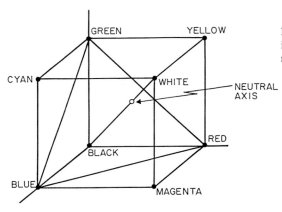

Fig. 7.2. The Maxwell Triangle shown in *RGB* space. CIE narrow-band primaries

$$R + B + G = 1 \ .$$

This plane includes the Maxwell Triangle, shown in RGB space in Fig. 7.2. Within this triangle, the amount of each ingredient increases the closer the point is to the particular vertex. In addition, the position of any chromaticity is at the center of gravity of the three components. Thus, any mixture of positive amounts of the primaries must be within the triangle having the primaries at the vertices. Since all real colors are composed of sums of spectral colors, they must lie within the locus of the spectral colors in the plane of the Maxwell Triangle as shown in Fig. 7.3. That this is so is evident from the construction of the triangle, but it can be demonstrated formally. Let $\hat{r}, \hat{g}, \hat{b}$ be unit vectors along the three axes. If we have two colors

$$C_1 = R_1\hat{r} + G_1\hat{g} + B_1\hat{b}$$

and

$$C_2 = R_2\hat{r} + G_2\hat{g} + B_2\hat{b} \ ,$$

then the mixture of p parts of one with q parts of the other is

$$pC_1 + qC_2 = (pR_1 + qR_2)\hat{r} + (pG_1 + qG_2)\hat{g} + (pB_1 + qB_2)\hat{b} \ .$$

Since in the Maxwell Triangle $R + B + G = 1$, the coordinates of the two colors and of the mixture in the plane of the triangle are

$$C_1 : \frac{R_1}{R_1 + G_1 + B_1} \ ; \ \frac{G_1}{R_1 + G_1 + B_1}$$

$$C_2 : \frac{R_2}{R_2 + G_2 + B_2} \ ; \ \frac{G_2}{R_2 + G_2 + B_2}$$

$$pC_1 + qC_2 : \frac{pR_1 + qR_2}{p(R_1 + G_1 + B_1) + q(R_2 + G_2 + B_2)} \ ;$$

$$\frac{pG_1 + qG_2}{p(R_1 + G_1 + B_1) + q(R_2 + G_2 + B_2)} \ .$$

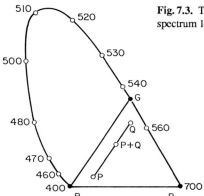

Thus, both coordinates of the mixture are the *same* linear combination of the coordinates of the original colors. If p and q are positive, the mixture thus lies in between and on a straight line joining the two points, as shown in Fig. 7.3.

7.3.2 Chromaticity Coordinates

The stratagem discussed above, in which the luminance in colored lights may be ignored so that their color only can be compared by means of the Maxwell Triangle, can be carried out more methodically by computing a normalized version of the tristimulus values. We define the "chromaticity coordinates:"

$$r = \frac{R}{R + G + B} \, , \; g = \frac{G}{R + G + B} \, , \; b = \frac{B}{R + G + B} \, .$$

Evidently $r + g + b = 1$. We have thus normalized with respect to luminance. It will be seen that all we have really done is to refer to colors by the coordinates of the intersections of their vector RGB representation with the plane of the Maxwell Triangle.

Since the trilinear display is awkward (and also redundant, since the third coordinate can be found from the other two) we can just as well project the triangle on the $R - G$, $B - G$, or $R - B$ planes to get a two-dimensional Cartesian chromaticity diagram. In Fig. 7.4 we have shown the spectrum locus on the $R - G$ plane. All real colors must plot in the shaded area.

7.3.3 Transformation of Coordinates

One of the most powerful results of Grassman's Laws is that the tristimulus values of any color with respect to one set of primaries can be transformed to a second set of primaries, provided that the tristimulus values of the first set of primaries with respect to the second set are known. Let $\hat{r}_1$, $\hat{g}_1$, and $\hat{b}_1$ be one set of primaries and $\hat{r}_2$, $\hat{g}_2$, and $\hat{b}_2$ be the second set. Then if

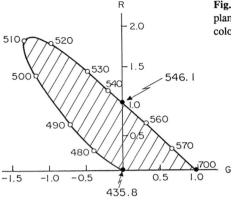

Fig. 7.4. The previous figure projected onto the *R-G* plane. The shaded area is the locus of all physical colors

$$\hat{r}_1 = a_{11}\hat{r}_2 + a_{12}\hat{g}_2 + a_{13}\hat{b}_2$$

$$\hat{g}_1 = a_{21}\hat{r}_2 + a_{22}\hat{g}_2 + a_{23}\hat{b}_2$$

$$\hat{b}_1 = a_{31}\hat{r}_2 + a_{32}\hat{g}_2 + a_{33}\hat{b}_2$$

and if a certain color is given by

$$C = R_1\hat{r}_1 + G_1\hat{g}_1 + B_1\hat{b}_1 = R_2\hat{r}_2 + G_2\hat{g}_2 + B_2\hat{b}_2 \ ,$$

then we can find R_2, G_2, and B_2 from the given data. By substitution

$$R_2 = R_1 a_{11} + G_1 a_{21} + B_1 a_{31}$$

$$G_2 = R_1 a_{12} + G_1 a_{22} + B_1 a_{32}$$

$$B_2 = R_1 a_{13} + G_1 a_{23} + B_1 a_{33}.$$

This is a very useful result, since we can now calculate (or measure) the tristimulus values with respect to any convenient set of primaries, and then convert them into the set actually used for display by a simple linear transformation. It is thus possible to achieve a colorimetrically accurate display, on a CRT, of any image described by tristimulus values, within the gamut of the CRT screen.

7.3.4 The CIE[5] Chromaticity Diagram

Referring to the chromaticity diagram in the $R - G$ plane (remember that the $\hat{r}$, $\hat{g}$, $\hat{b}$ primaries are spectral lines) it is seen that no *real* primaries can form a triangle that encompasses all real colors. This means that negative tristimulus values must result for some real colors for any real primaries. Even for colors that can be reproduced with positive amounts of the primaries, the computation of the tristimulus values involves multiplication of much of the spectrophotometric data

---

[5] Commission Internationale d'Eclairage (International Commission of Illumination)

by negative values,[6] since, as can be seen from the diagram, the red component of the tristimulus values of spectral colors below 545 nm is negative.

In picking a new set of primaries, the CIE was guided entirely by computational convenience. The following requirements on the new primaries were established:

1. The tristimulus values of all spectral colors shall be positive. This means that the primaries themselves cannot be real.
2. As far as possible, zero tristimulus values shall be obtained.
3. Two of the primaries shall have zero luminance. Thus the third will have all the luminance, since a corollary of Grassman's Laws is that the luminance of the sum of some colors is equal to the sum of their individual luminances.

Requirement number 1 means that the location of the new primaries on the $R - G$ plane must be at the vertices of a triangle that encompasses the spectrum locus. Requirement number 2 means that as much as possible of the spectrum locus should coincide with one side of this triangle. Both of these requirements can be met, as it happens, by locating two of the primaries along a straight line tangent to the spectrum locus in the red-to-yellow region where it is almost straight.

To handle the third requirement we should recall that the location of a color on the $R - G$ plane is the projection on that plane of the intersection of the vector representation of the color with the plane $R + B + G = 1$. As we shall see shortly, the locus of vectors of zero luminance is another plane in RGB space. Its intersection with the plane $R + B + G = 1$ is a straight line whose projection on the $R - G$ plane is also a straight line. If two of the new primaries are on this line, they have zero luminance.

When the spectrum primaries (453.8, 546.1, and 700 nm) are normalized at their match with equal energy white, their relative luminances are found to be in the ratio $.06 : 4.59 : 1.00$. Thus the plane of zero luminance is given by

$$R + 4.59G + .06B = 0 .$$

The projection of the intersection of this plane and the plane of the Maxwell Triangle onto the $R - G$ plane is found by substituting into the previous equation the relation

$$B = 1 - R - G$$

giving

$$0.94R + 4.53G + 0.06 = 0 .$$

It is on this line that two of the primaries must lie, as shown in Fig. 7.5.

---

[6] In 1931, this was disadvantageous.

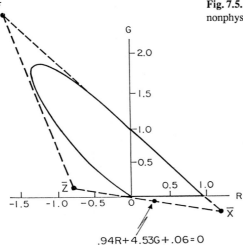

.94R+4.53G+.06=0

We have thus uniquely decided on two of the edges of the triangle of new primaries. The third is arbitrary and is selected to make the triangle as small as possible. We call the new primaries $\bar{X}$, $\bar{Y}$, and $\bar{Z}$, where $\bar{Y}$ is the primary associated with luminance.

7.3.5 The Nature of the CIE Primaries

$\bar{X}$, $\bar{Y}$, $\bar{Z}$ are supersaturated red, green, and blue colors, whose chromaticity coordinates can be found by projecting their coordinates in the $R - G$ plane onto the Maxwell Triangle. Thus

$$\bar{X} = 1.275\hat{r} - 0.2778\hat{g} + 0.0028\hat{b}$$
$$\bar{Y} = -1.7394\hat{r} + 2.7674\hat{g} - 0.028\hat{b}$$
$$\bar{Z} = -0.7429\hat{r} + 0.1409\hat{g} + 1.602\hat{b} \ .$$

Obviously we cannot physically generate these colors since they call for negative amounts of some real primaries. If we wanted to, however, we could make a visual colorimeter for experimentally determining tristimulus values directly in terms of these nonphysical primaries. It would use three pure spectral light sources. The knobs marked X, Y, Z would simultaneously direct the red, green, and blue lights in correct proportions onto the colorimeter screen, with positive values going on one side and negative values going on the side of the unknown color. Actually, this would be pointlessly difficult. It is much easier to use a real-primary colorimeter and then to convert the resulting tristimulus values by means of the expressions previously given.

It should be noted that in the physical realization of primary lights for additive colorimetry, the only important aspects of the light are their tristimulus values in terms of some known colors. In other words, only the *appearance* of the primaries (to the standard observer) affects how they mix. This is very much different from subtractive mixtures, in which colors that look alike may produce radically different results when used, for example, as pigments.

In order to give the CIE primaries more meaning, it is necessary to be able to calculate the tristimulus values of arbitrary colors in terms thereof. To do this, we must know the tristimulus values of the spectral primaries in terms of the CIE primaries. The derivation just given is not sufficient to do that, since we have so far only specified the chromaticity of the new primaries and have said nothing about their magnitude. This last point can be settled experimentally, of course, by normalizing the CIE primaries with respect to reference white. However, the transformation can also be done theoretically by the method of Wintringham, showing that if R, G, B are tristimulus values with respect to the previously used spectral primaries, and if X, Y, Z are tristimulus values with respect to the CIE primaries, then

$$X = 2.769R + 1.752G + 1.130B$$
$$Y = 1.000R + 4.591G + 0.060B$$
$$Z = 0.000R + 0.057G + 5.593B \,.$$

In particular, if this transformation is applied to the tristimulus values of unit energy spectral colors [$R(\lambda)$, $G(\lambda)$, and $B(\lambda)$ of Fig. 7.1], then we obtain the necessary data to calculate the tristimulus values of *any* color light in terms of the CIE primaries. These curves are usually labelled $\bar{x}$, $\bar{y}$, and $\bar{z}$ and are shown in Fig. 7.6.

In actual use, for any physical color specified by its spectrophotometric curve $E(\lambda)$, we find

$$X_E = \int \bar{x}(\lambda)E(\lambda)d\lambda$$

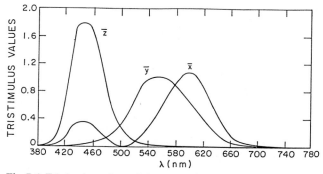

Fig. 7.6. Tristimulus values of the pure spectral colors. Like Fig. 7.1 except with respect to the CIE nonphysical primaries. White reference is still Illuminant E

$$Y_E = \int \bar{y}(\lambda) E(\lambda) d\lambda$$

$$Z_E = \int \bar{z}(\lambda) E(\lambda) d\lambda \,,$$

where Y_E is the luminance of the color. We then normalize as before, by defining the CIE chromaticity coordinates

$$x_E = \frac{X_E}{X_E + Y_E + Z_E}$$

$$y_E = \frac{Y_E}{X_E + Y_E + Z_E}$$

$$z_E = \frac{Z_E}{X_E + Y_E + Z_E} \,.$$

Colors are usually referred to in terms of Y_E, x_E, and y_E, the latter two being plotted in the CIE chromaticity plane of Fig. 7.7.

It will be seen that this plot is the same as that on the $R - G$ plane, except for translation, scaling, and rotation of coordinates. It is once again useful to plot the spectrum locus, noting that white is located at $x = 0.33$, $y = 0.33$.

By Grassman's laws, the mixture of two colors lies on a line between them. Thus, any color in the larger hatched area can be considered a mixture of white and a spectrum color, while any color in the smaller hatched area is a mixture of white and a nonspectral color. This leads to a very useful way to talk about colors, namely by their dominant wavelength and their purity. An example is shown in Fig. 7.7,

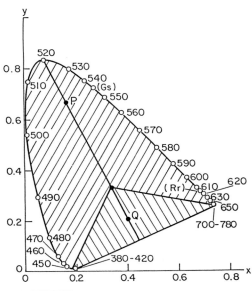

Fig. 7.7. The CIE chromaticity diagram, showing the spectrum locus. The *larger hatched area* is the locus of all physical colors resulting from mixing white with pure spectral colors. The *smaller hatched* area is the locus of nonspectral colors, consisting of mixtures of pure red, pure blue, (which together give the purples, which are nonspectral colors), and white

[7] The perceptual correlate of purity is saturation. The two words are often used interchangeably, to the dismay of purists.

where color Q has a dominant wavelength of 520 nm and a purity of 70 percent. Color P would be described the same way, with the additional specification that it is a nonspectral color. Another useful concept is that of complementarity. Two colors are complements with respect to some reference white if they can produce white when additively mixed.

7.4 Other Color Spaces

In the previous sections, we have discussed a number of graphical representations of color and of chromaticity. Most of these are obtained by various linear transformations from *RGB* space, which, in turn, is linear with light intensity. Each has a number of advantages, and most are valuable in explaining various aspects of color. They all have the profound disadvantage of being highly nonuniform from a perceptual viewpoint. The nonuniformity of the CIE chromaticity diagram was first described quantitatively by *MacAdam* [7.3]. At constant luminance, threshold chromaticity changes have more than 20:1 variation, being smallest for hue changes in the deep blue and largest for purity changes in bluish cyans, as shown in Fig. 7.8.

A perceptually uniform color space would be useful in color adjustment systems, to provide constant control sensitivity, and in transmission and processing systems, to provide optimum noise performance. Since the ability to discriminate visual stimuli depends markedly on the state of adaptation of the observer as well

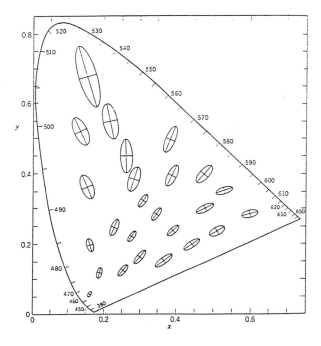

Fig. 7.8. MacAdam ellipses on the CIE chromaticity diagram

as related phenomena such as simultaneous contrast, it is impossible to have a perfectly uniform space. There is also a theoretical question as to whether color space can ever be perfectly Euclidean, which is necessary in order to have a valid distance metric. In spite of these questions, it is certainly possible to remap linear color space into one that is much more uniform than XYZ space.

7.4.1 CIE UCS (Uniform Chromaticity Scale) Diagram

In 1960 and 1964, the CIE adopted the first widely accepted remapping of the x-y chromaticity diagram, using a so-called projective transformation, i.e., a projection onto another plane at a certain angle. Such a projection is represented by a bilinear transformation of the type

$$u = \frac{ax + by + c}{dx + ey + f}$$
$$v = \frac{gx + hy + j}{kx + ly + m}.$$

Although such a projection cannot possibly turn MacAdam's ellipses into circles, the degree of nonuniformity is reduced to perhaps $5:1$, as shown in Fig. 7.9. The actual relations adopted were

$$U^* = 13W^*(u - u_0)$$
$$V^* = 13W^*(v - v_0)$$
$$W^* = 25\left(Y^{1/3}\right) - 17 \quad (1 \leq Y \leq 100),$$

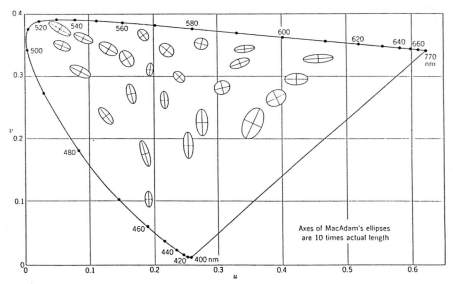

Fig. 7.9. MacAdam ellipses on the CIE 1960 UCS diagram

180

where u and v are defined as

$$u = \frac{U}{U+V+W} = \frac{4X}{X+15Y+3Z} = \frac{4x}{-2x+12y+3}$$

$$v = \frac{V}{U+V+W} = \frac{6Y}{X+15Y+3Z} = \frac{6y}{-2x+12y+3} \, ,$$

and where u_0 and v_0 are defined as

$$u_0 = \frac{4X_0}{X_0+15Y_0+3Z_0}$$

$$v_0 = \frac{6Y_0}{X_0+15Y_0+3Z_0}$$

and where X_0, Y_0, Z_0 defines reference white.

If u and v are considered normalized chromaticity coordinates like x and y, they are related to a UVW system which, in turn, is related to XYZ space by the transformations

$$U = \tfrac{2}{3}X$$
$$V = Y$$
$$W = -\tfrac{1}{2}X + \tfrac{3}{2}Y + \tfrac{1}{2}Z$$

and

$$X = 1.5U$$
$$Y = V$$
$$Z = 1.5U - 3V + 2W \, .$$

In 1976, the CIE adopted a lightness formula which, together with the UCS diagram, makes a complete color space that is substantially more uniform than XYZ space. Colors are usually referred to in terms of L, u, and v, and the space is normally called CIE LUV space. L has the same value as in Lab space in Sect. 7.4.3.

7.4.2 Munsell Space

A much older and much more uniform space is that of the empirically derived Munsell System. This system, which was eventually standardized in terms of CIE XYZ specifications, is in common use by artists and for the industrial specification of color. It was derived by having viewers select, from a large number of painted reflective samples, those that had equal perceptual increments from each other. The coordinates in this system are "value," corresponding to lightness, "chroma," corresponding to saturation, and hue. Since hue is a circular variable, Munsell space uses a cylindrical coordinate system. It is approximately uniform along its principal axes when the observer is adapted to a 50% reflective (gray) surface.

7.4.3 CIE $L^*a^*b^*$ Space

This is a closed-form approximation to the empirical Munsell space and is the most uniform space in common use. It also has provision for a variable white reference, brighter than any observed color, the assumption being made that the observer adapts completely to the reference, so that the corresponding tristimulus values can simply be linearly scaled. (Note that $L^*a^*b^*$ coordinates, unlike $L^*u^*v^*$, are nonlinear, and therefore do not represent amounts of primaries.) The color-distance metric, ΔE, is the most accurate simple such metric in common use.

$$L^* = 116 \left(\frac{Y}{Y_0}\right)^{1/3} - 16 \quad (.01 \le \frac{Y}{Y_0} \le 1)$$

$$a^* = 500 \left[\left(\frac{X}{X_0}\right)^{1/3} - \left(\frac{Y}{Y_0}\right)^{1/3} \right]$$

$$b^* = 200 \left[\left(\frac{Y}{Y_0}\right)^{1/3} - \left(\frac{Z}{Z_0}\right)^{1/3} \right]$$

$$\Delta E = (\Delta a^{*2} + \Delta b^{*2} + \Delta L^{*2})^{1/2} .$$

7.5 Additive Color Reproduction

Colorimetry provides a straightforward way to design a color image reproduction system, such as color TV, where additive color synthesis is used. For example, three component images might be displayed on three white CRTs and projected in register, through three filters, onto a white screen.[8] The filtered white lights are the display primaries of the system. If the intensities of the CRTs, at corresponding points, were proportional to the tristimulus values of the corresponding point in the original scene with respect to the three primary colors (the projection filters), then an exact colorimetric match would be achieved within the gamut of the primaries. Attempts to produce colors outside of the gamut would result in negative intensity. In practice, with present-day filters or red, green, and blue phosphors, highly satisfactory color rendition is possible.

To obtain the tristimulus values needed to display the separate images, the spectral intensity, $E(\lambda)$, of each point of the scene must be multiplied by $R(\lambda)$, $G(\lambda)$, and $B(\lambda)$, the tristimulus values of the spectrum lights with respect to the display primaries, and then integrated over the visible spectrum as discussed above. In order to implement these mathematical operations with simple filters, we could divide the incoming light into three images by means of beam splitters,

---

[8] TV tubes having closely spaced dots or lines of three colored phosphors also effect additive synthesis when viewed from a distance at which the component images merge. Another example is the projection of red, green, and blue images in rapid sequence.

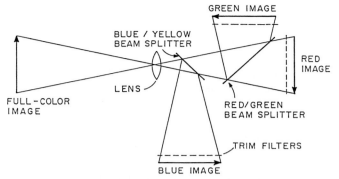

Fig. 7.10. An additive color head. This arrangement can be used as a camera or projector. In the former, the light travels from left to right and the colored images of the real world (on the *left*) are focussed on the targets of separate camera tubes. In the latter, the light travels from right to left, orginating on the faces of CRTs or the like and falling on a screen. Highest efficiency results when as much of the color separation as possible is done with nonabsorbing dichroic beamsplitters. Trim filters are almost always necessary to achieve the desired overall filter characteristics

and form separate video signals with three camera tubes whose spectral taking sensitivities were $R(\lambda)$, $G(\lambda)$, and $B(\lambda)$, as shown in Fig. 7.10. An obstacle to the physical realization of this scheme is that for all real primaries such as those used for the proposed receiver, these functions have negative lobes and are therefore unrealizable as passive filters. Only imaginary primaries[9] that define a triangle completely encompassing the spectrum locus on the chromaticity diagram always have non-negative color-mixture curves, which can be implemented as passive filters. To measure tristimulus values with respect to real projection primaries, the camera can be equipped with real color-separation filters that are color-mixture curves (CMC) – i.e., that are independent linear combinations of $R(\lambda)$, $G(\lambda)$, and $B(\lambda)$. Such CMCs must correspond to imaginary primaries. The three video signals from the camera can then be processed using a 3×3 linear matrix transformation such as in Sect. 7.3.3 to obtain the projection video signals.[10]

If the camera taking sensitivities are not true CMCs, (they almost never are, in practice), then there will be an error in the measured chromaticity coordinates. The magnitude of the error depends on the spectrophotometric curves of the colorants in the original. For color transparencies or prints, the error is usually small but quite noticeable. For arbitrary colored orignals, such as artists' paintings or the real world in front of a TV camera, the errors are often severe. Fortunately, precise color reproduction does not have the high priority in TV that it does in graphic arts.

CMCs are not usually used in cameras and scanners because, being very wide-band, they require larger off-axis coefficients in the correction matrix, which

---

[9] Such as the CIE primaries

[10] Note that the "taking sensitivities" $R(\lambda)$, $G(\lambda)$, and $B(\lambda)$ are never identical in shape to the display primaries, since if the latter are real the former *must* have negative lobes.

makes the process more error prone. In addition, there may well be a [cut] SNR in the matrix. The reasons why CMCs are not normally used will pr[cut] disappear in time, making possible far more accurate color reproduction [cut] present.

Naturally, there is more to designing a color TV system than covere[cut] but the colorimetric principles are as stated.

7.6 Subtractive Color Reproduction

In photography and printing, colors are generally mixed in such a way th[cut] spectral transmittance curves more nearly multiply, rather than add. As a [cut] the color of the mixture cannot be predicted solely on the basis of the appe[cut] of the components, but depends on the details of their spectrophotometric [cut] as well as the physics of the mixing process. For example, in the hypot[cut] case shown in Fig. 7.11 (which could be synthesized with multilayer interf[cut] filters), blue mixed with yellow-green #1 would yield neutral gray; with y[cut] green #2, green; with yellow-green #3, black; and with yellow-green #4 [cut] red. All four yellow-greens would appear, however, very nearly the same.[1] [cut] tractive systems are thus inherently more complicated than additive in pri[cut] and further complicated by imperfections of the colorants used.

Although children are often taught that the "primary colors" are yellov [cut] and blue, the widest gamut is produced by yellow, cyan (blue–green) and magenta (red–blue, also called purple or fuschia). Ideally, each of these dyes should be completely transparent over two thirds of the spectrum, different concentrations attenuating the remaining one third, more or less, as shown by the dotted lines

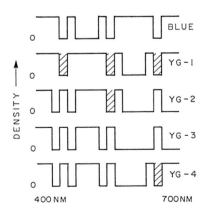

Fig. 7.11. Blue plus yellow-green = ? Spectrophotometric curves showing that the subtractive mixing rules are complex, and do not depend only on the appearance of the separate colorants. The net transmission bands of blue with the various yellow-greens are shown hatched

---

[1] These curves, as well as those to follow, show density, rather than transmittance, as a function of wavelength. Density is more convenient in subtractive systems since densities of perfectly subtractive multiple layers add, while transmittances multiply. Density: $-\log_{10}$ (transmittance)

Fig. 7.12. Ideal subtractive dyes. Each attenuates only one of the three principal wavelength bands, more or less, according to density

in Fig. 7.12. If the pass bands do not overlap and the dyes are transparent, they operate independently when mixed, each one controlling one primary light. For example, the yellow dye controls the blue luminance, because of which it is sometimes called "minus blue." Thus magenta is minus green and cyan, minus red. Such ideal dyes would work in much the same way as additive primaries. Note that $Y + M$ = red, $M + C$ = blue and $Y + C$ = green.[12] With ideal magenta and cyan dyes, the green obtained would be very much brighter than that resulting from the mixture of typical yellow and blue paints.

Real colorants, unfortunately, are far from perfect (yellow is best), as shown in Fig. 7.13. Even if they are perfectly transparent so that the density at any

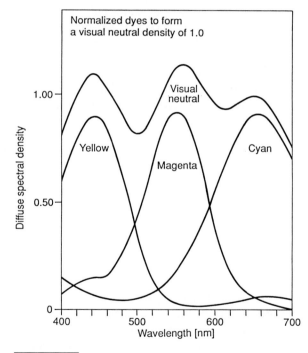

Fig. 7.13. Real subtractive dyes, in this case Kodak Ektachrome 200 Professional film. These dyes are typical of the better film dyes

---

[12] This last relationship gives the common rule of paint mixing that yellow plus blue produces green. This rule works only when the 'blue' paint actually reflects a substantial amount of green light, i.e., when it is really cyan. Deep blues give black when mixed with yellow. Arthur Hardy, one of the founders of colorimetry, once won a large bet when he found a blue that produced dark red when mixed with yellow. This occurs with blue colorants that have a small red 'leak.'

wavelength is the sum of the densities of the components, the unwanted absorptions outside their principal bands tend to make colors darker, desaturated, and of incorrect hue. Much of the theory and practice of high quality subtractive printing is concerned with overcoming the effects of these unwanted absorptions. What follows is a very brief account.

7.6.1 Elementary Color Correction in Subtractive Systems

Suppose we want to reproduce a colored original with three subtractive dyes or inks, having spectral density curves as shown in Fig. 7.14, each corresponding to

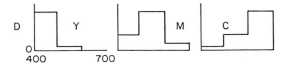

Fig. 7.14. Ideal block dyes with unwanted absorptions approximately as found in real dyes

unit concentration. We approximate the curves of real dyes with these peculiar-looking "block dyes" because we need consider their densities only in the blue, green and red regions of the spectrum, i.e., roughly 400–500, 500–600 and 600–700 nm, respectively. The degree of unwanted absorption shown here is typical of that encountered in practice, however. We first try to mix the dyes in proportion to get a neutral density (i.e., to white light) of 1.00. Defining "neutral" as the same density in each band requires a density in each color region of 1.00. Thus, since the density at each wavelength of several layers of perfectly subtractive material is the sum of the separate densities,

$$D_R = 1.00 = 1.0C + 0.2M + 0.0Y$$
$$D_G = 1.00 = 0.4C + 1.0M + 0.1Y$$
$$D_B = 1.00 = 0.1C + 0.4M + 1.0Y \ .$$

Solving these equations for the required dye concentrations, C, M, and Y, we have

$$C = 0.88 \ , \quad M = 0.58 \ , \quad \text{and} \quad Y = 0.68 \ .$$

These particular concentrations are said to be of unit equivalent neutral density (END) since they are the amounts of dye required to give unit neutral density.

It is easy to see that we can get correct rendition of a neutral grey scale,[13] from some minimum to some maximum density, by making the C, M, and Y ENDs equal to the density of the original. Therefore, we can make a reproduction system with these dyes by causing the density of each dye to be proportional to the density of the original in the corresponding band, i.e.,

---

[13] In printing, this is called adjusting the grey balance.

$$C = \gamma_c D'_R$$
$$M = \gamma_m D'_G$$
$$Y = \gamma_y D'_B \ ,$$

where the D's are the densities of the original in the separate bands.

In this case, to get correct reproduction of neutrals we simply make

$$\gamma_c = .88 \ , \quad \gamma_m = .58 \ , \quad \text{and} \quad \gamma_y = .68 \ .$$

Note that the individual gammas must be reduced from 1.00 to the extent that the unwanted absorptions of the other dyes contribute to the density in the relevant band.

If we had tried this procedure on colored originals, it would have worked very poorly. For example, if the original was just one of the reproduction dyes, we would still get all three dyes in the reproduction, even though only one was required. Instead, we can require

$$D'_R = D_R$$
$$D'_G = D_G$$
$$D'_B = D_B$$

for *all* combinations of densities and not just for the case where they are equal. We can then solve the first set of equations for the general case:

$$D_R = D'_R = 1.0C + 0.2M + 0.0Y$$
$$D_G = D'_G = 0.4C + 1.0M + 0.1Y$$
$$D_B = D'_B = 0.1C + 0.4M + 1.0Y$$

giving

$$C = 1.09 D'_R - 0.23 D'_G + 0.02 D'_B$$
$$M = -.44 D'_R + 1.13 D'_G - 0.11 D'_B$$
$$Y = 0.07 D'_R - 0.43 D'_G + 1.04 D'_B \ .$$

What kind of reproduction can we expect in this case? Clearly, any set of D'_B, D'_G, and D'_R in the original can be reproduced exactly, i.e., we can obtain precisely the same densities, as long as we do not call for physically unrealizable amounts of the reproduction dyes, C, M, and Y.[14]

The fact that negative dye densities may be called for means that some colors are nonreproducible. This is the result of the limited range, or gamut, of colors that can be reproduced with imperfect dyes such as in this example. If the dyes

---

[14] Strictly speaking, obtaining the same densities does not guarantee that the reproduction will look like the original, unless the reproduction colorants are the same as in the original. If the original is a photograph and the reproduction is with dyes, the errors will usually not be great.

were perfect, i.e., absorbed in only one band each, the off-diagonal terms in the last set of equations would be zero. In the case of real dyes, rather than block dyes, the procedure also works quite well. Of course, the coefficients would have somewhat different values.

In the general case, we can think of each of the three color-band images as a linear combination of three images, each derived from one color of the original. The images corresponding to the off-diagonal terms are "masks" that cancel, in each dye, the imagery due to the unwanted absorptions of the other dyes. The main masks control the green absorption of the cyan dye ($.44D'_R$), and the blue absorption of the magenta dye ($.43D'_G$), followed in importance by the red absorption of the magenta dye ($.23D'_G$). If the dyes are formulated appropriately, these last two masks can be identical, so that, in photographic systems, only two different masks must be made. The diagonal coefficients control the contrast in each color, taking account of the contrast-reducing effect of the off-diagonal masks. Of course, no amount of masking can extend the gamut of a reproduction process. Thus, to obtain the correct hue and saturation in prints or transparencies, it is often necessary to darken colors substantially, as in the dark blue skies often seen in Kodachrome. In a negative, of course, darkening by masking can readily be compensated for in printing the positive.

When making color reproductions photographically, the masking technique can produce highly acceptable results, since the main deficiency in that case consists of the unwanted absorptions. Two or three masks are usually used, with overall contrast often corrected by controlling development. Highlight and shadow masks may be used in addition for controlling both color and tone rendition in these areas. These complications result in high costs. On the other hand, the so-called "integral masks" that are incorporated in Kodacolor film (the orange cast is actually a colored positive image composed of two separate dyes) make possible high quality prints with no additional masking. In printing with inks, however, much poorer results are achieved since printing inks are much less well behaved than the dyes in color film. Ink actually behaves somewhat like paint. The upper layers tend to obscure the lower layers, rather than being transparent. As a result, a great deal of hand work is usually required to get good results by traditional photographic methods.

In recent years much color separation and correction have been done on electronic scanners, in which case no photographic masks are made, the equivalent operation being performed electronically. In addition, colors may be selectively adjusted to achieve a more precise color-by-color correction. Much less hand work is required. Nevertheless, a set of corrected separations, with proofs, may cost as much as $500 to $800. Systems currently under development in which ink correction is done by look-up table and aesthetic correction is done interactively while observing a TV display that mimics the final output, show promise of substantial reduction in cost and improvement of quality.

7.6.2 Problems in Color Reproduction with Printed Inks

It is evident from the previous section that subtractive reproduction is considerably more complicated than additive since the physical definition of each primary colorant (i.e., the spectral curve rather than just the appearance) must be considered. In printing, the situation is even more difficult, since, when inks are overprinted, their spectral densities do not simply add as they do, for example, in photographic transparencies. This is partly due to the fact that the inks are not completely transparent, so that some light is scattered within the ink layers. Some is also reflected from the ink/ink and ink/paper boundaries. In addition, the halftone process, particularly in light tones, introduces an element of additive mixing when dots fall side by side. The net result is that the previous equations are not completely valid.

These departures from exact multiplication of spectral reflectivity of the individual inks are grouped by specialists into two categories – proportionality failure and additivity failure. The first class includes errors due to the fact that the ratio of spectral density at different wavelengths may change with the overall density of a single ink. This tends to desaturate midtone and shadow colors. The second includes errors resulting from the fact that the density of several ink layers is less than the sum of the densities of the separate layers. This may lead to severe errors, since the upper inks tend to 'hide' the lower inks, thus reducing the effect of the latter on the color of the combination.

These imperfections, in addition to unwanted changes in the colors of combinations of inks, also restrict the color gamut and dynamic range of press prints as compared with transparencies, and even more so as compared with original subjects. The result is than many colors (the word as used here including the intensity aspect as well as the chromatic aspect) of originals may be nonreproducible with any given set of inks and paper. This requires a compression of input gamut to fit within the output gamut. A convenient way to think of this process is as a mapping of the input color space into the space of output colors. For all of these reasons, the achievement of high quality reproduction requires much more elaborate color correction than described in Sect. 7.6.1.

7.6.3 Color Correction in Electronic Scanners

The masking techniques discussed previously can be extended to deal with the problems mentioned by means of altering the masking factors and using nonlinear masks. These techniques require a great deal of skill and time, and thus are very costly. The advent of electronic scanners made the job much easier, although still not simple. At present, a very high proportion of high quality color printing is done by means of these scanners, even when computer-based prepress systems are used. What follows is a brief account of color correction in Hell scanners, which are representative of modern practice.

The scanner always has a meter that indicates the predicted output in all three colors (dot percentage or separation density) with the current settings of the controls and the current point being scanned. In adjusting the many controls on the scanner, the operator consults a book of printed color patches so that he knows what color will be printed with the current predicted output, and he adjusts the controls to achieve the desired color for each of a large number of input colors.

Light reflected or transmitted by the original is separated into red, green, and blue components and sensed by photomultiplier tubes, forming the three (RGB) colorhead signals.[15] The filters, usually Kodak 25, 58, and 47B, are not color-mixture curves, but rather a compromise between the wide-band filters needed in colorimeters and the narrow-band filters that would be used if an attempt were being made to read the dye densities of the original image. The signals are normalized to equal values while scanning a neutral highlight, and then are compressed to a "quasilog" tone scale of the general form

$$y = \log(x + a) .$$

If a is small compared with the maximum value of x, the relationship is nearly logarithmic, while if a is a substantial fraction of x_{max}, then the relationship is nearly linear. With the values normally used, y is approximately a "lightness" signal.

The function of this compression is two-fold. One purpose is to compress the dynamic range to fit the output process in a perceptually acceptable manner. The other is to work with the signal in a form in which computation noise is minimized. This is particularly important if part of the processing is digital.

The next step is the gradation adjustment, in which all three signals are subjected to the same nonlinear transformation. Following that is "basic" color correction, which emulates photographic masking. This is usually carried out in an analog "color computer." For each of the three input signals, a masking signal is formed consisting of the difference between the main signal and a variable mixture of the other two. The gains are chosen so that for neutral colors (for which the main signals are equal), the masking signals are zero. Thus, when the masking signals are added to the main signals, no change is made to the gradation of neutrals. There are nine controls in this procedure, giving the same number of degrees of freedom as in the 3×3 photographic masking equations. Many scanners, however, have two sets of masking controls, one for dark colors and the other for light colors.

Initial adjustment of the masking controls is usually done while scanning a chart consisting of the six principal colors – cyan, magenta, yellow, red, blue, and green, made by printing all combinations of one and two printing inks. These particular colors are used because the correct outputs take a very simple form. Final adjustment is made with normal originals, paying special attention to flesh

---

[15] If we were printing with ideal dyes, as in Fig. 7.10, each dye could simply be controlled by one of the three signals, and with the appropriate nonlinear transformation, good reproduction would result. With real inks, the images would be of the wrong color balance and very desaturated.

tones, grass, sky, etc. Since there are literally hundreds of colors to be corrected, basic color correction cannot do the entire job since there are not enough degrees of freedom.

Further improvement in rendition is provided by the selective color correction controls, which have labels such as 'the amount of yellow in red'. Additional masking signals are formed that are not only zero for neutral colors, but are also zero for all colors except those close in hue to one of the six principal colors. The controls can be used to adjust the hue and brightness of these six colors, one by one. There may also be similar selective controls for light red (flesh tones) and dark brown.

The skill, time, and patience required to set these controls is substantial. Even so, it is virtually impossible to get all colors correct at the same time, especially with a variety of copy. In many cases, local correction is used, in which the color of individual objects is adjusted by using geometrical masks to limit the correction to a given area.

In some scanners, after the controls of the color computer are set, the information is transferred to a digital lookup table (LUT), which is then used to process the colorhead signals during actual scanning. One way to transfer the information is to cycle the RGB signals through all combinations of a number of steps from zero to the maximum value, to use the RGB signals as addresses to the LUT, and then to enter the color computer output into the corresponding memory locations in the LUT. As few as 4096 entries may be used. The LUT is provided with an interpolator to handle RGB values that are not exact addresses. This method has some important advantages over the analog color computer, principally for the ease of saving and down-loading settings, and for greater repeatability. In some cases, higher speed can be reached than readily achievable in the analog circuitry.

In spite of the cost, most high-quality printing is done by these methods at the present time, with excellent results, although quite variable between operators and between plants.

7.6.4 Color Correction by Computer Processing

Granted that an economical size LUT, with the appropriate contents, can be used to represent the correct dot percentages given the RGB signals obtained by scanning the original, it is clear that color correction consists 'simply' of finding the table contents. It is not necessary to use only the methods found in electronic scanners, which are, in essence, the electronic implementation of the masking equations. Here we think of the LUT as effectively mapping RGB space into $CMYK$ space.

A straightforward way to find the LUT contents is to print color patches with all combinations of a sufficient number of steps of each ink. Experience shows that 5 or 6 levels of each is enough with normal inks. The tristimulus values of each patch are measured, giving an empirical relationship between the ink and the resulting color. For each RGB (colorhead signals or tristimulus values) address, the LUT contents (the CMY values that match the color) are found by interpolating in the color data base.

If all the colors of the input image are reproducible, and if colorimetrically accurate reproduction is desired (as would usually be the case in this situation), this method is so easy and so accurate, it is hard to see why any other method should be used at the present time. If some colors are nonprintable (Sect 7.6.5) they must be changed to make them printable. Provision also must be made for editorial correction of input colors. These additional changes may be made by additional processing units, or they may be incorporated into the LUT.

However the correction is determined, an enormous improvement can be made if the corrected output image is displayed, preferably instantly, in a manner that accurately indicates the final appearance of the press print. It is fair to say that no system in commercial use at the present time is accurate enough, so that operators continue to rely on their color books and paper proofs. However, progress is being made, and sufficiently accurate "soft proofs" will very likely eventually become available. Even if the TV predicts the output only approximately, it may still be very useful. The direction and approximate degree of change can be seen. Pictures can be compared on the screen for color balance, brightness, contrast, hue, and saturation (the principal components of image quality) with other pictures known to print correctly or which are to appear on the same page. Finally, the operator eventually becomes 'calibrated' to the display so that he can infer the output in spite of the deficiencies of the display.

Interactive color editing is most accurate when the conditions under which the CRT is viewed are the same as those used in examining the original copy and the reproduction. This takes a good deal of trouble – in particular, keeping the room lights very low – so that attempts are often made to use more convenient viewing conditions and to correct for the difference. This may eventually be possible, but this author has yet to see a convincing demonstration.

One approach for computer color correction is to use the system in a manner analogous to the scanner, but to make the process easier and faster. For example, if a given color is to be changed to another color, the computer can be instructed to do so, and can quickly calculate and make the change. The chromatic neighborhood to be affected can easily be specified. Local color correction can be implemented with the aid of a tablet-drawn outline, and the display tube can be used to check the outline accuracy. Gradations can be displayed and quickly changed.

On the other hand, it is also possible to use computer power to carry out some operations very quickly and easily that are slow, difficult, or even impossible on the scanner. For example, it is possible to develop methods for editing LUTs if they require correction in just a portion of color space. Another technique is to select colors to be changed as well as the colors to which they are to be changed by picking them from the displayed image. Particularly valuable is to change color coordinates from RGB to LHS, so that changes can be made independently in luminance, hue, and saturation, parameters whose effects are much easier to predict.

7.6.5 Nonprintable Colors

Most color reproduction systems suffer from the fact that the gamut of reproducible colors is less than the typical gamut of subject matter. Photographs of outdoor scenes in sunlight can only show a small fraction of the dynamic range of the focal-plane image, and printed reproductions of transparencies must be reduced in both brightness and chromaticity range. Long experience in photography and graphic arts indicates that the S-shaped D-log$E$ curve of transparency film is quite appropriate to natural scenes, while the quasilogarithmic curve of Fig. 7.15 is best for printed reproductions. Even after this lightness compression, there may still be some chromatic nonprintability, as evidenced by excess saturation. Since saturation is an important element of overall quality, it must be reduced judiciously. Because of the fact that the chromatic gamut of most reproduction systems is greatest in the midtones and less in both highlights and shadows, some colors can be made reproducible by making them darker or lighter, respectively. When saturation must be reduced, it is generally better to reduce the more saturated colors proportionately more than the less saturated ones, using a relationship such as that in Fig. 7.16.

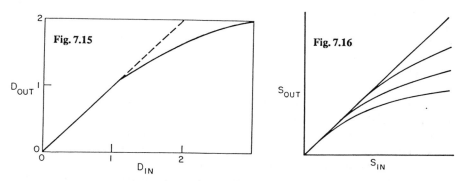

Fig. 7.15. Quasilogarithmic curve showing the appropriate nonlinear transformation when the input is a photographic transparency, rather than a signal from a camera looking at the real world

Fig. 7.16. A nonlinear transformation suitable for manipulating saturation

7.6.6 The Black Printer

In spite of the fact that good reproductions can be made using only cyan, magenta, and yellow ink, black ink is also used in printing for the following reasons:

1. to save money by replacing some expensive colored ink with cheaper black ink,
2. to improve the lay-down and drying of ink by reducing the total amount used,
3. to reduce the effect of misregistration and color balance errors, and
4. to increase the density of certain dark colors.

The process of replacing colored ink by black is called under-color removal (UCR). The black image is called the "black printer" or "key", and, if very light, is sometimes called the "skeleton key." If the inks were transparent so that they obeyed the additivity rule, it would be simple to calculate the amount of black to be added and colored ink to be removed so that the same appearance would be achieved. If C, M, and Y were the amounts of ink required to produce a certain appearance, and T_{min} were the least of the three ink densities, we could simply remove a fraction, k, of this amount from all three inks and replace it with K, an equal quantity of black.

$$K = kT_{min}$$
$$C' = C - K$$
$$M' = M - K$$
$$Y' = Y - K .$$

These formulas are accurate only for small amounts of black. In conventional systems, either photographic or using the "color computer" of drum scanners, the operator manually selects the amount of black and the reduction in the other inks, either relying on his experience or consulting his book of printed patches. Generally, no black at all is used in light tones. The key is made to start in the midtones and is increased toward the shadows, being made equal to a fraction of the lightest colored ink. In some cases, where the 3-color image is not dense enough in the shadows, a light key may be added without removing any colored ink at all.

In recent years, most suppliers of computer-based prepress systems have produced proprietary systems for much more accurate UCR, often called by such names as GCR, or grey component removal. In these systems, it is, in principle, possible to print, with only two colors plus black, anything that normally would require all three colored inks. It is believed that most such systems are based on fairly accurate models of ink-mixing performance.

In systems using a LUT, the contents of which are found from measurements of a color-patch printing test, precise UCR is possible on a completely empirical basis, even with arbitrarily large amounts of black. Of course, it is necessary to make a 4-ink printing test with enough samples to characterize the ink/paper performance adequately – perhaps 500 to 1000 patches. When such a test is made and the resulting color data base analyzed, it is found that there is one section of printable color space that can be matched only without black. (The colors produced with only one ink are examples.) Another section, mainly the darker colors, requires at least some black. (Many of these colors would not be printed at all with conventional methods of UCR.) The largest section can be matched with a continuum of 4-ink combinations, with the black ranging from none to some maximum value. This is to be expected, since the color has only three degrees of freedom while the ink combination has four.

Given a LUT with RGB input (colorhead signals or tristimulus values of the desired color), and with $CMYK$ contents (the matching ink combination), the first step is to search the color data base for ink combinations that match each LUT address. Interpolation is used to find a series of matching ink combinations in which the black ranges from minimum to maximum. The minimum amount is called 0% and the maximum 100%. This procedure is carried out for all LUT addresses. A LUT can now be made for any desired percentage. Actually, there is little point in using any black at all in light areas. The nonlinear relationship between black added and colored ink removed makes it impossible to save much colored ink except at high density. It is quite practical to make the percentage vary with overall density, so that a great deal of black is used in the shadows and very little in the light tones. A saving of 10% to 15% of colored ink is possible.

7.6.7 Making Accurate Color Copies

There are many occasions when it is desirable to make accurate copies of a color image in a different medium. One such case is taking pictures of color CRTs. Another, and very important one, is proofing, where it is desired to see, on paper, an accurate prediction of the output of a printing press, but without the expense and time required to make plates and operate a proofing press. In all such cases, the LUT method mentioned above can be made to give excellent results.

The procedure to be followed is similar to that previously indicated for finding the contents of an RGB to CMY or to $CMYK$ LUT for printing. It is first necessary to make a color-patch test on the output material, using all combinations of sufficient numbers of steps of each of the input variables to the output process. For example, in photographing a CRT, these variables would be the red, green, and blue brightness of the image on the faceplate of the tube. The resulting patches can then be measured on a colorimeter to find their tristimulus values. A LUT is now made by interpolating in this color data base, to find, for each output color desired, the input variables to the output process that will produce that color.

A common complication in these systems is that the image to be copied may not be specified in tristimulus values, but rather in RGB colorhead values, where the taking sensitivities of the scanner are not true CMCs. In that case, a second LUT can be used to transform the signals into tristimulus values. This LUT can be calculated from a second color-patch test involving patches to be scanned, which are also measured on a colorimeter. In some cases, the colorimetric errors of the scanner are not very large, in which case adequate correction can be made with a 3×3 linear matrix, rather than a LUT.

Of course, the same problem of nonprintable colors may occur in copying as discussed in Sect. 7.6.5 with respect to printing. Again, a similar solution must be used. First the printable colors must be shifted so as to be printable in the copying medium, and then the conversion must be made from the original color coordinates to those of the copying medium, typically in a LUT.

7.7 Color Editing

A pervasive problem in many of the systems discussed above is the adjustment of colors from what they are to what they should be. In the case of operators with many years of experience with photographic correction or with analog scanner correction, it is quite feasible to think entirely in terms of the three color separations, i.e., the three (or four, if black is used) ink densities or dot percentages. This is so much the case that it is not unusual for proofs to be returned with notations such as '-10% yellow.' Such experienced practitioners may or may not be willing to change to more modern methods that are simpler and more accurate. However, there is no question at all that users without such experience, or those who must adjust color in computer graphics, in-plant and desk-top printing, etc., would be severely handicapped if forced to adopt traditional methods. It is therefore of some interest to explore methods that are easier.

It has long been recognized that the specification of color requires three degrees of freedom.[16] Furthermore, laymen display a remarkable degree of unanimity as to which three aspects of color to use in identification, when the color is being described independently of the material used to produce it.[17] First comes the color name, or hue. (Of course, we do not all use exactly the same names for a particular hue.) Nearly everyone is familiar with the color circle, in which the colors are arranged as in the rainbow, the opposite ends of the spectrum being brought together through the purples. The other two aspects most often used are lightness/darkness and purity, or saturation. Hue and saturation are polar coordinates in the chromatic plane. Again, although many different names are used for these two aspects, the concepts are well known. Everyone has a good understanding of what is meant by dark red and light red, and of the continuous scale from white through pink to red.

These common understandings are the basis for most so-called 'color-order' systems, such as those of Munsell and Ostwald. There are, however, some that do not use hue and saturation, but use more-or-less rectangular chromatic components, such as 'blue-yellow' and 'red-green'.[18] In all such systems, laymen with no special color experience have little difficulty in adjusting colors. In most cases, they can quickly achieve whatever color is desired, as long as they can see the results of any adjustment immediately.

Imagine the much more difficult task facing the operator of a conventional correction system. Not only must he use a color identification scheme (CMY)

---

[16] Evans [7.1] makes a convincing argument for five parameters in the perception of color, but that need not concern us here.

[17] Painters often refer to the specific colorant itself, such as 'Prussian blue' or 'scarlet madder,' but they certainly understand that they are using a shorthand notation (and method) to get the color they want by employing a certain pigment.

[18] These 'opponent-color' schemes have a significant advantage over the hue/saturation schemes in that the sensitivity of control in each coordinate of the former is much more independent than in the latter. They have a disadvantage in that the color effects produced by each control depend very much on the setting of the other, and they probably take more training to use effectively.

whose characteristics are quite outside normal experience and therefore must be learned, but, in most cases, he must wait for a proof, hours or days later, to see the results of an adjustment. In the case of scanner operators who are using the predicted dot density shown on a meter and then checking by looking at patches in a color book, feedback is much quicker. However, this arrangement has two important disadvantages. One is that the color patch in the book never produces the same subjective effect as the colored object in the picture. The other is that he cannot simultaneously see what is happening to the other parts of the image. Indeed, facing these handicaps, it is quite remarkable that such good results can be achieved.

7.7.1 Appearance Editing

In conventional color correction, the operator must compensate for the characteristics of the output process – the ink and paper or the photographic material. Additionally, he must adjust for any changes desired in the input image, for errors due to the fact that the scanner filters are not true CMCs, and for the preferred reproduction of nonreproducible colors. It is very desirable to separate these several corrections, particularly with respect to whether they need operator judgment or can be done entirely automatically by the computer. It is evident from the foregoing that interactive editing in terms of the appearance of the image, rather than in terms of the physical constituents of the output, should be much easier. Likewise, it is clear that, to the extent to which the performance of ink, paper, and press are stable and can be specified, such correction is a matter of 'number-crunching' and need not, indeed should not, be left to the operator.

The essentials of an appearance-editing system are a scanner with CMC filters so that it measures the tristimulus values of the input image, a colorimetrically accurate display, a system of interactively adjusting the image by altering its luminance, hue, and saturation (or possibly luminance and two opponent-color coordinates) to achieve the desired appearance of the output, and a means for automatically converting between the desired appearance and the physical constituents of the output.

This arrangement, which is more fully described in [7.4], still leaves nonprintable colors to be accommodated. The usual quasilogarithmic luminance compression deals quite well with excess dynamic range, while interactive hue-selective saturation compression and brightness adjustment takes care of most remaining problems.

Interactive systems of this type facilitate the achievement of correct color balance and tone scale, the most important elements of color image quality, along with saturation. Whereas adjustment of the latter requires a 3-dimensional mapping of color space, the first two can be accomplished by independent gradation adjustments in R, G, and B. Since the main cause of incorrect color balance is the illumination of the scene in front of the camera, correction can be made by processing the signals so that they become what they would have been, had the red, green, and blue exposures of the film been different. This requires knowl-

197

edge of the $D - \log E$ film characteristics. With such a control, it is also possible to correct a certain amount of under- and overexposure. Gradation adjustments are made as in monochrome systems to achieve optimum tone scale with the particular subject.

7.8 Color Coding

With three degrees of freedom and using a simple-minded approach, color images require three times the storage space or channel capacity as monochrome images. The three image components can be processed independently by any of the methods developed for monochrome. This approach ignores both the statistical relationships among the three components and their different fidelity requirements. Depending on the purpose of the processing – storage, transmission, editing, etc. – other representations may be more convenient, and quite different efficiencies can be achieved. In this section, we deal primarily with data compression. The appropriate metric of efficiency is thus the image quality achieved per unit storage or transmission capacity. Because of the very large amount of work done in this area, this section cannot even be a summary. What is presented here is a short explanation of the fundamental principles that are used in most color coding systems.

7.8.1 Color Representation in Digital Systems

The greatest possibility for data compression obviously occurs when we have complete freedom to encode colors arbitrarily, considering both the required accuracy of representation with respect to colorimetric factors, i.e., those dependent on the colors themselves, and to image factors, i.e., those dependent on the distribution of colors with a scene or sequence of frames. Taking full advantage of all the factors is not feasible except in the digital domain, and even there, practical matters such as cost and processing speed often preclude the use of the most efficient methods.

a) Spatial Resolution Requirements. An interesting early experiment that clearly indicated the great differences in required resolution of the different color components was done by *Baldwin* in 1951 [7.5]. Using an optical projector that superimposed red, green, and blue color separations to form a full-color image, he defocussed the components one by one to measure the just-noticeable differences in resolution. He found that the green component required maximum resolution, the red component required about half as much, and the blue component could be defocussed the full amount built into the projector without causing any visible degradation of the final picture. When viewed alone, the blue image was so blurred that it could hardly be deciphered as a picture, but it was needed nevertheless, since without it, the image was very yellow. There is an object lesson here; image components cannot be discarded solely because they are small!

Much of the difference in resolution is the result of differences in brightness. For the pure spectral primaries of Sect. 7.3.4, the relative brightnesses of red, green, and blue are 1.00 : 4.59 : 0.06. The relative resolutions, however, cannot be selected simply in accordance with the variation of acuity with brightness, as the required resolution also depends on the particular image. In general, color detail corresponds to luminance detail, as at the edges of colored objects. Where there is a luminance edge, the chromatic resolution can be very low without ill effect. However, at edges with little or no luminance difference, substantially higher chromatic resolution is required to avoid obvious defocus. In practice, as in the NTSC color television system described below, a compromise is struck between a good overall compression ratio and some loss of quality in unusual situations. In many cases, it is found that the red resolution can be half and the blue resolution one quarter of the green resolution without perceptible effect. Highly satisfactory results can often be obtained when no more than 10% of the data is chromatic.

Color images are judged substantially superior to monochrome in virtually every case, regardless of whether the display is for information or entertainment. Since it takes so little additional channel capacity to add color, doing so can realistically be considered a capacity conservation technique; color is not a luxury – using it is an obvious way to *improve* channel utilization. For example, when luminance resolution is slightly reduced to make room for color, the quality is nearly always increased [7.6].

In digital systems, it is common to keep the representation in RGB form throughout the system. Since scanners and displays always operate in these terms, this procedure avoids matrix transformations, which are rather costly in hardware unless special chips are used. However, even greater compression is possible with some kind of luminance/chrominance representation such as Y, x, y or L, a, b since the chromatic components have zero luminance.

An alternative to using high luminance and low chrominance resolution is the "mixed highs" representation due to *Bedford* [7.7]. This involves low-resolution RGB plus high resolution spatial high-frequency luminance components. One advantage of this scheme is that if color editing is required, it can be performed on the low-resolution RGB image only.

b) Amplitude Resolution Requirements. The number of levels needed for each color component depends both on the actual dynamic range of the variables and on the visual sensitivity to color increments. In TV displays, 7 or 8 bits is usually used for green or luminance signals, while 5 or 6 suffice for the others. Roberts' randomization process works as well for chrominance as for luminance. The same kind of nonlinear transformation to lightness discussed in Chap. 3 also is appropriate for luminance in color images, and, in fact, is the transformation used in L, a, b space.

A complication in quantizing color components is that most color spaces are not scaled cubes, so that the range of each component usually depends on the values of the others. The boundaries of the space are sometimes defined by physical variables. One example is the color space defined by all the colors that can be produced with a given set of inks on a given paper. Another is the color space of a cathode-ray tube. Each phosphor brightness has a certain range, independent of the other two. Thus the space is a scaled cube. Each component can be quantized separately without any unused codes, although this space is highly nonuniform perceptually. If each component is nonlinearly compressed, approximate uniformity can be achieved for each separate color, but the 3-dimensional space is still nonuniform.

When the CRT RGB color space is transformed to any other, such as a linear $L, C1, C2$ space, the latter will no longer be a scaled cube. There will be unused codes as well as perceptual nonuniformity.

Boundaries of other spaces may be set in part by psychophysical factors. For example, we can talk about the space of all perceptible colors. Since these must be real, they are formed by combinations of pure spectral lights. They are bounded by the spectrum locus as well as the minimum and maximum perceptible luminance at each chromaticity. This space is very far from a scaled cube.

In cases where such highly irregular spaces are to be coded efficiently, the variables, no matter what they are, must not be quantized independently. What can be done is to divide the space into a large number of small 3-dimensional cells and to give each cell a code. If the cell lengths in each direction all represent approximately equal perceptual increments, then this coding is quite efficient, particularly if the mixed-highs scheme is used and the colors are those of the lower spatial frequencies only [7.8]. Of course, for particular images, the various cells occur with nonuniform frequency, so that a statistical code can be used for further compression.

7.8.2 Color Representation in Analog Systems

Analog systems must use simpler color representations than digital systems for practical and not theoretical reasons. The most common technique is simply to proportion the bandwidth of the components so as to get the highest quality for a given total bandwidth using typical pictures. This scheme makes use only of the lower permissible resolution in the direction of scanning of two of the components compared to the third. It gives a compression ratio of 1.5 to 2 as compared with using equal bandwidth for all three signals.

a) The NTSC System. The most important application of these ideas has been in the color system proposed by the second National Television System Committee[19] in 1953 [7.9]. The most important guiding principle of this system was that it was to be compatible with existing monochrome receivers. This feat, which many

---

[19] The first NTSC, in 1941, set the 525-line monochrome standard. The color system was introduced in 1953 as a modification of the 1941 standard. PAL and SECAM, different systems based on similar principles, are used in 50-Hz countries of Europe, Asia, and Africa. NTSC is used in the US, Canada, Japan, and Latin America and is the standard for a little less than half of all the TV receivers in the world.

thought was impossible, was achieved by using two narrowband chrominance components modulated in quadrature on a subcarrier placed near the upper end of the video band. The subcarrier and its sidebands cancel out, more or less, on old monochrome receivers, thus achieving compatibility. There is, unfortunately, some crosstalk between luminance and chrominance that degrades the image on color receivers. In addition, there was an effective decrease in luminance resolution as compared with the 1941 system. Most of the defects can be eliminated and part of the lost resolution can be regained by sufficiently complicated spatiotemporal filtering at both transmitter and receiver, either adaptive or nonadaptive [7.10]. Some of these processes are beginning to be used commercially.

b) The Vectorscope. The signals that are transmitted in the NTSC system are called Y, I, and Q, and are found from the RGB camera signals by the following transformation:

$$Y = .299R + .587G + .114B$$
$$I = .540R - .274G + .322B$$
$$Q = .211R - .523G + .312B .$$

A common diagnostic tool used in NTSC television, and one that can quickly tell us a lot about color signal statistics, is the vectorscope. This is an oscilloscope in which the horizontal and vertical deflections are made proportional to the two chrominance signals, I and Q. Statisticians call such a display a scatter diagram. As shown in Fig. 7.17, this display usually comprises a fairly small number of 'spokes' radiating from the origin together with a 'cloud' of low amplitude but little structure. The structure that appears in the figure indicates a high degree of

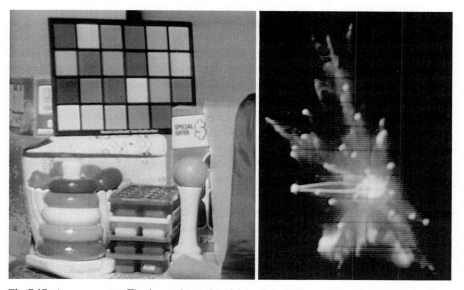

Fig. 7.17. A vectorscope. The image is produced by using R-I and B-Q for the horizontal and vertical deflections of an x-y oscillocope

correlation between the two signals. Note that many possible colors, especially those of high saturation, are absent in any one image.

The cloud of data points near the origin indicates the preponderance of image areas of low saturation, i.e., that are nearly neutral.[20] Each spoke corresponds to a group of colors of one chromaticity but different luminances. Usually, each spoke is from one object in the image. Many objects, both man-made and natural, are thus shown to be of a single chromaticity. The factor that varies from point to point is the luminance. This variation is due to the illumination. As discussed in Chap. 2, when the illumination has a directional component, the light flux incident per unit area depends on the angle between the normal to the surface and the direction of the incident light. Surface inclination and texture combine to produce most of the point-to-point variations in luminance.

7.8.3 Vector Coding

Since a color is specified by a triad of numbers, it is natural to apply vector coding [7.11]. We effectively did this when we divided color space into small cells and gave each a distinct code group. Lossless statistical coding can be applied to the vector representation to achieve significant compression, since it is clear that most possible colors do not appear in any one image. Much more compression is possible in this case, however, by using lossy coding.

Suppose we want to describe a color image using only eight bits per pel. This allows a palette of only 256 colors rather than the 2^{24} (16 million) colors that are possible with the more usual eight bits per component per pel. (Since there are only about 200,000 pels per frame in NTSC, note that no more than 18 bits can possibly be needed, even on a lossless basis, for any one picture.) A way to do this is to choose the 256 most 'popular' colors in the scene and to map all the others to the nearest one used. These pictures are surprisingly good, but do have some visible color quantization, especially on smooth curved surfaces. Nearly perfect pictures are possible in most cases with more judicious choice of the 256 colors based on location as well as the histogram, together with a certain amount of randomization [7.12]. In this procedure, as in other forms of vector coding, encoding is rather complicated but decoding is quite simple. Here decoding merely requires an 8- to 24-bit codebook, sometimes called a "color map" or lookup table. This codebook, which is different for each image, must be transmitted and considered part of the picture code.

An interesting question that arises from the successful representation of images using only a very small number of different colors is how many perceptibly different colors actually exist. Clearly it is less than 16 million, since a properly calibrated color picture tube contains all of them. From the results achieved in a

---

[20] In an RGB system, the three variables are nearly identical in such areas and are thus highly correlated. One of the benefits of the luminance/chrominance representation is that for the very common near-neutral areas, two of the signals are made near zero, rather than merely being correlated.

variety of experiments, a reasonable guess is about 2 million. This large number applies only to the very best conditions for seeing small color differences, which means using a target such as in Fig. 3.3. In practical image-viewing conditions, our judgment of absolute color is very poor. It is color differences to which we are most senstive, and these are somewhat obscured by noise and texture. It is such phenomena that make it possible to produce acceptable results with so few different colors.

7.9 Some Experiments in Color Perception

Although we all experience color whenever our eyes are open, some of the most elementary phenomena are far from obvious and are best revealed by simple experiments. These experiments cannot be reproduced in a book or on a television screen, but must be done 'live.' The equipment required is quite simple and the results are revealing. The following items are needed:

1. Two slide projectors, preferably with variable transformers for dimming.
2. A good-quality white projection screen.
3. Kodak gelatin filters: red (29) green (58) blue (47b) cyan (44a) magenta (32) yellow (12). Theatrical "gels" that look like the real thing are not good enough, but an orange filter (21) or gel is useful.
4. A spectroscope (highly dispersive) prism at least 1 inch high, such as Edmund Scientific catalog no. N31,801.
5. Special slides as shown in Fig. 7.18.

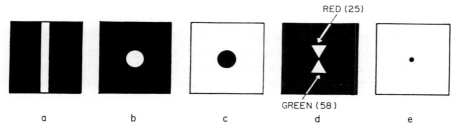

Fig. 7.18a-e. Special slides used for the color experiments. See the description in the text

7.9.1 The Spectrum

The room should be very dark. Project the slit slide (Fig. 7.18a) on the screen. Place the prism in front of the lens and turn the projector as shown in Fig. 7.19 to center the image on the screen. The familiar rainbow will appear, with each point in the spectrum corresponding to a different wavelength.[21] Rotate the prism

---

[21] The experiment itself does not prove that.

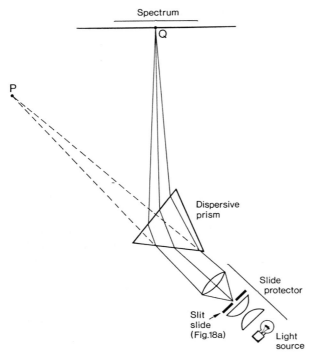

Fig. 7.19. Arrangement for displaying the spectrum. The slit slide of Fig. 7.18a is placed in the slide projector and focussed at P. A dispersive prism is placed in front of the projector, moving the slit image to Q on the screen. Dispersion in the prism causes the image to be spread out on the screen into the typical rainbow. The slit must be narrow enough to make the yellow region of the rainbow easily visible

on its vertical axis to make the spectrum as wide as possible.[22] Newton drew several conclusions from observing the spectrum, the most important of which is that all of these colors combine to produce white light. Note that the experiment does *not* prove that white light invariably consists of these spectral components. (It doesn't.) All of the common colors except the purples appear.[23] Note that yellow occupies a very small portion of the spectrum and that the blues are very dark. In the region near yellow, the hue changes very rapidly with wavelength. Although not evident from this experiment, the colors in the red-orange-yellow range also change apparent hue with brightness. All become more-or-less brown when dark enough.

---

[22] The image on the screen is the convolution of the image without the prism with the rainbow produced by an infinitesimal slit. A narrow slit gives good wavelength resolution, while a wide slit makes a brighter image. Use a slit narrow enough to show clearly the yellow portion of the spectrum.

[23] Newton called the colors red, orange, yellow, green, blue, indigo, and violet. Nowadays, Newton's 'blue' is called 'cyan' and the combination of his 'indigo' and 'violet' is called 'blue' by color scientists. 'Magenta' is now the scientific name for 'purple.'

7.9.2 Spectral Composition of the Common Colors

Red, green, and blue are the "primary" colors of additive mixing systems while cyan, magenta, and yellow are the "primary" colors of subtractive mixing systems. Place all the filters in the slide projector, one by one, and note which portions of the spectrum remain. Note that the red filter passes only the red portion, the blue filter the blue portion, but that the green filter permits some adjacent colors to pass as well. Each filter passes about one-third of the spectrum. The blue and green filters are not perfectly transparent even to their own colors, so the image will be quite dark.

The subtractive primaries each pass about two-thirds of the spectrum. They can be labeled by the colors they eliminate, so that magenta, e.g., is sometimes called 'minus green'. Cyan is shown to be a combination of blue and green, while magenta (or purple), the color missing from the spectrum, is the combination of red and blue. These mixtures seem reasonable. Yellow, however, is the sum of red and green. The mixture of red and green produces virtually the same color as the monochromatic yellow of the spectrum. Thus two colors that appear alike can have very different physical constituents. Those who have not seen this before are usually surprised. Another way to demonstrate it is to add up red and green light from the two projectors.

Subtractive mixing, which is defined as multiplication of spectral transmittances, can be demonstrated by placing two filters in tandem in front of a projector. The additive primaries yield black – no light at all – since they have no common pass band. The subtractive primaries do have common pass bands. Thus yellow and magenta give red,[24] magenta plus cyan give blue, and yellow plus cyan give green. This last combination is also somewhat surprising, since yellow, cyan, and green are distinctly different colors. The fact that yellow and blue paint usually produce green is because 'blue' paint is usually cyan, i.e., it also passes some green light.

Finally, we can try the subtractive mixture of one additive primary and its complementary subtractive primary. All yield black since there are no common pass bands. Note that yellow plus real blue do not give green; they give black. The rules of additive mixing depend only on the appearance of the lights to be added, but the rules of subtractive mixing depend on the details of the spectral curves of the colorants. A good demonstration of this (see footnote 11) can be made by experimenting with blue theatrical gels. Some of them pass enough red light so that yellow plus 'blue' gives dark red.

7.9.3 Simultaneous Contrast

Colorimetry is based on color *matching*, not color *appearance*. Matching is nearly independent of conditions of viewing and the state of adaptation of the observer,

---

[24] putting to lie a rule taught to me in elementary school: "A primary color is one that cannot be produced by mixing two other colors." There is no such 'primary' color. There are simply better and worse combinations of primaries.

whereas appearance depends very much on both. This can readily be shown using slides 7.18b,c, mounted in separate projectors, registered on the screen to produce the arrangement of Fig. 3.3, with provision for varying the two luminances independently. With the center patch adjusted to a middle intensity, its appearance can be made to change from nearly black to nearly white by changing the illumination of the surround. If the room is truly dark and the surround is changed slowly and smoothly enough, the eye adapts to the latter and it is the center patch that seems to be changing and the surround that appears to be constant.

If this experiment is now done with separate colors in the two areas, the central patch will be changed in color toward the complement of the surround. Again, if the surround is changed slowly, it is the patch that seems to be changing. This phenomenon is ever-present and so strong so that it is a very important element in the normal viewing of pictures. Clever use of simultaneous contrast enables painters to give the impression of bright sunlight even when using a medium as dark and as lacking in dynamic range as oil paints. It is what makes television pictures so life-like with similar limitations. For example, the shadows of a TV image seem to be darker than the tube itself when it is turned off. Yellows seem quite saturated even though TV yellows are really very 'muddy' from the printer's standpoint.

7.9.4 Sequential Contrast and After-Images

Place slide 7.18d in one projector and 7.18e in the other, and register the two so that the dark spot falls on the touching tips of the triangles. Set the brightness of the second projector at a fairly dim level. Now project the triangles only and stare at the joining tips for 20 to 30 seconds. Turn off the first projector and turn on the second, keeping the eyes fixed on the spot. Very clear complementary-color after-images will be seen. They slowly fade away but can temporarily be restored by changing the brightness of the second projector. Any bright, contrasty color slide can be used to show after-images. The phenomenon is explained by assuming that the retina is adapting to the first image, reducing its red sensitivity in the area illuminated by red light and the green sensitivity in the green-illuminated area. When exposed to white light, white is shifted to the complement of the color to which the retina adapted and to which it therefore has a lower response. Like simultaneous contrast, after-images are continuously present in normal vision, although not so obvious. They have a strong effect on perception at scene changes in movies and television.

7.10 Lessons for the System Designer

Undoubtedly the single most important principle that emerges from this study is that colorimetry 'works' in the sense that when a reproduction matches an original according to the laws of colorimetry, nearly everyone agrees that the result is satisfactory. This takes much of the assumed 'art' out of color work and makes it possible to replace it with methods that can be developed and applied in an objective manner that does not depend on the presumed skill of the operator. When physical limitations prevent a match, and therefore the output color space must be constricted, judgment is required and skill is very useful. Further developments will surely solve this problem in part, but it is likely that the ultimate decision as to when a picture is entirely satisfactory will properly remain in the hands of operators for many years to come.

In addition to this kind of subjective judgment, the principal role for the observer, as in monochrome work, is in the field of data compression. Where colors are changed deliberately to achieve a reduction in transmission or storage capacity, the effects can only be measured, at present, by psychophysical testing. There are some lines of research that may lead to methods to replace human judgment in this area by sophisticated distortion measures, but results so far have not been very promising.

Finally, the additional data required to add color to a monochrome image, when the addition is done correctly, is very small. In view of the importance of color in increasing the sense of reality that we get when viewing reproductions, it is almost always better to devote system resources to color than to resolution, no matter what the purpose of the imaging system.

8. The Design of Improved Television Systems

In this chapter, we apply the principles and methods previously discussed to a topic of great current interest. TV broadcasting systems are likely to be revamped in the coming years so as to deliver substantially higher picture and sound quality to home receivers. Today's US monochrome standards were established in 1941 and color was added in a compatible manner in 1953. Since then, a true revolution in electronics technology has taken place, and, in addition, our knowledge of visual psychophysics and of TV signal processing has advanced a great deal. It would be unreasonable to believe that, in spite of all these changes, the design decisions made so long ago would still be appropriate. In fact, a newly designed system, taking advantage of increased knowledge and improved technology, can give much better performance than today's systems. A much more difficult problem is devising a transition scenario by which such a totally new and necessarily incompatible system can be put in place without putting an unacceptable financial burden on industry and viewers alike. Such matters are discussed in further detail in the Appendix.

This treatment is concerned principally with design methods based on the properties of images, the characteristics of human vision, and the nature of the physical channels that are used to transmit video signals to the home. It is not meant to be a comprehensive treatment of the coding of moving images. Much of the recent work is reviewed, including the various methods of hiding enhancement information within the NTSC signal so as to make possible a single-channel receiver-compatible system. The NTSC constraints have proved so formidable, however, that the emphasis is on completely new systems, free of these constraints except the need to fit into the same 6-MHz channel. We discuss the recent work we have done at MIT and the conclusions we have reached as to the appropriate design methods for our times. We also deal with JPEG and MPEG, systems undergoing international standardization for application to still and moving images, respectively. Most of the treatment is concerned with source coding, i.e., the search for more efficient image descriptions based on a knowledge of images and vision. We also deal to some extent with channel coding, since that has become such an issue with the proposals for digital terrestrial broadcasting. The other analog media, such as cable and satellite transmission, are dealt with only briefly as the technical difficulties are much less.

An issue that emerges with remarkable clarity from the coding discussion is spectrum efficiency – the provision of the most television service of a given quality within a given overall spectrum allocation. This is related, not only to the quality that can be delivered within a certain bandwidth, but to the number of channels available in each locality as a function of the overall spectrum allocation. Today's efficiency is very low, not only because NTSC gets rather poor quality in 6 MHz, but also because it is extremely sensitive to interference. To bring into being a very efficient system requires making progress in both aspects, a necessity that has not yet fully penetrated the thinking of system designers.

8.1 A Review of TV Basics

In order to provide a coherent and self-contained treatment of the subject, a certain amount of duplication of material from earlier chapters is unavoidable. Of course, we shall concentrate on moving color images, and of levels of spatial and temporal resolution that are appropriate for broadcast television. A more complete treatment of some topics is found in [8.1] and the bibliography.

8.1.1 The TV Chain as a Generalized Linear System

Considerable insight into the fundamental operation of a television system can be developed by thinking about the problem on the basis of the block diagram of Fig. 8.1. Light from the scene before the camera is caused to form an image $i(x, y, t)$ in the focal plane. We call this the "video function," and describe the purpose of the entire chain as the production of a modified version $i'(x, y, t)$ on the display device. The video function is converted to a "video signal" $v(t)$ by a scanning process operating on the charge image developed by the camera. In typical cameras, the signal associated with each point in the focal plane is proportional to the integrated light that had fallen on the point since the previous scan. A version $v'(t)$ of the video signal is delivered to the display device, invariably somewhat altered by channel processes such as modulation, filtering,

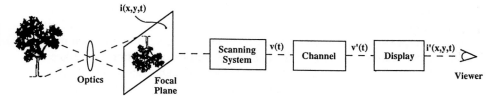

Fig. 8.1. The TV Chain as a Linear System. An image of a scene, continuous in space and time, is formed on the focal plane of the camera. A scanning system samples the image periodically, forming a video signal that is transmitted through the channel. The received signal is used to modulate the display, making the luminance of each point on the display proportional to the illuminance of the corresponding point of the focal-plane image. The relationship between the two images is that of the input and output of a linear system characterized by a three-dimensional frequency response

digitization, transmission, etc. The display process can be thought of as tracing out, on the viewing surface, a scanning pattern (raster) related to that in the camera. At each point, the viewing surface emits a total amount of light proportional to that collected at the corresponding point of the focal plane, usually in an interval considerably shorter than the frame time.[1]

8.1.2 Sampling

Scanning is a kind of sampling process, to which we can apply standard methods from linear systems analysis, as discussed in Chap. 4. Partly because of physical limitations and partly because television technology originated long before systems analysis was well understood, the pre- and postfiltering operations usually used are far from ideal. As a result, the displayed image is nearly always quite different from that on the focal plane. The latter is continuous in space and time while the former is discontinuous, particularly in the temporal domain. Postfiltering is carried out, for the most part, in the human visual system. Since our visual apparatus evolved to its present state by looking at the natural world rather than TV screens, it is anything but ideal for this new purpose.

As a practical matter, the spatial bandwidth usually encountered is not so high that it is impossible to render it adequately in television systems. However, the velocity of objects in front of the TV camera is quite often high enough to generate temporal bandwidths that cannot be properly rendered, even at hundreds of frames per second. The degree of temporal bandlimiting required to prevent aliasing with practical frame rates would often produce unacceptable blurring. Temporal aliasing, manifested as defective motion rendition, is thus the rule, rather than the exception, in both motion pictures and television. A completely satisfactory solution to the dilemma posed by the tradeoff between smooth motion and the sharpness of moving objects still eludes us, although recent developments in adaptive selection of subbands seem quite promising, as indicated in Sect. 8.12.1.

8.1.3 Interlace

All current TV systems use a system of interlace in which alternate lines are traced out on alternate vertical scans of the screen. This can be thought of either as an attempt to double the large-area flicker rate while keeping the total number of lines per frame constant, or to double the vertical resolution while keeping the large-area flicker rate constant. Actual measurements show that the factors that are supposed to be doubled are increased only marginally, the percentage improvement falling rapidly with screen luminance [8.2]. In addition, a number of troubling artifacts are produced such as interline flicker in finely detailed areas and defective rendition of vertical motion. Finally, interlace greatly complicates transcoding from one format to another. One beneficial effect of interlace is that,

---

[1] In storage-type displays, light is emitted at a steady rate until changed by new data.

210

provided the appropriate prefilters are used, it is possible to provide the diamond-shaped vertical-temporal frequency response that is often advocated [8.3].

As is shown below, the objectives of interlace can be achieved very well by separating the display scanning parameters from those used in the channel. This requires a frame store in the receiver, but such stores are expected to be used in all advanced TV systems. Similarly, the camera scanning parameters can also be made independent of those in the channel by using a frame store at the transmitter. In view of the fact that interlace does not perform very well, introduces artifacts, and complicates transcoding, its use in new TV systems does not appear to be advisable.[2]

8.1.4 The Camera

All of today's television cameras that use interlaced scanning discharge the target nearly completely on each vertical scan [8.4]. This reduces the vertical resolution and raises the temporal resolution so much that NTSC cameras act much more like 262.5-line, 60-frame progressively scanned devices than 525-line, 30-frame interlaced devices. This improves the motion rendition at the expense of resolution, and greatly reduces the interline flicker from what it would have been had the vertical resolution been the full 525 lines. (This accounts for the very large difference in motion rendition between film and NTSC television. The latter effectively has 2.5 times the frame rate, not just 25% more.)

The typically low vertical resolution of cameras gives an incorrect impression of the relationship between line number and maximum possible picture quality. This is an important issue in the design of new systems, since modern methods permit the attainment of very nearly the full theoretical resolution without defects [8.5].

A point worth noting about HDTV cameras is that, for a given lens diameter, the number of photons collected per picture element goes down as the square of the linear resolution. If the extra resolution is used to increase the observer's viewing angle, the subjective SNR goes down rapidly with resolution. (If the overall viewing angle remains the same, the SNR is unchanged, and the redundancy rises.) There is also a loss of SNR with target capacitance, a problem made more serious at higher bandwidths. These effects are equivalent to a loss of camera sensitivity, and are problems endemic to HDTV.

Television cameras normally integrate for the full field time, resulting in both high sensitivity and motion blur. The latter can be reduced by employing a shutter, but this reduces the sensitivity and produces less smooth motion. There are no simple solutions to this quandary. With respect to sensitivity, CCD cameras are already close to the theoretical limits, although tube cameras still have some way to go. Further improvements would require using physically larger cameras and lenses to collect more light from the scene.

In addition to sensitivity and resolution, tone reproduction is an important characteristic of cameras. Especially when TV systems are proposed as replace-

---

[2] In spite of this, television systems are still being designed with interlace!

211

ments for conventional cinematography, this factor must be carefully considered. Generally speaking, the ability of current electronic cameras to render shadow detail is inferior to that of photography.

8.1.5 The Display

Displays are characterized by luminance, spatial resolution, temporal effects, and, of course, cost in relation to size. Cathode-ray tube technology has persisted a good deal longer than expected, and flat-panel displays have been very slow in coming. Most observers believe that HDTV requires displays of 1 meter or more, which makes CRTs impractical for many viewing situations. If this is true, and no economical, relatively thin display device is developed within the next few years, the growth of HDTV may be inhibited. There is a good possibility that CRT projection systems will soon meet all the requirements except brightness. If viewers become content to watch HDTV in dimly illuminated rooms, and save their daylight viewing for lower-resolution programs, this may provide a solution of sorts. Solid-state light valves show promise of meeting the brightness requirement, but require more development.

8.1.6 The Channel

Digital television transmission is appropriate in fiber-optic cables and could be used in satellite channels as well, although FM has been the normal choice. In principle, digital transmission could also be used in cable, but the entire cable plant in the US, which serves more than half of all households, would probably have to be rebuilt. Until quite recently, it had generally been thought that over-the-air transmission (in regulatory jargon, this is referred to as "terrestrial broadcasting," while cable service, which is also on the surface of the earth, is simple called "cable") in which the SNR at the receiver varies over a very wide range, would remain analog for many years to come. Recent proposals to use digital terrestrial broadcasting have attracted an enormous amount of attention. This matter is discussed in Sect. 8.13.

Digital channels are simply characterized by data rate and error rate. Traditionally, analog channels are correspondingly characterized by bandwidth and SNR. This is inadequate, since in terrestrial transmission, image quality is also affected by multipath (ghosts), interference from other TV stations, and distortion due to improper frequency response. In fact, the spectrum problem of television broadcasting is related, not only to the bandwidth used by each signal, but rather to the total amount of spectrum devoted to TV service in relation to the number of channels and picture quality available in typical homes. At the present time, the overall efficiency of utilization of spectrum is very poor. In each locality, less than one-third of the channels can be used, primarily because of the very poor interference performance of current TV systems.

8.2 Relevant Psychophysics

In this Section, those aspects of visual psychophysics that most closely relate to television coding are briefly reviewed. Further details are found in Chap. 3 and the references.

8.2.1 Normal Seeing

As a result of evolution, the human visual system (HVS) is remarkably effective in rapidly extracting a large amount of useful information from the scene before the observer. We have sharp vision only in the central few degrees of the field of view, but the eye is in constant voluntary (e.g., tracking) and involuntary motion so that we usually are able to see what we wish or need. The cone cells, concentrated in the fovea around the optic axis, are responsible for fine detail and color, while the rod cells, which are absent from the fovea, provide vision under low light levels. We can see well over the very wide range of commonly encountered lighting conditions partly because of changes in iris size but mostly because of adaptation. The latter process is photochemical, taking place in the retinal receptors.

8.2.2 Contrast Sensitivity

The term contrast sensitivity relates to the ability to see fractional differences in luminance between neighboring points in an image. It is essential for perceiving detail. With adaptation, about a 1% difference can be seen over more than a 10,000:1 range of luminance, depending on the exact character of the image. The enormous dynamic range resulting from adaptation is greatly reduced while looking at any one scene – the instantaneous dynamic range at any one point on the retina is only about 100:1. In one sense, this simplifies the task of man-made imaging systems, which do not have to reproduce the full dynamic range found in nature under all conditions. Under outdoor sunlight however, we may encounter as much as 10,000:1 dynamic range, which the HVS deals with by rapid adaptation as the eye moves. Imaging systems cannot do as well, so that various stratagems must be used to get good pictures under those circumstances. When possible, the dynamic range is deliberately reduced by supplementary lighting. No electronic display device has the necessary four decades of dynamic range, so the realistic reproduction of outdoor scenes requires that the inevitable compression of large-area dynamic range be accompanied by an expansion of small-area contrast. This gives a result similar to what the eye perceives when it is differentially adapted to the light and dark areas in a particular image.

Tone reproduction is associated with contrast sensitivity, and under many circumstances, it is the most important element of picture quality. Full adaptation implies a logarithmic relationship between stimulus and response. The partial adaptation that occurs nearly instantaneously while looking at a scene results in a quasilogarithmic relationship, which is therefore a good characteristic for imaging systems.

8.2.3 Temporal Frequency Response

Flicker and motion rendition are associated with this factor, which therefore has received a great deal of attention. When measured using a large, featureless field so as to reduce the effect of spatial variation, sensitivity is seen to peak at 5 to 25 Hz, and to remain significant up to 80 Hz or higher [8.6]. That means that systems such as TV or motion pictures that simulate temporally continuous stimuli by a sequence of images must use a very high display rate to avoid flicker completely. To avoid a concomitant increase in bandwidth, modern systems separate the frame rates of transmission and display [8.3].

A somewhat surprising result of temporal response measurements is the fall-off in response at low frequencies. The HVS acts as a temporal differentiator at low frequencies, which some have associated with the heightened sensitivity to change and to reduced sensitivity to unchanging parts of the image.

8.2.4 Spatial Frequency Response

Visual acuity – the ability to see small details – is one of the most obvious aspects of vision, and the one most often measured. If we assume linearity, then all aspects of spatial behavior can be elucidated by measuring the spatial frequency response using sine-wave test patterns [8.7]. Step response can also be measured [8.8]. The results are remarkably similar to the corresponding temporal measurements, showing a peak at about 5 cycles/degree subtended at the eye, going to zero at about 50 cycles/degree. Differentiation at low frequencies is associated with heightened sensitivity to edges, and rather low sensitivity to the absolute luminance level in broad areas lacking detail.

For systems that use spatial sampling (such as the scan lines in television) it is highly desirable to eliminate the sampling structure from the display. The sampling theorem tells us that the baseband may extend as high as one-half the sampling frequency, but the HVS simply does not provide a sufficiently sharp-cutting low-pass filter to separate the baseband from the sampling frequency. The visibility of sampling structure, whether spatial or temporal, must reduce the sensitivity to the real information being displayed.

8.2.5 Spatiotemporal Interactions

Assuming that the eye has a fixed spatiotemporal frequency response, it is clear that any relative motion of image and retina effectively changes the spectrum of the stimulus. For the fixated eye, off-axis response, i.e., the response to diagonal spatial frequencies as well as to stimuli that are simultaneously high in the spatial and temporal frequency dimensions, is rather low. Thus, detail of moving objects is blurred. However, when we track moving objects, holding their images stationary on the retina, then the detail is nearly as clear as if the object were stationary. This presents a serious problem for imaging systems, as the bandwidth cost of rendering the full three-dimensional spectrum is very high.

A possible solution to this problem is found in the fact that, in most cases, important detail is far from isotropic.[3] For example, edges, either stationary or moving, generally have highly nonuniform spectra. If we can find some way to deal only with that portion of the spectrum that is required for each small area of the image, a very large reduction in channel capacity may be possible without substantial loss of information.

Two points to be kept in mind, especially if some kind of adaptivity is to be used, is that all the bandwidth limits of human vision are gradual, and that a small amount of energy may be associated with some image element of great visual importance. In general, no picture information is entirely *irrelevant* in the sense that it can be removed without any harm at all. What usually is true is that some information is more important visually and some less so. Within a given overall channel capacity, it is therefore often possible to improve image quality by replacing less important information with more important information.

8.2.6 Noise Visibility and Masking

The channel capacity required for picture transmission is directly proportional to the SNR, in dB, in the resulting image. High capacity is required if we insist on high SNR in all image areas, at all brightnesses, and throughout the spectrum. Therefore, one of the most fruitful approaches to TV compression is associated with masking, i.e., the tendency of picture detail to reduce the noise visibility in the spatial, temporal, or spectral neighborhood [8.9]. This is not a small effect. Some of our results indicate that suppression of noise in highly detailed image areas is as much as 25 dB [8.10]. This is over and above the variation of noise visibility due to the spatiotemporal frequency response of the HVS. This phenomenon is undoubtedly the main reason why SNR, even weighted by frequency response, is a very poor indicator of image quality.

Masking aside, noise visibility depends not only on its spectrum, but its degree of randomness. One of the most significant advances in image processing was the discovery by *Roberts* that quantizing noise could be made much less visible if it were randomized [8.11]. This result is so striking that an appropriate conclusion is that quantizers should always be randomized unless there is a compelling reason to the contrary.

Another important phenomenon is the visibility of noise throughout the tone scale. Noise is least visible if it is uniformly visible.[4] If the noise is added in the channel, this can be controlled by the tone scale of the transmitted signal. The quasilogarithmic scale mentioned previously, which is very common in graphic arts, is optimum. The "gamma-corrected" tone scale, common in television, is very nearly optimum. A linear tone scale results in more shadow noise and a truly logarithmic scale results in most noise in highlights.

---

[3] Picture areas that have uniform spectra generally consist of texture rather than detail, in which case accurate sample-by-sample reproduction is not required.

[4] This is because, in the case of nonuniform visibility, the most visible portion can be made less visible by rearranging the noise.

8.3 The 1941 NTSC System

The scanning parameters and modulation method of today's television system were set by the first National Television System Committee and accepted by the FCC in 1941 [8.12]. They were predicated on receiver technology that was appropriate at that time, and so resulted in a system that is quite inefficient in spectrum utilization and is extremely vulnerable to transmission impairments such as echoes and interference.

8.3.1 Audio System

One excellent decision was to use a high-quality FM audio system and to specify the relative power of the sound and picture carriers so that at CNRs where the picture is quite noisy, the sound quality is still excellent. This was quite appropriate considering the reaction of viewers, who have more tolerance for picture degradation than for audio degradation. With guard bands, however, about 0.5 MHz, or nearly 10% of the channel bandwidth, is devoted to sound. Another problem is that with "intercarrier" receivers, the audio cannot be properly recovered unless the picture carrier amplitude is quite high and generally free of phase distortion. This inhibits the use of certain types of recently proposed video enhancement methods that involve quadrature modulation of the picture carrier.

8.3.2 Modulation Methods

Double-sideband amplitude modulation requires an rf bandwidth twice that of the basebandwidth. Since television signals require a much larger bandwidth than audio, it is always desirable to conserve bandwidth, if possible. Therefore, vestigial-sideband AM (VSB) was used, in which the carrier is placed away from the center of the band, as shown in Fig. 8.2.[5] VSB was a very unfortunate choice. With an rf bandwidth of 5.5 MHz, the video bandwidth is only 4.2 MHz. Double-sideband quadrature modulation (DSBQM) would have given a basebandwidth of the full 5.5 MHz. In addition, very careful filtering is required to avoid transient distortion in VSB, while no such distortion occurs in DSBQM.

8.3.3 Scanning Parameters

The 1941 NTSC system employed no storage, so that the camera, channel, and display all used the same scanning standards. To avoid flicker as well as to reduce the visibility of interference from stray signals related to the power line frequency, 60 Hz was chosen as the vertical frequency.[6] The choice of horizontal frequency

---

[5] It is said that VSB was discovered accidentally during experiments with transmission of very low-resolution pictures in radio channels during the 1920s. It was found that detuning increased the resolution! It should be remembered that the "sideband theory" was not generally believed by radio engineers at that time, in spite of work by Fourier 100 years earlier.

[6] Similar power-line considerations led to a choice of 50 Hz in Europe, which gives substantially more flicker. However, Europeans have apparently gotten used to it.

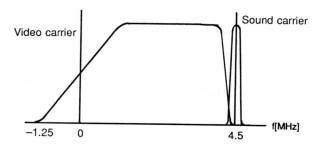

Fig. 8.2. The 1941 NTSC Spectrum. The video signal is impressed on one carrier using vestigial-sideband amplitude modulation, while the audio signal frequency-modulates a second carrier. The video carrier is located 1.25 MHz above the lower band edge and the audio carrier 4.5 MHz above the video. The video bandwidth is 4.2 MHz and the channel width is 6 MHz, which includes small guard bands

trades off vertical and horizontal resolution, second-order interlace being used in an attempt to improve the tradeoff. The choice of 525 scan lines per full frame every 1/30 second, together with generous retrace intervals (at least 16% horizontal and 5-8% vertical) completed the selection of parameters. This results in about 480 active scan lines with about 420 resolvable elements per line. It was thought that the effects of interlace and vertical sampling, as expressed by the Kell factor [8.5], would result in approximately equal vertical and horizontal resolution for the 4:3 aspect ratio, chosen to be the same as motion pictures.

As discussed above, interlace doesn't work very well and also leads to artifacts such as interline flicker. With the camera tubes and display tubes of the time, which had rather low resolution, it would not have made much difference if 262 lines with progressive scanning had been chosen. The use of interlace in both the NTSC system and the European system did considerably complicate transcoding between the two.

8.4 Adding Color

The 1941 standards did not come into commercial use until 1949, but as soon as television broadcasting was introduced after World War II, it became very popular and engineers turned immediately to the question of color. *Kodachrome* film and *Technicolor* motion pictures were already in use, so that the public was quite accustomed to high-quality color images. Work began and eventually a second NTSC was formed. Adding color to the monochrome television system seemed like a formidable task. The first important problem was the color picture tube, and the second was how to squeeze the three signals required for color into the 6-MHz channel that was being used for mediocre black-and-white pictures.

8.4.1 The CBS Field-Sequential System

Nearly everyone thought at the time that it would be desirable to add color in a compatible way so that existing monochrome receivers could continue to be used.[7] However, by 1951, the FCC had concluded (incorrectly) that compatible color was not on the horizon, and selected the CBS field-sequential system, shown in Fig. 8.3. The Korean war prevented its commercialization, and, within two years, the FCC reversed itself and selected the NTSC compatible system, shown in Fig. 8.4. In this system, a composite video signal, using the same parameters as the earlier monochrome signal, is formed by adding a color subcarrier on which the color information is impressed [8.13].

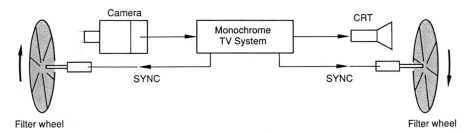

Fig. 8.3. The CBS Field-Sequential Color System. This system, which was the US standard from 1951 to 1953, transmitted the red, green, and blue components of the signal sequentially. To accommodate the three signals within the same 4.2 MHz bandwidth, the scanning parameters were 24 fps, 405 lines/frame instead of NTSC's 30 fps, 525 lines/frame. The horizontal resolution was reduced about 50%, but the sharpness was partly restored by "crispening." The usual implementation was with color filter wheels in front of both the camera and the display, although other methods are also possible

Field-sequential color had been demonstrated much earlier and had even been used at one time in motion pictures. It is very simple, and the color quality is excellent, requiring no adjustments at all. All that is necessary is to transmit and display the red, green, and blue components in sufficiently rapid sequence. The components can be produced simply by using rotating filter wheels with red, green, and blue segments in front of the camera and display.[8] Any short-persistence white picture tube can be used. There is, of course, a bandwidth problem, since many more frames per second are required. CBS dealt with this by reducing the frame rate and resolution. They used 24 full color frames per second, 405 lines/frame, 2:1 interlace, giving 144 fields/sec. With a horizontal

---

[7] In retrospect, this may not have been such a good idea. There were about 10 million receivers in use when compatible color was adopted in 1953, and there are 180 million now, all with defects due to the decision to keep the relatively much smaller number of receivers in use for a few more years. In addition, the incentive to buy color receivers was greatly reduced since the old receivers could be used to watch the new color programs. This may well have been one of the reasons for the slow growth of color. It is worth thinking about this now, since the compatibility issue is an important one for HDTV.

[8] The two sets of filters must be different. See Chap. 7.

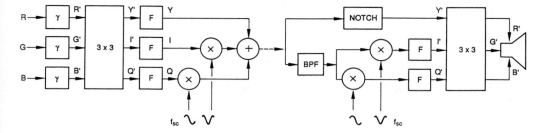

Fig. 8.4. The NTSC Color System. The red, green, and blue signals from a color camera are gamma-corrected and then subjected to a 3 × 3 linear matrix transformation. The resulting signals are band-limited to 4.2 MHz (luminance), 1.6 MHz (I), and 0.6 MHz (Q) before being combined to form the composite signal that is transmitted. I and Q are quadrature-modulated onto the color subcarrier of 3.58 MHz, audio is frequency-modulated onto a 4.5 MHz carrier, the three signals are added, and the resulting composite signal is applied to the picture carrier using vestigial-sideband modulation. The means shown in the figure, by which the signals are separated and combined, has been the usual method, but is now being replaced by comb filters

scan rate of nearly twice that of monochrome TV, the horizontal resolution was about half that of the black-and-white system. The addition of color more than made up for the lower resolution, however. There is a color fringing problem with motion since the three components are not then in geometrical register.[9] In spite of these important defects, it was the color wheel that was the subject of the most derision. It did limit the picture size. However, it worked quite well, although the brightness was much lower than the same tube could produce with monochrome pictures.[10] Of course, the field-sequential system could also have used a color picture tube, but if the latter had been available, higher bandwidth efficiency could have been achieved with a simultaneous system and the color fringing problem easily avoided. A simultaneous color system, such as later adopted, *requires* a full-color display, and cannot use the simpler color wheel.

With storage, the field-sequential system could have been made compatible by using a simultaneous camera. By taking a portion of the height of the mono-chrome field for the chrominance components as shown in Fig. 8.5, a wide aspect-ratio image could have been produced. The monochrome receiver would simply mask off the color components, while the color receiver would resynchronize the three components to derive the *RGB* signals, then retime them to drive the sequential display.

---

[9] The severity of this effect, which is similar to that caused by showing each frame twice in motion pictures, depends on the degree of motion and the extent to which the viewer tracks the moving object. It can be avoided to some extent with a simultaneous (3-tube) camera and storage of the successive color fields. No receiver change is required.

[10] Another method that does not require a color CRT is to use separate red, green, and blue tubes, and to register the three images, either by projection or direct viewing. This is used in most of today's projection systems with great success. See Chap. 7.

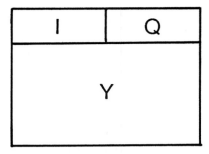

Fig. 8.5. A Compatible Field-Sequential System. With modern methods, it would be possible to make a compatible field-sequential system with much less loss of resolution by taking a portion of the monochrome frame for two luminance components, as shown. The chrominance area would be masked off on the monochrome receivers, and frame stores would be used to get precise synchronism among the three components in the color receivers

8.4.2 Luminance/Chrominance and Mixed Highs

The bandwidth inefficiency of the field-sequential system comes from the fact that equal resolution for the red, green, and blue images, which is inherent in the color wheel, is far from optimum. Much higher overall resolution is attainable within the same bandwidth if channel capacity can be apportioned appropriately among the three components. Even better results are attainable if the *RGB* signals are matrixed to produce one luminance and two chrominance components, since the required resolution for the chrominance signals is even lower than for red and blue. If this process is carried out two-dimensionally, then the luminance resolution must be reduced only about 10% to add color to a monochrome system without any increase in bandwidth [8.14]. In NTSC color, since storage is not used, all components must have the same vertical resolution. Horizontal luminance resolution would have had to be reduced about 30% to add color, were it not for band sharing (Fig. 8.6).

An alternative to luminance/chrominance is the "mixed highs" system, originally proposed by *Bedford* [8.15]. The luminance signal is divided into a low-frequency component, equal in bandwidth to the chrominance signals, and a high-frequency component, which can be considered to be a combination of the

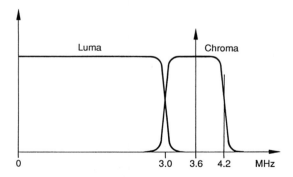

Fig. 8.6. A Compatible System Without Band-Sharing. If enough of the luminance spectrum were given over to chrominance, band-sharing with its inevitable cross-effects could be avoided completely. The actual reduction in luminance resolution would not be very high and the decoding circuitry would be quite simple

high-frequency portion of the three original wideband *RGB* signals. The three low-frequency signals can be dealt with as *RGB* or L/C lows, as desired. This representation is somewhat less convenient for compatible systems, but much more convenient for color-only systems, since most of the bandwidth is used for the single-component luminance highs. As we shall show below, keeping the dc information entirely within a narrow-band component also has advantages.

8.4.3 The Color Subcarrier and Quadrature Modulation

Aside from the reduction in resolution of chrominance, the most important element in compatible color – the element that made compatibility feasible – is the ingenious band-sharing principle, in which the color information is impressed on a subcarrier near the upper end of the monochrome video band, as shown in Fig. 8.7. The scheme does not, however, work perfectly with the kind of processing envisaged by its advocates in the 1950s. In saturated areas, the color subcarrier produces a highly visible cross-hatch pattern on old monochrome receivers, even though the subcarrier is out of phase on successive frames, since the eye does not integrate exactly over 1/15 second. In addition, considerable crosstalk between luminance and chrominance is often seen on color receivers. Since the luminance and chrominance bands overlap, if no steps are taken to separate the signals, fine luminance detail produces "cross-color" and sharp color edges produce "cross-luminance."

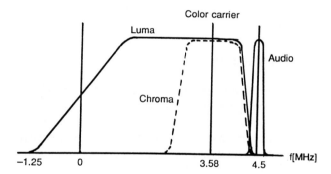

Fig. 8.7. The 1953 NTSC Spectrum. Color was added compatibly to the 1941 monochrome system by impressing two chrominance signals onto a color subcarrier located about 3.58 MHz above the picture carrier. The visibility of the subcarrier and its interference with the luminance signal are reduced by making the subcarrier frequency an odd multiple of half the line scanning rate. This causes the luminance and chrominance harmonics to be interleaved. Separation of the two signals has never been perfect

Failure to appreciate this problem properly is puzzling, since comb filters can separate the signals rather well. Such filters were known for other purposes at that time and had been discussed by *Gray* in his 1929 patent, the earliest mention of frequency interleaving [8.16].

Receivers with notch filters (formerly the most common kind, but now being replaced by comb filters) eliminate cross-luminance only and substantially reduce the luminance resolution. It is the author's opinion that a much more satisfactory result would have been obtained if the original luminance bandwidth of 4.2 MHz had been reduced to 3.0 MHz and band-sharing not used at all. While the right kind of multidimensional filtering [8.17] can recoup much of the quality lost when NTSC color was introduced, these improvements come only after decades of defective images, and not very long before NTSC itself is likely to be abandoned for a much more efficient HDTV system.

One element of the NTSC color system – quadrature modulation – does work very well. This technique permits the bandwidths of the baseband and rf signals to be the same, and has proven simple and reliable. What should have been clear from its success was that the earlier VSB technique for the luminance signal ought to have been abandoned then and there, rather than keeping it and thus greatly reducing picture quality on all color receivers for the next 40 years.

8.5 NTSC System-Design Limitations

A list of NTSC defects and limitations is a list of its principal design choices, since every single choice, made in 1941 and 1953 based on then-existing receiver technology, would now be made differently. It should be remembered, while cataloging all these shortcomings, that the NTSC system has nevertheless been very successful, thus testifying to the importance of color as one aspect of image representation.

8.5.1 Limitations Due to Use of a Simple Receiver

The lack of storage required a rather high frame rate to eliminate flicker, which, while beneficial for motion rendition, gives very low spatial resolution. Many years' experience with motion pictures clearly demonstrates the acceptability, if not the preference for, lower temporal and higher spatial resolution. Nearly 20% of the frame time is lost to retrace intervals, and very large synchronization pulses cause higher interference than necessary between stations.

8.5.2 Interlace

Although there is not yet universal agreement on this point, the evidence is that interlace should not be used in the camera or display, but might be used in the channel. The main purpose of interlace, which is to improve the tradeoff between vertical resolution and large-area flicker rate, is achieved only to a small extent at the high luminance levels actually used. Specific defects, such as interline flicker, the disappearance of half the scan lines during vertical motion, and the complications inflicted on transcoding, all indicate that this idea, if ever justified, ought now to be abandoned. Especially in an era when all receivers will have frame stores that permit nearly perfect achievement of the original purpose, the argument for interlace is no longer persuasive. Nevertheless, as discussed in the Appendix, interlace is still being specified for some of the new HDTV systems.

8.5.3 Vestigial-Sideband Modulation

Use of VSB in new systems is clearly an error. Double-sideband quadrature modulation has now been used successfully for nearly 40 years for the color subcarrier and its chrominance information. This permits the basebandwidth to be the same as the rf bandwidth, rather than 30% smaller. It also eliminates the defective transient response of VSB without particular attention to the filters since there is no inherent problem to be corrected. There is still some use of VSB in digital transmission systems.

8.5.4 The Separate Sound Carrier

No new system would use frequency multiplexing for the audio signal, since time multiplexing is more efficient, and the required storage can be provided easily and cheaply. In addition, the great popularity of compact disks make it likely that any new system would incorporate digital audio, which would use an even higher proportion of the channel capacity if impressed on a separate carrier.

8.5.5 Chrominance Limitations

In NTSC color, the relative vertical chrominance resolution is too high and the horizontal chrominance resolution is too low. While a certain amount of anisotropy is permissible [8.18] the 7:1 disparity is a noticeable defect. The chrominance frame rate is actually higher than necessary, especially in view of the very low spatial resolution. A much more satisfactory color signal would result from a better balance of horizontal, vertical, and temporal resolution.

8.5.6 Violation of the Constant-Luminance Principle

A more subtle problem with NTSC is the result of the way the overall tone scale is controlled. Because of the near-square-law characteristic of picture tubes, a corresponding compression is required somewhere in the system. In order to simplify the receiver, this is done at the transmitter, as shown in Fig. 8.4. Since matrixing to obtain Y, I, Q from R, G, B is done on the compressed, or "gamma-corrected" signals, there is some unwanted luminance mixed with chrominance and vice versa. This would cause little harm except that the three contaminated signals are differently band-limited. Thus the portion of luminance that is improperly carried in the chrominance signals is reduced in bandwidth. This leads to very low resolution of certain saturated colors, which is particularly noticeable in red areas.

There have been some proposals to eliminate this problem by using linear, rather than gamma-corrected signals in the channel. This would produce a worse problem, however, since the compressed signals are much better with respect to performance in the presence of channel noise. Thus, there is no simple solution to the problem. Compression for dealing with channel noise must be carried out separately from linearization of the display characteristic. Neither is difficult in future systems that will use digital processing at both encoder and decoder. Most American proposals for improved NTSC have disregarded this problem; a partial solution is included in Japanese plans.

8.6 Compatible Improvements to NTSC

Because of the very large installed base of NTSC receivers and VCRs, there has been considerable sentiment to improve NTSC rather than to replace it [8.19]. Elimination of its perceived defects is also deemed essential to any receiver-compatible HDTV system that adds enhancement information to the current NTSC image to produce a new image with higher resolution, whether the extra information is hidden within the NTSC signal or transmitted in a separate channel.[11] In considering such improvements, it is important to distinguish between quality judgments made by the mass audience and those made by experts. There is considerable evidence that the NTSC shortcomings most noticed by experts (who, for the most part, look at studio-quality images) are ignored by ordinary viewers, watching in their homes. These include interlace effects, cross-color, cross-luminance, and poor motion rendition. Some evidence that even broadcasters realize this is provided by the fact that most cross-color, for example, could be eliminated simply by banning the offending clothing from the studio, a simple step that is rarely taken.

Our own audience tests, as well as much anecdotal evidence, indicate that the principal limitation to perceived quality in the home is that due to channel impairments such as noise, ghosts, and interference [8.22]. These defects could, in principle, be eliminated by better channels such as may eventually be provided by digital fiber-optic cable. However, making these improvements in today's analog channels is much more difficult. This is partly a question of cost, but it is the extreme vulnerability of NTSC to channel impairments such as interference that is the root cause. This cannot be corrected except by replacing NTSC with a system that has better interference performance. Realization that this is the case developed very slowly in the broadcast industry. It was the FCC that decided that the new US HDTV would be noncompatible; the industry seems to have accepted the decision, for the most part reluctantly.

---

[11] It goes without saying that any TV system can be improved by providing additional bandwidth. The two most prominent proposals for doing that were those of North American Philips [8.20] and Glenn [8.21]. In both these systems, an unaltered NTSC signal was to be transmitted in one channel and enhancement information in a second channel, either 3 or 6 MHz wide. Eventually, the FCC removed all augmentation systems from consideration on account of poor spectrum efficiency. It is quite remarkable that it should have taken an FCC decision to show the disadvantages of augmentation, which necessarily retains all the defects of NTSC and permanently uses more bandwidth.

8.6.1 Cross Effects

Cross-color and cross-luminance are by far the most often cited problems with NTSC, even though they do not seem to trouble the mass audience. At some cost, they can be removed completely by three-dimensional filtering at both encoder and decoder. However, there are several theoretical and practical difficulties with this solution.

Since the video signal is a function of x, y, and t, it has a three-dimensional spectrum. The color subcarrier occupies a portion of four octants in frequency space, as shown in Fig. 8.8. Complete separation of luminance and chrominance requires that they be appropriately bandlimited (three-dimensionally) before being combined to form the composite video signal. Such filters are called comb filters because their equivalent one-dimensional characteristics are periodic, like the teeth of a comb. Similar filters are needed at the receiver, where they add some cost. A problem in most studios is that all cameras and other signal sources produce composite signals. Thus the filters would be needed in hundreds of places. This can only be avoided in component studios, which are growing in popularity for other reasons, but are still unusual.

An increasing number of home receivers are now equipped with two-dimensional comb filters. It is not clear that much improvement in perceived

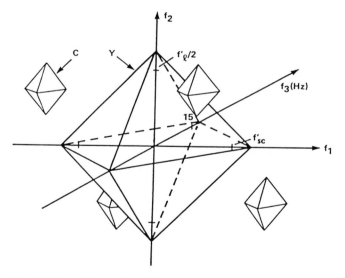

Fig. 8.8. The 3-Dimensional Spectrum of the NTSC Signal. Although the color subcarrier is electrically generated, it can also be produced by scanning, in effect, a theoretical spatiotemporal frequency component located in four of the octants of 3-d frequency space as shown. This method of thinking about the spectrum shows that chrominance and luminance can be separated completely by 3-d filtering. The required deletions from the luminance spectrum reduce the sharpness of moving objects and also reduce both the spatial and temporal resolution of color. Although experts usually prefer the alias-free result of 3-d filtering, it is not clear that the mass audience would feel the same way. (Diagram courtesy of Eric Dubois)

quality is achieved, partly because substantial improvement requires filtering at both encoder and decoder. Very little encoder filtering is being done at present. However, there is another, more basic problem. These cross effects constitute a form of aliasing. As pointed out in Chap. 4, aliasing is not a crime – it is simply a defect that must be traded off against the blurring that results from eliminating it. In the case of three-dimensional filtering, which is not yet done commercially, it appears that the blurring is excessive [8.23]. In the case of today's 2-d comb filters, my own impression is that there seems to be little, if any, net overall improvement. There appear to be no credible tests of the mass audience to support the use of such filters.

8.6.2 Upconverted Displays

Another possibility often proposed for improving NTSC, and now implemented in some improved-definition (IDTV) receivers, is the use of progressive-scan displays that are upconverted from the normal interlaced transmission. Such displays do eliminate interline flicker, if present, and could be used, in principle, to obtain much higher vertical resolution than seen today by raising the Kell factor. Again, there are some obstacles that stand in the way of such improvement, aside from cost.

If the camera has full vertical resolution, then an interlaced display has unacceptable interline flicker. This flicker disappears on the progressive display, but not on existing standard receivers. Hence this route to higher vertical resolution presents a backward compatibility problem, although not as severe as that presented by completely noncompatible systems. If the higher resolution information is not present in the signal, there will be no improvement in resolution on the so-called improved display. Actually, most interpolation methods used for upconversion result in *lower* sharpness, albeit a smoother looking (structureless) picture. There is no credible evidence that visible line structure, by itself, degrades perceived quality, at least for the mass audience, while viewing today's pictures that have low vertical resolution.

Another problem with upconversion is that, in order to preserve the sharpness of moving objects, motion-compensated interpolation, or at least motion-adaptive interpolation is required.[12] Nonadaptive interpolation noticeably degrades sharpness. This effect, combined with the typically low vertical resolution of today's transmissions, results in very little improvement, if any. In one test carried out in our Audience Research Facility, the viewers actually preferred the interlaced display [8.25].

---

[12] A simple but rather effective method of motion-adaptive interpolation is used in MUSE. In stationary areas, interpolation is temporal, while in moving areas, it is vertical [8.24].

8.6.3 Hiding Augmentation Information in the NTSC Signal

Having eliminated its defects, further improvements in NTSC require transmitting additional information. If sufficient information could be hidden within the NTSC signal in such a way that performance on today's receivers were not substantially impaired, then broadcasters would have achieved their goal of HDTV or EDTV[13] within a single channel in a completely compatible manner. If one accepts the premise that NTSC quality is limited primarily by channel impairments, then this goal is clearly impossible. However, under good transmission conditions, it does appear to be possible to hide a worthwhile amount of information. In all the methods cited here, the visibility on the screen of a given amount of hidden enhancement data can be minimized by adaptive modulation and scrambling, as discussed in Sect. 8.12.4.

The Fukinuki Hole. A good principle of efficient image representation is to allot channel capacity according to the psychophysical importance of the data to be transmitted. In three-dimensional frequency space, information in the corners, which corresponds to small moving detail, is of less importance than the components that are low in at least one frequency dimension. This is the unstated premise that was relied on in adding color, since the low-frequency chrominance information replaced diagonal luminance information. *Fukinuki* has proposed using the four octants of frequency space not now used for color for a second subcarrier to transmit additional luminance data [8.26]. This method was incorporated into the ACTV proposal of the Sarnoff Laboratories [8.27]. There is little question but that some picture improvement can be obtained in this manner. The method produces additional cross-color flickering at 30 Hz, which is nearly invisible. The loss of diagonal luminance resolution may produce some small artifacts.

Quadrature Modulation of the Picture Carrier. In vestigial-sideband modulation, both upper and lower sidebands are present in the region around the picture carrier. Therefore quadrature modulation is possible within this region, which is about 2.5 MHz edge-to-edge [8.28]. The difficulty in utilizing this essentially unused spectrum space is that there are two kinds of demodulators used in current receivers. One is an envelope detector and the other is a synchronous detector. The result is that, no matter how the extra information is impressed on the carrier, it causes some crosstalk on one or both types of receivers. In addition, the resulting phase modulation of the carrier may interfere with audio reception in intercarrier receivers. For this reason, the amplitude of the impressed signal, and therefore the degree of enhancement that can be achieved, cannot be very large.

The Noise-Margin Method. It is well known that the required SNR in video goes down with spatial frequency [8.29]. With white noise, if the SNR is satisfactory,

---

[13] Enhanced-definition (EDTV) TV usually means resolution between 500 and 1000 lines, HDTV means resolution of at least 1000 lines, and improved-definition TV (IDTV) refers to better pictures that do not require additional transmitted information, but rely primarily on receiver processing.

then it is probably higher than needed in the higher frequency bands. Information can therefore be added nearly invisibly in these higher bands if it looks like noise [8.30]. This process is symbolized in Fig. 8.9. In Fig. 8.10, the simplest possible version is shown. The two-dimensional spectrum is divided into the dc component and one group each of middle-frequency and high-frequency bands. The dc component and the middle bands comprise the "NTSC" image and all three comprise the "EDTV" image. The middle bands are quantized into 16 to 32 levels, and the high bands reduced in amplitude so as to fit within a single quantum step. Decoding is obvious. In this method, the middle bands have added quantization noise, and the high bands have reduced SNR due to the relatively higher channel noise. To bring the second effect into an acceptable range requires adaptive modulation of the high bands, as discussed in Sect. 8.12.2.

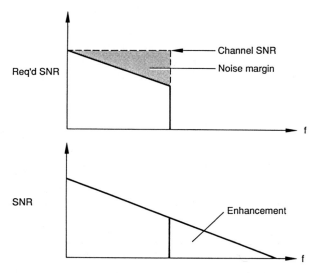

Fig. 8.9. Conceptual Basis of the Noise-Margin Method. In a channel with white noise, if the SNR of a video signal is satisfactory, it is usually higher than necessary at high frequencies, since the required SNR decreases with frequency. Therefore, there is a "noise margin" within which it is possible to add enhancement information invisibly and compatibly as long as it looks like noise. This extra data can be extracted in a special receiver to raise the resolution

Adding Information to Chrominance. The capacity of the NTSC chrominance channel is not well defined, since it overlaps with luminance, producing cross effects. In Fukinuki's method, chrominance is confined to a very small volume in frequency space, and no additional data can be added. When two-dimensional comb filters are used, half of the chrominance channel can be used for additional data if the chrominance frame rate is reduced to 15 fps [8.31]. Even more space is available for additional data if luminance and chrominance are separated by one-dimensional filters, because then the vertical resolution of chrominance is higher than optimum. In both of these cases, in order to use part of the chrominance

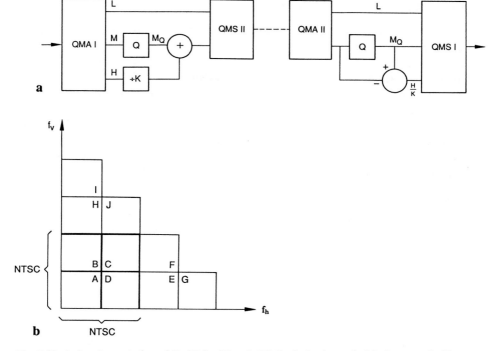

Fig. 8.10a,b. Implementation of the Noise-Margin Method. As shown in (a), the normal video signal is divided into a dc component, L, and three middle-frequency components, symbolized by the single signal, M. Six high-frequency components, symbolized by H, are hidden, three per frame, in the three M components. To do this, the M signals are quantized and the frequency-shifted H signals are added after being reduced in amplitude sufficiently to remain within one quantum step. Separation at the receiver is obvious. In order to have high enough SNR in the hidden components, they are adaptively modulated, after which they are scrambled to reduce their visibility. In (b), the effect on the spectrum is shown. A high-definition image is divided into 16 components and the highest 6 discarded completely. A,B,C,and D comprise the NTSC signal. On one frame, I, J, and H are hidden in B, D, and C, while F, G, and E are hidden on the next frame. In this manner, the static vertical and horizontal resolutions are doubled. The frame rate of the 6 highs components is halved

channel for other data, it is important that, when decoded on all types of existing receivers, the additional data does not cause significant degradation.

Extra data added in the spectral region between 15 and 30 fps can be made to appear as 30-Hz chrominance flicker, which is essentially invisible. This is accomplished by sending each of 15 chrominance frames per second twice, alternately adding and subtracting the same extra information. At the receiver, chrominance is found by adding pairs of frames, and the extra information is found by subtracting pairs, a process known as frame combing. The extra information might well be digital audio. An analogous process can be carried out to utilize the excess vertical resolution, in which case the extra data can be used to double the horizontal chrominance resolution.

8.6.4 PALplus and Wide-Screen NTSC

Another method of adding information to PAL or NTSC in a nearly compatible manner (in the sense that old receivers can operate with new signals, but with a somewhat altered display) is to usurp 25% of the screen height for enhancement information, leaving a 16×9 window for the image. New receivers have a 16×9 aspect ratio and use the enhancement information to improve the picture quality. If the video signal format in the window interval is normal NTSC, and if the signal in the enhancement interval contains standard blanking and color burst signals, then all existing receivers operate properly. The appearance of the screen on these receivers, with bars at the top and bottom, is quite similar to the "letterbox" format often used to show wide-screen motion pictures on television in continental Europe. If the enhancement information is of very small amplitude, solid-color bars are seen. However, if the enhancement information is allowed to be of a large enough amplitude, a much higher degree of improvement is possible by this method than any of the others. With appropriate processing, the appearance in the bar areas can probably be made acceptable.

There once was a very strong opinion among broadcasters in the US and the UK that letterbox format was unacceptable to viewers. (That opinion may well have been incorrect.) Views on this matter are slowly changing along with the growth of the idea that a wide-screen presentation may be desirable for its own sake. Until the recent enthusiasm for digital broadcasting (see below) there had been a possibility that PALplus, which operates in this manner, would actually go on the air in Europe. There is also some support for a similar system in the US. Wide-screen receivers are already on the market in Europe in which the bar areas are invisible whether they contain enhancement data or not.

Since the enhancement information itself need not be in a compatible format, efficient data-packing methods developed for HDTV can be used. With scrambling and adaptive modulation, as described in Sect. 8.12.4, the vertical resolution could probably be raised from the 360 active lines in the 16×9 frame to 720 lines, using 15 frames/sec for the extra material, while at the same time minimizing its visibility. Using an effective ghost canceller, this method would make excellent pictures, although not as good as can be produced with entirely new systems designed without the compatibility constraint.

8.7 Designing from Scratch

It is hardly necessary to point out that a completely new system design is bound to be able to provide much better performance than NTSC, whose fundamental parameters were set in 1941 and whose color capabilities were added in 1953 under the heavy burden of compatibility. The intervening years have seen the invention of the transistor and the integrated circuit, while our understanding of video signal processing and the relevant aspects of visual psychophysics have also improved markedly. We therefore have a much better idea of what processing should be done and we have spectacularly better methods of implementation.[14] In this section, principles are given that would be used in any new system design, regardless of its performance specifications [8.32].

8.7.1 Receiver Processing Power

There is a strong tradeoff between receiver processing power and bandwidth efficiency. Use of a frame store alone permits more than a factor of two reduction in bandwidth. The most sophisticated methods, such as vector coding of subbands, can give a factor of at least ten, and probably much higher. The very first step in system design, therefore, is to choose the level of receiver processing power. This is a cost issue. At present, signal processing accounts for 10-20% of receiver cost, depending heavily on the size of the display and the degree of integration. Most cost is in the display, display-related electronics, cabinet, power supply, and tuner. *A fundamental lesson of the semiconductor revolution is that complexity of signal processing no longer necessarily equates with high cost.* For large-screen HDTV receivers to be made by the millions, we can therefore postulate a very high level of signal-processing capability with only a small effect on cost.

Of course, signal processing has not yet become quite cost-free. Current proposals to use motion-compensated prediction plus the discrete cosine transform for HDTV require receivers with some 10 billion operations per second, which is comparable to the computing power of present-day supercomputers. It remains to be seen whether near-term developments in integrated-circuit manufacture will be able to produce such powerful processors at acceptable cost.

Even the simplest HDTV receivers will require substantially more processing power than NTSC receivers just because they will have to transcode NTSC transmissions into the HDTV format for display. They are highly likely to use frame stores to provide a progressively scanned display. The electronics required for these two functions already provides processing power far in excess of that used in today's receivers. At very little additional cost, this circuitry can be rearranged and augmented for added power and flexibility.

A highly advantageous configuration is the open-architecture receiver, or OAR [8.33]. A possible configuration is shown in Fig. 8.11. Such a receiver could be made adaptable to a certain range of transmission standards by programmability or by addition of hardware or software modules. It could readily be interfaced with various peripherals and communications links such as fiber. It would permit the system to be improved over time as better components become available and our knowledge of signal processing develops. An open architecture would encourage third parties to develop add-ons, as has proved so successful in the case of the IBM PC.

The OAR proposal was uniformly denounced by traditional manufacturers who had become accustomed to commodity-type single-purpose receivers, but it

---

[14] When I came to teach at MIT in 1959, my mentor, Peter Elias, in talking about bachelor's theses, said "Don't let them use more than one or two tubes or else they'll never get finished." Current hardware BS theses may involve *dozens* of integrated circuits and *millions* of tube-equivalents.

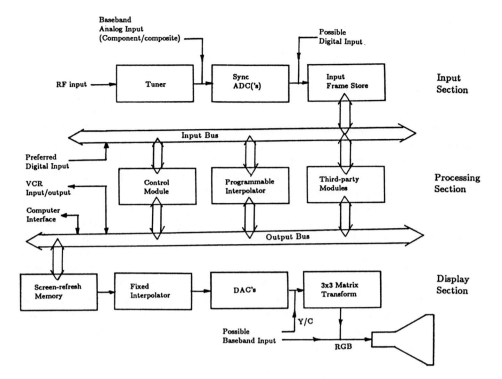

Fig. 8.11. The Open-Architecture Receiver. This is a proposed receiver configuration that uses computer and digital signal-processing techniques to permit decoding a range of signal formats. Two busses are provided, the first dealing with digitized representations of the input signal, and the second dealing with the signal in the output (display) format. In the input section, the rf signal is shifted to as low a frequency range as practical, digitized, and stored. After passing through the input bus, the signal is decoded with the aid of control and computation elements, put into a raster format, and passed through the output bus to the display section. Here the signal is stored and then interpolated to the display format. Third-party modules can be added to the processing section. Such a configuration is only feasible with highly integrated semiconductor circuitry

was very congenial to computer companies and those involved with modern communication systems. Programmability and open systems have become accepted principles with these groups. Recently, as the computer and imaging industries have become aware of the imminence of HDTV broadcasting, they have become very concerned about interoperability between TV standards and those to be used elsewhere. For example, progressive scanning and "square pixels" are very attractive to nonbroadcasters because transcoding is thereby simplified. As a result of agitation by these groups, the Federal Communications Commission has now included interoperability in its list of desired features for the forthcoming American HDTV broadcasting standard and has appointed a subcommittee of the Advisory Committee on Advanced Television Services to deal with the issue. These matters are further discussed in Sect. 8.8.4.

232

8.7.2 Scanning Standards

One very useful principle that should be followed is to separate completely the scanning standards of camera, display, and channel. Although this does require scan conversion at both transmitter and receiver, an operation not needed in NTSC, the advantages are overwhelming [8.3]. It allows the question of flicker to be separated from that of motion rendition. Depending on the sophistication of the interpolation between channel signal and display, it permits a substantially lower transmission frame rate (and correspondingly higher spatial resolution for the same bandwidth) with no loss of output quality. Almost surely, the camera and display would use progressive scanning. The scan pattern in the channel would be related to the desired spatiotemporal frequency response, and it probably would be offset in some manner akin to three-dimensional interlace.

One proposal often heard is to accommodate different transmission standards by having separate ports on the receiver [8.34]. This is proposed as an alternative to the OAR, which deals with the problem by programmability. The multiport approach is effective only when the display scanning parameters are standardized. Otherwise, a separate scan converter is needed for each port. For the highest-quality output from whatever signal is transmitted, the display line and frame rate should be as high as possible. This is a useful way for receiver manufacturers to distinguish between their less expensive and more expensive models, and therefore the display format should *not* be standardized.

8.7.3 Modulation Methods

Certain NTSC signal-design decisions should be changed in any new system because they are inefficient, out of date, or cause defects. This includes the use of subcarriers for audio and color, vestigial-sideband modulation, and the band-sharing principle.

Audio Subcarrier. Time multiplexing is more efficient than frequency multiplexing. It does require storage, but all new receivers will have sufficient storage for other reasons. The subcarrier approach also makes audio quality dependent somewhat on the video modulation. This is unacceptable in future systems, which will be required to provide so-called compact-disk quality.

Band Sharing. With multidimensional filters, we can now make band sharing work in the sense of eliminating luminance/chrominance crosstalk. However, it has obscured the actual frequency response of the various components. We cannot steal bandwidth from luminance and use it for chrominance without noticing it. Like audio, color information must be assigned a baseband spectrum appropriate to its importance. This is also made easier and more straightforward, as well as more efficient, by time multiplexing.

Vestigial-Sideband Modulation. This is an idea whose time has come and gone. Thirty-five years' experience with double-sideband quadrature modulation for the color subcarrier proves that for analog amplitude-modulation systems used in cable and over the air, DSBQM is the method of choice. It gives a base bandwidth as wide as the rf bandwidth, has no transient distortion, and is easy to implement.

8.7.4 Vision-Based Design

Although even the 1941 NTSC standard appealed to psychophysics for authority for some of its parameters, our knowledge of visual psychophysics is now more complete. Two related visual principles that have not previously seen their way into TV standards are masking and the variation of noise sensitivity with frequency.

Masking. The visibility of noise depends on the image activity in the spatiotemporal neighborhood. This is not a small effect. Some of our earlier work indirectly indicated that noise visibility was suppressed by as much as 25 dB in areas of high detail content [8.10]. I believe that this is the main reason why rms error is not a good quality criterion. Noise should be equally visible to be minimally visible. The masking phenomenon teaches us that noise should be redistributed from blank areas to busy areas. One way to accomplish this is to separate low- and high-frequency signal components and to use nonlinear processing of the latter, as is sometimes done in VCR noise-reduction systems [8.35].

Frequency Dependence of Noise Visibility. Although this phenomenon is not new [8.29] there are new and more effective ways to exploit it. With quadrature-mirror filters [8.36] the video spectrum can readily be divided into three-dimensional components. Each component can be allocated the appropriate transmission capacity. In particular, the higher frequencies require much lower SNR than the lower frequencies, providing a good opportunity, in both digital and analog systems, to achieve a higher overall resolution. The technique of separately dealing with the components, now called subband coding, may have originated with a paper by *Kretzmer* [8.37]. It has been found to be effective with both video and audio.

Appropriate Spatiotemporal Frequency Response. Except for the inroads of the color subcarrier and its harmonics, NTSC has a scaled cubical frequency response, whereas, as discussed in Sect. 8.2, information at the corners of frequency space is relatively less important than that close to any of the axes. At least for the fixated eye, a diamond-shaped response makes better use of the available spectrum. The preferred response can be provided by using appropriate pre- and postfiltering together with a nonorthogonal sampling lattice. Some of these possible sampling patterns and corresponding frequency responses are shown in Fig. 8.12.

Alternatively, the spectrum can be divided into a rather large number of relatively small three-dimensional blocks as in subband coding. The desired response can then be built up by choosing which blocks are to be transmitted [8.38]. This method also lends itself both to the assignment of SNR to each subband as needed and to the adaptive selection of subbands according to the degree of motion in

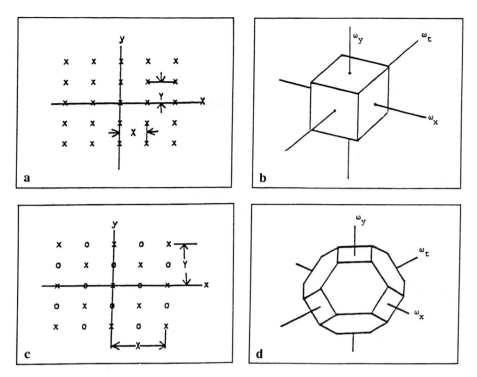

Fig. 8.12a–d. Sampling Patterns. If a moving image is sampled on the same Cartesian grid on every frame, as in **(a)**, the most natural alias-free baseband is as shown in **(b)**. To obtain the desired diamond-shaped spatial-frequency response in still pictures, the image is sampled diagonally. To extend this concept to moving images, the spatial sampling pattern is interleaved on successive frames, as shown in **(c)**, resulting in a 3-d response, shown in **(d)**, that is approximately ellipsoidal, with the highest cut-off frequencies being found along the principal axes

the image. Note that eye-tracking keeps the images of tracked objects almost stationary on the retina, effectively rotating the 3-d spatiotemporal frequency response. There is therefore no fixed frequency response that is always suitable. Dealing with this effectively requires adaptive adjustment of the response of the system.

8.7.5 Providing for Evolutionary Improvements

One thing that we surely have learned in the current period of rapid developments in electronics is that this process is not going to stop suddenly when new TV standards are adopted. On the contrary, it is going to continue, motivated by the competitive drive for market advantages in the electronics industry as well as by military research and development. In the past, the timing of standards has always been problematical. A premature standard was necessarily obsolete by the time it reached the market, while postponed standards run the danger of being subverted by several different *de facto* standards getting to market beforehand.

235

A principal feature of new electronics design is programmability. While this principle originated with computers,[15] it is now used in many different electronic products. With programmability in TV receivers, we can adopt standards without fearing that we are foreclosing the good ideas that will surely develop in the coming years.

Another way to introduce flexibility into standards is to provide for the addition of hardware modules at a later date. This is another computer idea without which the rapid progress that we have seen in recent years would not have been possible. A bus-structured design is ideal for this purpose. Video games, which are expected to use modules not even thought of at the time of original design, would be impractical without this principle.

A general approach to nondisruptive improvement over time, which the FCC desires (but has not insisted on) in new systems is representation in the frequency domain with some form of subband coding. If the selected information is appropriately packed into the encoded data stream, and if it is possible for various encoders to extract only a portion of the data to make a lower-quality picture, ignoring other information, the problem is solved. This scheme also would provide a high level of interoperability, since encoders of a range of quality could communicate with decoders of a range of quality. Since there are known ways to accomplish this, the real question concerns the penalty in transmission efficiency that must be paid.

8.8 Desirable Characteristics of a New System

To exploit to the fullest extent the luxury of designing from scratch, it is helpful to make a list of desirable attributes before thinking about the details. This is more complicated than it may seem, particularly if we try to take into account the interests of viewers, broadcasters, other claimants to spectrum, the economy as a whole, and perhaps even the national security. The following list is therefore quite personal, reflecting, as it does, the priorities that the author assigns to the various concerns. However, the list is not totally idiosyncratic – a fair amount of agreement will be found within the TV community on most of these points.

8.8.1 Spectrum Efficiency

Although broadcasters have not yet had to pay for it, spectrum is valuable, since many wealth-creating services can be devised requiring its utilization. Society as a whole thus pays for wasteful use of spectrum. Since such a large proportion of the most easily used spectrum has been allocated for television, there are constant demands for its reallocation for services such as cellular telephones and other mobile applications.

---

[15] Actually, computers borrowed the idea from textile mills. For nearly 200 years, no one has thought it a good idea to build a weaving machine that could only produce one pattern.

Looked at from this perspective, the spectrum problem goes beyond minimizing the bandwidth needed for a single transmission of a given quality; it also involves fitting as many transmissions as possible into whatever spectrum is allocated. A qualitative way to describe spectrum efficiency is therefore the attainment of the maximum service in the least bandwidth. Quantitatively, we can speak of the number of distinct programs of a given technical quality available to each viewer within a given overall allocation, and actually attained under practical operating conditions.

By these measures, NTSC is very inefficient, and not just because of low picture quality in 6 MHz. Of the 68 channels allocated for TV and therefore not available for anything else, only from 10 to 20 are usable in any one locality. Adjacent channels cannot be used in the same area, and stations on the same channel must be at least 155 miles apart. Some of these "taboos" are due to assumed poor performance of receiver tuners and antennas, but most are due the very poor interference characteristics of NTSC. It is just as important to raise the ratio 20:68 as it is to maximize the quality obtainable within one channel.

The systems discussed later in this chapter and in the Appendix get much better quality in one channel and also have superior interference performance. The highest possible spectrum efficiency is reached with single-frequency networks, discussed in Sect. 8.9.6, a scheme that completely eliminates taboo channels (at least those due to cochannel interference), so that the ratio 20:68 becomes 20:20.

8.8.2 Picture and Sound Quality

While people watch TV primarily for the programs and not for the aesthetic pleasure of the visual experience, it is nearly always desirable to have higher quality. Since present-day picture quality appears to be limited primarily by transmission impairments, any new system should do much better than NTSC in this respect. One piece of evidence that this is possible is that today's TV sound quality is not limited by transmission, but by receiver quality. The audio channel uses FM, a method well known for its good noise performance.[16]

Our own work indicates that much better performance is possible with new designs and that excellent picture quality can be obtained, even with very poor channel conditions. The same methods that deal with noise can also deal with interference, so that much higher spectrum efficiency is also possible with new systems.

Once transmission impairments are overcome, it is possible to decide on the goals for spatial resolution and quality of motion rendition. These are generally selected on an economic basis, perfection always being too expensive. The term "theater quality" is often used, which seems to mean about twice the horizontal and vertical resolution as we have now. This is a very inexact term, since today's movie quality is highly variable, and most technically trained people would like

---

[16] In spite of this, most HDTV system proposals include digital audio, primarily because of the popularity of compact disks. Our audience studies indicate that most listeners cannot tell the difference, but market pressures probably will dictate the use of digital audio anyway.

to improve on the motion rendition obtained at 24 fps. As a practical matter, it is likely that the quality will be limited by the channel capacity, with 6 MHz by far the most likely value to be used.

8.8.3 Production, Distribution, and Transcoding

Television programs are assembled in a complex postproduction process that often involves numerous signal sources, several generations of recording, and a good deal of signal processing in special-purpose equipment. Programs are distributed to local stations and cable companies, often by satellite, and sometimes by tape or film. Cable companies and terrestrial broadcasters retransmit these programs to viewers along with locally originated material and often exchange programs among themselves. At present, all of these steps use the NTSC signal format, but may vary the modulation method in accordance with the characteristics of the transmission medium. If programs are exchanged with 50-Hz countries, they must also be converted to or from PAL or SECAM.

Even after 25 years' experience in PAL/NTSC transcoding, that process is still not entirely satisfactory in the case of moving images. With HDTV, it will become even more complex because of the appearance of additional standards, including several different production standards and several different transmission standards. The introduction of digital transmission on fiber-optic cable will result in at least one new digital transmission standard. Broadcasters using the various media all want to select optimum transmission formats. It is clear that transcoding will become an even more important issue than it is now, and it will be complicated by the greater need to optimize the format for each medium [8.39].

In principle, high quality transcoding should be possible between any two formats. In practice, defect-free transcoding is likely to be very expensive for a long time to come, particularly between interlaced formats. It is therefore highly desirable that this issue be carefully considered in the design of new systems. As will be seen below, it appears to be possible to simplify transcoding greatly and at the same permit accurate optimization if the appropriate transmission formats are selected [8.40].

8.8.4 Interoperability

This term refers to the easy interconnectibility of TV systems of different spatiotemporal resolutions, different applications, different industries, and different epochs. It is an expansion of the idea of transcodability. In its ideal form, signals from any source and quality could be used by any application without special equipment. It also implies that systems could be upgraded over time without making any existing equipment obsolete, as has been proposed by the FCC for any new TV system. It is related to the highly desirable capability of designing receivers of low price and performance, and for easy conversion to NTSC. This ideal form is probably impossible to achieve except by placing an uneconomic burden on all new systems. However, there are many ways interoperability can be facilitated at low cost if the new systems are designed with this property in mind.

The issue of interoperability, and its related properties of scalability and extensibility, have been raised primarily by computer interests, which hope that any new TV system will be interoperable with multimedia workstations. The use of progressive scan and "square pixels" (equal horizontal and vertical resolution) are often advocated for this reason. While these measures are helpful, they do not go far enough. Conceptually, interoperability can be achieved by representing video information in the spatial or spatiotemporal frequency domain, for example by using the DCT. The more coefficients used, the higher the quality. If a system could be devised in which different encoders transmitted more or fewer DCT coefficients, and different decoders used more or fewer of the received coefficients, ignoring those not used, a very high degree of interoperability would be attained.

8.9 Channel Considerations

Of the three kinds of analog channels now in use, the satellite channel is the best, being simply characterized by bandwidth and CNR. The terrestrial channel is the worst, since it suffers noise, ghosts due to multipath transmission, interference, and widely varying signal levels at different receiving sites. Cable often has lower CNR than one would expect, suffers from leakage and some nonlinearity, and is plagued, in the US, by low-level reflections due to unterminated connectors. Nevertheless, it is generally superior to terrestrial transmission. All channels have some distortion due to imperfect frequency response. When digital channels are eventually provided by fiber, the channel capacity will be governed strictly by data rate, since the telephone company or other service provider will supply the terminal equipment and guarantee the performance.

8.9.1 Variable Signal Levels

Since the capacity of an analog terrestrial channel is proportional to the product of bandwidth and SNR (in dB), it necessarily varies enormously from close-in to far-out receivers.[17] This manifests itself in analog systems as a smooth fall-off in quality with distance. Due to the use of FM audio, the sound quality remains high if the picture is at all viewable. Unlike many other kinds of data, pictures are useful over a very wide range of quality. To get the highest average quality, the latter must be maximized at each receiver. There is thus no virtue at all in

---

[17] Due to the use of very tall antennas and rather narrow vertical beam widths, as well as the practice of aiming the beam at the horizon, the field strength at receiving antenna height for the first ten miles is nearly constant, while in the fringe area, it goes down at least as the fourth power of range. The variation of channel capacity over the service area may be as high as six to one.

delivering the same quality to all viewers. This is not the situation with NTSC and it has never been a stated goal of the FCC.

The soft threshold that is inherent in analog transmission is helpful in attaining high spectrum efficiency. (Of course, one must also examine the picture quality attained with a given transmission rate.) In spite of this, digital transmission is being seriously considered for terrestrial broadcasting. The first systems proposed delivered the same data rate to all receivers, thus achieving high spectrum efficiency only in the fringe area. These matters are discussed in more detail in Sect. 8.13 and the Appendix.

8.9.2 Noise

In principle, noise should set the limit to transmission quality over the air. At the high transmitter power levels now used, this is only true if there are no other limitations, such as interference. In the absence of interference, a large but affordable antenna can produce acceptable images at the boundary of line-of-sight transmission, and quite primitive antennas give adequate signal levels in the city. Studies indicate that viewers pay just enough for antennas to get barely viewable pictures – an interesting illustration of the degree to which the audience may be willing to pay for image quality alone.

8.9.3 Multipath

Ghosts due to reflections from buildings are the main problem within cities. They are partially correctable with good antennas. If all transmitters are not co-located,[18] multiple antennas or steerable antenna mounts are needed. Echo cancellers have been demonstrated, and can probably be incorporated into new digital receivers at acceptable cost. In general, SNR will suffer with ghost cancellation. Variable ghosts due to moving trees and airplanes are harder to correct.

8.9.4 Frequency Distortion

The over-the-air path is distortionless. There is some frequency-dependent loss in cable, especially at the higher carrier frequencies, and tuned antennas may also have an effect. However, the main cause of this problem is the filters used at transmitter and receiver, which are often far from perfect. They must have a high level of discrimination against out-of-channel signals. The failure to set tight specifications for these channel-defining filters, by law or regulation, leads to a noticeable loss in horizontal resolution.

Frequency distortion can be corrected, in large part, by automatic equalizers similar to those used for digital data transmission in telephone lines. If the equalizers have a long enough impulse response, they may also correct for echoes, which can be modeled as a linear effect. The degree of equalization that can be accomplished is limited by the accompanying reduction in SNR. Moving echoes are more troublesome than fixed echoes.

---

[18] Another piece of jargon, which means that all antennas are close together.

8.9.5 Orthogonal Frequency Division Multiplex (OFDM)

The basic principle of this system, developed in Europe initially for digital audio broadcasting (DAB), involves dividing the spectrum of the signal into a large number (typically hundreds) of components, each delivered in a separate sub-channel, so that the symbol length of the components is longer than the temporal spread of the echoes [8.41]. In the DAB experiments, for example, the symbol duration is 80 μs, of which 64 are active and 16 are ignored [8.42].

This absolutely guarantees that, up to a time spread of 16 μs, all echoes arrive within one symbol period, where they are added, rather than interfering with each other. Rapidly changing multipath is suppressed as effectively as fixed echoes; no channel equalizer is required. Certain frequency ranges can be left unused to gain higher immunity to cochannel interference from PAL or NTSC. OFDM can be used equally well for digital, analog, or hybrid transmission, although in the tests so far, digital transmission has been used exclusively.

The implementation of the system is quite practical. Signals in adjacent sub-channels are orthogonal to each other, so that no guard bands are required. The hundreds of separate modulated carriers need not be generated individually. Instead, the signal to be transmitted is produced by sampling the coefficients produced by a single Fourier transformation of a group (typically hundreds) of signal samples. In a similar fashion, the inverse Fourier transform is used to separate the samples at the receiver. It is important to note that OFDM and its theoretically expected operation have been fully proven in field tests in Europe, Canada, and the US. It is a thoroughly practical system and will almost surely be used for DAB in Europe, to replace FM broadcasting.

The main reason for the use of OFDM for DAB is the multipath immunity and the resultant high reliability without resort to automatic channel equalizers, [19] particularly when the receivers are themselves mobile. The multipath immunity has proved so effective that it is possible to use fill-in retransmitters in areas shadowed from the main signal by buildings or mountains. These retransmitters are fed from nearby receiving antennas and operate on the same frequency. At the boundary of reception areas of the main and retransmitted signals, the two appear to be echoes and are successfully dealt with by the receiver. In the test in Montreal, such a retransmitter was installed in a vehicular tunnel with excellent results [8.43].

OFDM is not a new idea. Some earlier papers are listed in [8.44] and some recent papers in [8.45].

---

[19] Actually, equalizers can be completely eliminated only for binary transmission. For multilevel transmission, the gain of each channel must be known accurately at the receiver. This is accomplished by a simple equalizer using a staircase test signal.

8.9.6 Single-Frequency Networks (SFN)

In an extension to the idea of using fill-in transmitters, with OFDM, all or a part of the reception area of a station can be served by a cellular array of low-power transmitters, all operating on the same frequency [8.46].

Another, centralized, portion of the area can be served by a medium-power transmitter or even by a satellite with a small footprint. The cellular transmitters derive their signals from each other, from the main signal (where there is one), or by wire. The size of the cells is limited by the temporal spread for which the system was designed. The need for soft thresholds has much less urgency with SFN, since the various receivers have much more nearly the same signal level.

This arrangement has the potential to raise the above-mentioned ratio, 20:68, to 20:20, as it completely eliminates cochannel interference. Service areas for stations can be of arbitrary shape. Colocation for stations on adjacent channels is simplified since they all can use the same cellular sites, allowing such adjacent channels to be used in the same city.[20] SFN is such a powerful method of raising spectrum efficiency that, at the very least, it requires very careful examination by regulatory authorities.

The degree of multipath immunity required for SFN is far beyond that attainable with conventional automatic channel equalizers. "Ghosts" may be of the same amplitude as the "main" signal. They cannot be eliminated by the use of highly directional antennas, since nearby transmitters can be in the same direction; they are rarely in opposite directions as is the case in the region in between two conventional high-power transmitters. OFDM adds up all the echoes constructively, while equalizers cancel them out using a filter with many taps.

SFN requires some additional capital expenditures by current broadcasters. (New broadcasters would probably find it cheaper to install than a conventional high-power system.) However, the total transmitted power is much less than at present, and it has many other benefits, such as precisely defined service areas. It can be introduced station-by-station and city-by-city. Cheaper omnidirectional receiving antennas can be used at most locations. In rural areas, or in small cities located far from other cities, it need never be implemented. Its overwhelming advantage is the fact that we can have at least as much TV service as today and still give back hundreds of Megahertz for other wealth-producing activities.

8.10 Interference Considerations

Interference from other TV transmissions degrades image quality in the countryside, but its most important effect is to reduce the efficiency with which the allocated spectrum can be utilized.

---

[20] It is possible that the receiver-to-receiver variation in signal strength will be low enough so that colocation will not be required. This depends on the size of the cells and the adjacent-channel performance required of the receivers.

8.10.1 Taboo Channels

With NTSC, stations on the same channel must be at least 155 miles apart, which is dictated partly by antenna considerations and partly by the required signal-to-interference ratio. In the VHF band, adjacent channels are not used in the same city because a low level of adjacent-channel rejection on the part of the receiver is postulated and because the unwanted signal may be much stronger than the wanted signal. (Adjacent channels can be used in cable, where all the signals are of about the same level.) In UHF, there are many more "taboos" associated with an even lower level of receiver performance. Many of these taboos could be removed with modern receivers, but that would create a problem for the large installed base of old receivers.

No set of taboos can guarantee good reception on all channels at all receiving sites. The goal is to have up to seven VHF channels and and a somewhat higher number of UHF channels available to all receivers. There are many locations, particularly between large cities, where many more stations[21] can be received than are guaranteed. In such cases where more channels can be viewed, the FCC does not protect the service. Reassignment or relocation of stations may cause the loss of some reception.

The net result of all the taboos is that only about 20 of the 68 allocated channels are available to viewers at any one location, even if all the usable channels are on the air. This inefficiency is very costly, in view of the other services that might be provided were spectrum available.

8.10.2 Desired/Undesired Ratio

Looking forward to a time when NTSC is replaced by a new system, it is clear that it is highly desirable to use less spectrum while, at the same time, providing at least as many channels at each location as at present. There are two elements in this. One is to eliminate some or all of the UHF taboos by requiring better receiver design and possibly to require colocation of transmitting antennas in certain cases. The other is to achieve good reception with a lower ratio of desired to undesired signals so as to allow physically closer station location on the same channels.[22] With NTSC, the required desired/undesired (D/U) ratio is 28 dB with precision offset carriers and 45 dB with unsynchronized carriers. A goal of 12 dB would allow hundreds of MHz to be released for other purposes while accommodating all of today's licensees. To operate at such low D/U ratios, the interference must *not* take the character of a coherent undesired image that is superimposed on the desired image. Coherent interference of -12 dB (25% of the amplitude of the desired signal) is totally unacceptable. With a 12-dB D/U ratio, cochannel spacing could be reduced to about 100 miles and nearly every station

---

[21] Usually, there are duplications of network transmissions in these locations, so that the number of different *programs* available is not as large as it might seem.

[22] The same system design that permits closer cochannel spacing also permits much stronger signals in adjacent channels, given a certain level of receiver selectivity.

in the country could be provided with an extra channel to use for simulcasting HDTV.

8.10.3 The Zenith System

In 1988, Zenith proposed a system of simulcasting in which a completely independent HDTV signal was transmitted in a single 6-MHz channel at such low power that today's taboo channels could be used without undue interference to existing NTSC transmissions [8.47]. The system used nonadaptive subband coding and hybrid analog-digital transmission.

In accordance with an earlier FCC decision that, at least for an initial period, all HDTV programs should be made available to NTSC receivers, Zenith proposed that the new transmissions use the same programs as are transmitted on the associated NTSC channel, a practice called simulcasting.[23] Of course, this is an operational, not a technical, decision and could be reversed at a later date.

The ability to achieve HDTV in the same bandwidth used for NTSC depends on superior efficiency of signal design, based on the principles discussed above. Reduced interference requires the recognition of the source of NTSC's vulnerability in this respect, which is the lack of any threshold, the large synchronization pulses, the nonuniform spectrum common to all raster-scanned video signals, and the coherence of interference, which makes even a very low-level unwanted signal quite visible.

One way of operating successfully under high levels of interference is to use a modulation method with a sharp threshold, such as FM or even digital transmission. Unfortunately sharp thresholds go along with poor channel efficiency and, until recently, were thought unsuited to terrestrial broadcasting with its wide range of CNR (and thus channel capacity) at various receiving locations. A way out of this is to use a modulation system with a sharp threshold only for a small part of the spectrum – the dc component and very low frequencies – and to take other steps to mitigate interference at the higher frequencies, which do not require such a high SNR. Zenith accomplished this with subband coding, using digital modulation for the dc component. The coherency of the highs was somewhat reduced using time-dispersive filters [8.48]. The SNR in blank areas was raised by companding.

Although the Zenith system was eventually abandoned in favor of an adaptive all-digital scheme, the company made an important contribution to the development of HDTV by showing that enough of the taboo channels could be used so that every broadcaster could get an extra channel for HDTV. This made it possible for the FCC to choose noncompatibility, a previously unthinkable idea in the United States.

---

[23] Simulcasting was successfully used in France and the UK when PAL was adopted in 1965. PAL was not compatible with the monochrome systems in those two countries. Some transmissions on the old standards were continued for 20 years.

8.11 Video Function Characteristics Useful for System Design

In addition to the principles of Sect. 8.7, it is necessary, in designing from scratch, to examine carefully the characteristics of the information to be transmitted. The focal-plane image, although itself an abstraction of the real three-dimensional world in front of the camera, is a good place to look for these characteristics. The transformations of reality involved in forming this image, the so-called central projection, are well understood, and a faithful rendition of it on the display would be deemed fully satisfactory.

8.11.1 Spatiotemporal Redundancy

An important feature of the "video function," our name for the focal-plane image, which is a function of x, y, and t, is that it is mostly continuous. Generally, there is very high correlation in at least one direction at almost every point in x, y, t space. In stationary areas, this direction is parallel to the t axis. In moving areas, it is in a direction determined by the velocity,[24] which itself is a vector point function, continuous except at the edges of objects.

The video function can be visualized by drawing lines parallel to the velocity, in much the same way we visualize an electric field. These "optical flow" lines originate at the edge of the field or when objects that were occluded emerge from behind foreground objects, and they terminate when objects leave the field or move behind other objects [8.49]. Discontinuities can also occur due to shadows. Flow lines usually persist for at least several seconds (a hundred or more conventional frames) and often very much longer.

This method of visualizing optical flow shows that correlation in the video function *must* be very high if the function relates to a real image.[25] The redundancy must be higher and the entropy must be correspondingly lower than that of a random function of three variables in the same space.

This property enables noise to be distinguished from signal much more easily than in still images. Low-pass filtering along the flow lines can remove noise without removing fine detail [8.50] except in the very unusual case that the detail is essentially random. Even fine-grain texture on the surface of objects, giving rise to signal excursions lower than the noise level, can be separated from noise as long as it is in a region of space with well structured flow lines.

In addition to noise removal, this phenomenon also permits a substantially more efficient signal representation by reducing the frame rate and recreating the intermediate frames by interpolating along the flow lines, rather than parallel to the time axis. Implementing either noise removal or interpolation by what is

---

[24] In the case of transparent objects, the velocity field can be two-valued.

[25] Temporal aliasing is very common in TV, since the frame rate is often too low to give continuous output in the case of rapidly moving objects. This reduces temporal correlation and makes plotting the optical flow lines more difficult.

usually called "motion compensation" requires accurate knowledge of the velocity field and sophisticated receiver signal processing [8.51]. Velocity information, if used as part of the signal description, should not be thought of as undesirable overhead. There are many ways to represent signals. Taylor series and linear transforms are two. Sampling is traditional in video, and it is probably not the best method. Since optical flow is such a fundamental characteristic of moving images, its direct representation is likely to increase the efficiency of the representation. An alternative way to make use of temporal redundancy was discussed above in Sect. 8.9.6.

As we saw in Chap. 5, noise sets a limit to the effectiveness of entropy coding in still pictures. In moving pictures, the partial ability to distinguish noise from image information on the basis of optical flow may well make it possible to get useful results with this method.

8.11.2 The Preponderance of Relatively Blank Areas

Meaningful still images comprise representations of objects that are, on average, much larger than picture elements in the focal plane, as discussed in Sect. 2.5.5. Away from the edges of these objects[26] in x, y, t space, there are fairly large volumes with low entropy, but in which noise is highly visible. In the complex regions, however, there is much higher signal activity, and noise visibility is much lower. The information content in such regions is higher than in the blank areas but not as high as it seems because of the lower required SNR. The rather high SNR usually thought to be required for video[27] is related to the signal in the complex areas and the noise in the simple areas. The local-area SNR can be much lower in these areas while the image can remain apparently noise-free.

To take advantage of this characteristic, it is clear that the information in these two distinct types of regions must be rendered differently.

8.12 Some Specialized Video Signal-Processing Techniques

In this section, a number of techniques are discussed that have, until recently, been little used in the design of TV broadcasting systems. These methods, in combination, have shown themselves to have good potential for the design of spectrum-efficient systems. The remarkable resistance that can be achieved to interference and other analog channel impairments promises substantially improved picture quality in the home, under typically imperfect conditions, and much better utilization of spectrum.

---

[26] I shall use the word "object" here to mean the focal-plane image of a real object; confusion is unlikely to result.

[27] Similar considerations hold for audio. The short-time SNR required for apparently noiseless audio transmission is no more than 30 or 35 dB [8.52].

8.12.1 Subband Coding

The first use of this technique for picture coding was by Kretzmer in 1952, who found that fewer bits/sample could be used for the higher-frequency than for the lower-frequency components of a standard video signal [8.37]. It has also been used for some time in audio coding [8.53]. The procedure is to separate the two- or three-dimensional spectrum of the signal from a high-quality camera into a number, typically 8 to 64 or more, of spatial or spatiotemporal components. Separable filters are preferred for simplicity, with quadrature-mirror analysis banks currently the most popular type. The resulting components are time-multiplexed for transmission, each component being independently processed, for example by a static nonlinear amplifier and an adaptive modulator. At the receiver, the components are demodulated and then interpolated and combined to produce the output signal, usually with a quadrature-mirror synthesis bank. This was the basis for the Zenith proposal mentioned above.

The advantage of this approach is that the treatment of the various subbands can be tailored to how they are perceived and to their importance to overall quality. The higher-frequency components are mostly very small and therefore suitable for adaptive modulation. They do not require as high an SNR as the lower-frequency components. The single component (three if the image is in color) containing dc can have any value anywhere and cannot be adaptively modulated. It must be rendered with high SNR. A point not fully understood at present is the minimum bandwidth of the noise-free dc component required for good overall noise performance.

The very highest diagonal frequencies are characteristically of low power and low perceptual importance and can be omitted everywhere with little degradation. There are many subbands that are negligible in some image regions and significant in others. *It appears to be a fundamental property of sensible images that energy is distributed in a highly nonuniform and variable manner among the subbands.* Energy is found to be concentrated in just a few of the bands at any one point in x, y, t space. This property makes advantageous the adaptive selection of subbands, either on a scene-by-scene, frame-by-frame, or block-by-block basis. In the latter procedure, x, y, t space is divided into blocks (the block division is needed for adaptive modulation anyway) and only the most important subbands used for each block. A very significant improvement in sharpness of moving objects results from this method. The theoretical basis for the improvement is the rotation of the plane of high spectral content with motion, as discussed in Chap. 3.[28]

Although its philosophy is quite different, 3-dimensional block transform coding is actually a special case of subband coding [8.54]. If the filters used in

---

[28] The subbands produced by 3-d separable filters do not distinguish between the energy distributions caused by motion in opposite directions. When adaptive selection of components is used, the efficiency is therefore lower than what it would be if nonseparable filters were used. In addition, when the velocity is more than one pel per frame, the tilted-plane model breaks down and the energy is not as concentrated into a small number of subbands as with slower motion. As a result, 3-d subband coding is not as efficient with rapid motion as might be thought.

subband coding have a region of support identical to the block dimensions, then the transform coefficients are samples of the subband signals. Such filters tend to have higher sharpness but more blocking effects than occur with those having a larger region of support.

In both cases, the signal is represented by elements that are confined in both the space-time and frequency domains. The filters that are typically used to divide the spectrum in subband coding are chosen specifically to produce a smooth roll-off in both domains and thereby avoid the artifacts that are typical of block transform coding. In general, their region of support is larger, in pels, than the number of subbands in each dimension.

8.12.2 Adaptive Modulation

The spectrum of conventional video signals is highly nonuniform because of the large average object size, as discussed previously. The high-frequency SNR actually attained is thereby reduced. This property, by itself, guarantees that the capacity of normal channels is not fully utilized. Some improvement is possible simply by filtering, but the degree of preemphasis that can be used is limited by clipping at high-contrast edges. Adaptive modulation applied to the subbands permits much more of this lost capacity to be regained.

To apply this method, x, y, t space is divided into small blocks and the maximum adaptation factor that produces a given small amount of distortion is found for each block. This factor is transmitted as part of the signal description. The factor used at each picture element is found from the block factors by interpolation. It therefore varies smoothly from pel to pel and block effects are avoided. The highest efficiency is achieved when separate factors are used for each subband, but good results can also be achieved with as few as three different factors. Even a single factor produces worthwhile results [8.10, 38].

In the blank image regions where noise would be most visible, the subband signals are very small and can therefore be adaptively increased. When the corresponding reduction is applied at the receiver, noise added in the channel, from whatever source, is decreased by the adaptation factor. This is not a small effect. Maximum adaptation factors of 32 (30 dB) and even more have been used successfully. The effectiveness of noise suppression in the blank areas is usually higher than needed, as a result of which a better overall effect is achieved by using reverse companding in tandem to shift some noise from busy to blank areas.

A limitation on the effectiveness of adaptive modulation in controlling additive noise is that it leaves a narrow strip of noise around sharp edges. The width of this strip depends on the block size (i.e., the data rate of the adaptation

Fig. 8.13a–c. Adaptive Modulation in the Subbands. The effectiveness of adaptive modulation in suppressing channel noise is shown here. The original image is at (**a**), the result when the signal is transmitted in the normal manner at a CNR of 20 dB is shown at (**b**), and the result when adaptive modulation is used is shown at (**c**). In this case, the spectrum was decomposed into 16 components, the dc component being transmitted digitally without any channel degradation

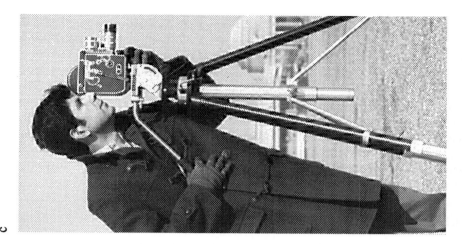

c

b

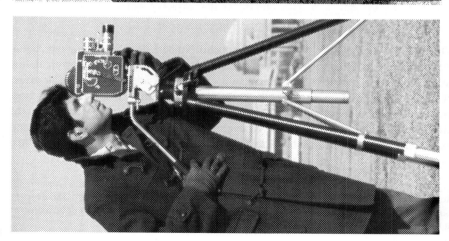

a

Fig. 8.13a–c. For caption see opposite page

249

information) and on the bandwidth of the frequency components, which must be adequate. Masking is then relied on to reduce the visibility of the strip of noise to an acceptable level. An example is shown in Fig. 8.13.

The division of both signal space and frequency space into blocks causes the signal to be represented by elements that are localized in both domains. The success of this approach depends both on the characteristics of moving images and on the way we see. These elements are fundamental to the organization of the visual system as well as to the physical properties of the real world of moving objects. Since the visual apparatus has developed in response to the need to extract useful information from the world around us, it is perhaps not surprising that this should be so.

8.12.3 Scrambling

All new television systems will make use of frame stores at encoder and decoder, at least for converting between the higher line and frame rates used at camera and display and the lower rates used in the channel. Transmission can be thought of as replicating, for each frame, the samples of each subband in the encoder store at corresponding locations in the decoder store. The samples can be transmitted in any order, and if transmitted in pseudorandom order, several useful effects are produced.[29]

The Signal Is Encrypted. The pseudorandom sequence, which can be changed on every frame if desired, must be known to reconstruct the image. Such sequences are described by their starting state and generating function [8.55]. If a suitably long sequence is used, a single generating function can be used and the starting state can be chosen at both transmitter and receiver according to a previously agreed-upon scheme, such as using the output of a second, known, sequence generator. If conditional access is desired, then both starting state and generating function can be chosen independently at the encoder and made known to the decoder by some kind of separate privileged transmission.

The Spectrum Is Made More Uniform and Noise-Like. Shuffling the signal samples leaves the average power unchanged, but decreases the power at low frequency and raises it at high frequency. This effect adds to the similar effect of adaptive modulation. If the noise spectrum is uniform, as is the case in most analog channels, the signal spectrum must also be uniform to maximize the efficiency of channel utilization. Conventional video signals, because of their extremely nonuniform spectra, use the channel inefficiently in the sense that

$\longrightarrow$

Fig. 8.14a,b. Scrambling. In this example, the same picture is shown at (a) after transmission through a channel with a 40% ghost and distorted by 40% intersymbol interference. The degradations are dispersed into random noise with the same energy by scrambling, the result being shown at **(b)**

---

[29] Some kinds of scrambling are often called interleaving or shuffling.

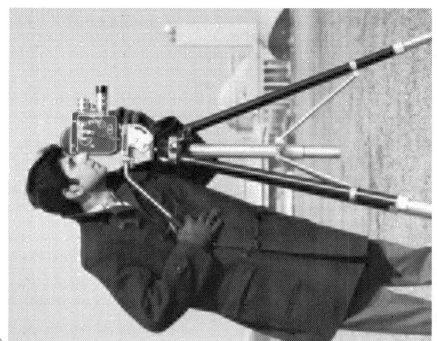

b

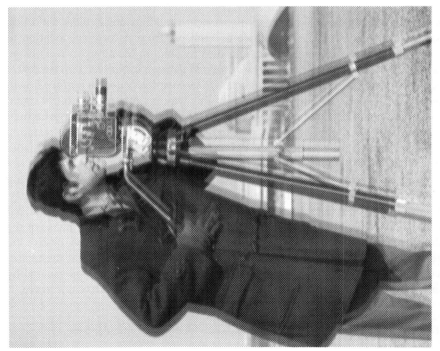

a

Fig. 8.14a,b. For caption see opposite page

251

the narrow-band SNR is very low at high frequencies if it is adequate at low frequencies. At no extra cost in channel capacity, much higher SNR can be achieved at high frequency by using these two techniques. If this higher SNR is not needed on account of perceptual considerations, it then becomes available for other signal components.

Interference and Echoes Are Dispersed as Random Noise. Since pels that are adjacent in the image are randomly located in the channel signal, unwanted coherent signals are transformed into random signals of the same power. These random signals are much less visible, both because of their more uniform spectrum and because the HVS is less sensitive to noise than to organized patterns.[30] An example is shown in Fig. 8.14.

The reduction in the visibility of interference is mutual for different signals of a given relative power that are scrambled with respect to each other. Thus two signals of a given relative power, one or both of which are scrambled, interfere with each other much less than two conventional signals transmitted in raster order. This is extremely important for video transmission, since it means that stations on both the same and adjacent channels can be located much closer to each other, thus permitting many more stations to be accommodated within a given overall spectrum allocation, or a given number of stations to be accommodated within a smaller spectrum.

Frequency Distortion Is Dispersed as Random Noise. One of the more surprising results of scrambling is that intersymbol interference caused by imperfect channel frequency response is also dispersed as random noise. This distortion distributes a fraction of the amplitude of each pel from its original location to its neighbors in the channel signal, and thus to random locations in the image. *The effect of scrambling on a signal transmitted through a channel with imperfect frequency response is to eliminate the distortion completely and to replace it with random noise.* The total noise produced in this way is proportional to the signal power, and would be excessive for conventional video signals and a large amount of distortion. When applied to the subband signals, however, the noise resulting from scrambling in the presence of a given degree of distortion is much smaller because the average power in these signals is quite low.

When a signal is scrambled in this manner and then transmitted in an analog channel, the received signal must be resampled to recover the pel values correctly. The resampling phase must be accurate to substantially less than one signal sample. This is a similar problem to that of recovering the reference phase of the color subcarrier in NTSC. While not difficult, it must not be ignored, or else the intersymbol interference will be greatly increased. It is essential that the phase-recovery method be immune to the effects of multipath. This can be assured by using a pseudorandom sequence for synchronization, rather than a sine wave.

---

[30] This is the principle used by Roberts to decrease the visibility of quantization noise, as discussed in Sect. 4.7.2.

8.12.4 Scrambling and Adaptive Modulation Combined

When a signal is first adaptively modulated and then scrambled, with the process reversed after transmission through a channel that has interference, echoes, and frequency distortion, the noise produced by scrambling of these unwanted signals is divided by the adaptation factor at the receiver. In the relatively blank areas of the image where noise would be most visible, the factor is large and the noise is suppressed by a very large amount, as shown in Fig. 8.15. Conventional video signals, as measured by their peak power, are very weak interferers with adaptively modulated and scrambled signals because the former have such low power at high frequencies. Except for scrambling, the latter interferes with the former to about the same degree as conventional signals because, for a given quality, the high-frequency power of the two types of signals is about the same. For a given power in the interfering signal, scrambling reduces its visibility considerably simply because incoherent signals are less visible than coherent signals. In one test made in our laboratory, this effect was about 6 dB.

Interference between a pair of signals that are adaptively modulated and scrambled (or between such a signal and its echo) is much less than between conventional signals, for two reasons. One is that the resulting noise that appears in the blank areas is greatly reduced. The other is that the undesired signal is random rather than picture-like. Our tests have shown that a 12 dB ratio between desired and undesired signals gives essentially perfect pictures, with 9 dB giving high quality, as compared with 28 dB for conventional transmission. Even two signals of the *same* amplitude, but with different scrambling patterns, can be independently recovered with quality better than many people see in their homes today under typically poor reception conditions! When two different signals are deliberately transmitted in a single channel, one can be enhancement information for the other, with the relative SNR controlled by the relative transmission power.

In NTSC-compatible systems in which enhancement information is hidden within the normal signal, adaptive modulation combined with scrambling is effective in reducing the visibility of the hidden data.

8.12.5 Motion Adaptation

The very high correlation in the video function corresponding to real scenes, discussed earlier, suggests a true information content much lower than that of random functions. There are several ways to take advantage of this, one of which is statistical coding. As pointed out in Chap. 5, noise generally sets a limit to the efficacy of this method. It is possible that motion-compensated noise reduction could produce a signal of high enough SNR so that this method would work well, but this approach has not yet been attempted. Another scheme is to transmit at a lower-than-normal frame rate and use knowledge of the optical flow to interpolate intermediate frames. This method seems to hold a great deal of promise. It requires very accurate motion estimation at either the encoder

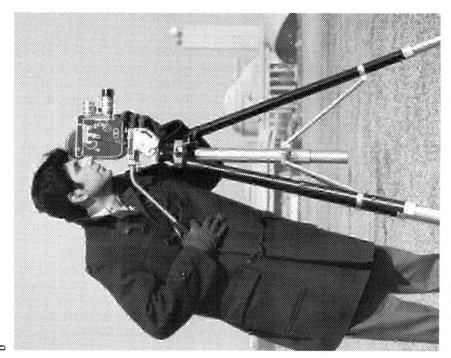

b

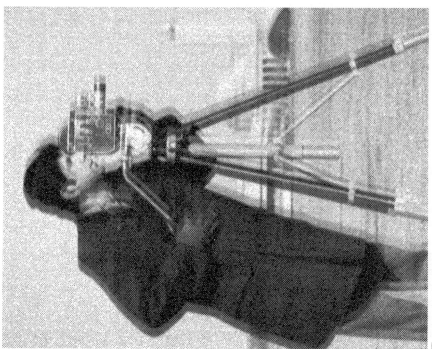

a

Fig. 8.15a,b. For caption see opposite page

or decoder[31] and motion-compensated interpolation at the receiver [8.51]. To make the latter operation practical requires the development of special integrated circuits. A method that is, at the moment, cheaper to implement in receivers is discussed here.

The three-dimensional spectrum of a stationary image is confined to the f_h, f_v plane. Motion rotates the spectrum about the origin, the axis and sense of rotation depending on the direction of motion. This changes the distribution of power among the subbands, emptying some of them and moving the power to others that were previously vacant. The fixated eye has a roughly diamond-shaped frequency response sometimes called the "window of visibility" [8.56]. Some higher-frequency components may well become invisible on rotation, a roundabout way of saying that images moving on the retina become *spatially* blurred by the *temporal* response limitations of the HVS. In real life, our eyes attempt to track interesting moving objects so as to hold them stationary on the retina. This effectively counter-rotates the spectrum. The fixated object then remains sharp and the stationary background becomes blurred.

To cause a television system to mimic the HVS, making use of limited resources to optimize the transmission process, we can adaptively choose the components to be transmitted. We can do this on a scene-by-scene, frame-by-frame, or even on a block-by-block basis within the frame. The last option actually improves on the operation of the eye (a very rare possibility indeed) which can only rotate as a whole. The principal limitations on this process, which is quite easy to implement, are block artifacts that may be produced and the uncancelled aliasing that may remain when some, but not all, of the components are used. Of course, additional data must be transmitted to indicate which components are used in each block. Like adaptation information, this data is highly correlated and therefore amenable to statistical coding [8.57].

8.12.6 Motion Compensation

The use of motion-compensated interpolation to reduce the frame rate, and therefore the data rate, requires exceptionally accurate motion estimation, since any errors have a direct effect on the quality of the interpolated frames. An alternative method is to transmit the error, in coded form, between the predicted and actual intermediate frames. Errors in motion estimation no longer have a direct effect on image quality, but they do increase the amount of data that must be transmitted, and therefore have an indirect effect.

◀ ——————————————————————————————————————

Fig. 8.15a,b. Adaptive Modulation and Scrambling Combined. In this example, all of the degradations of the previous two figures have been combined, and the result of normal transmission is shown at (a). At (b) we show the received image when both adaptive modulation and scrambling are used. Adaptive modulation suppresses the noise in blank areas, both that added in the channel and that due to dispersion of ghosts and intersymbol interference by scrambling.

---

[31] The estimation of motion at the encoder and its transmission as part of the signal description seems to us to be the preferred method.

In this method, which is described in further detail in Sect. 8.14.6, the image is divided into blocks and one velocity vector is transmitted for each block. There is a tradeoff between the block size and prediction error. Since the prediction error tends to zero in the absence of motion, the data rate depends on both the complexity of the images and the amount of motion, and is therefore highly irregular. On scene changes, correlation drops to zero, and a great deal of data is needed to build up the new picture. Fortunately, perceptual limitations permit the buildup to be rather gradual (up to about .3 seconds) so that the peak data rate can be limited. For rapid motion, there is also some permissible delay. However, there are bound to be some sequences for which resolution or SNR must be visibly reduced in the presence of excessive motion and/or complexity, in order to maintain a constant channel rate. It remains to be seen how serious this problem is for practical use.

8.12.7 Data-Packing Methods

One source of inefficiency in utilizing the analog channel for video transmission is that the various spectral components require different SNR. Amplitude modulation in the presence of white noise results in equal SNR for all components.[32] Frequency modulation, with its triangular noise spectrum, achieves a better distribution of SNR, at least in the horizontal direction in the frequency plane. With subband coding, it is possible to distribute channel capacity among the components in a more optimal manner by using less capacity for the subbands that require lower SNR.

In digital channels, bandwidth and SNR can readily be interchanged in strict accord with Shannon's rule [8.58]. This is also possible, but not so easy, in analog channels. For example, with a CNR of 100, there are about 100 discernible levels. If these are numbered 0 through 99, the tens digit can represent one 10-level signal and the units digit a second 10-level signal. We have thus doubled the bandwidth and halved the logarithmic SNR. Channel noise causes crosstalk between the two channels, but this can be alleviated by using reflected decimal code,[33] in which alternate groups of ten numbers are inverted, so that, for example, 39 follows 29 and 41 follows 31.

Using this scheme, which we have called the "data-under" method, two subband components can be combined for transmission, distributing the CNR between them as desired. A first component, called the "digital-under" signal, is quantized to n levels, typically 2 to 5 bits/sample, and the second, called the "analog-over" signal, is reduced in amplitude by a factor somewhat larger than n, and then added to the first quantized signal. Either or both of the signals can be scrambled and/or adaptively modulated. The two signals can readily be separated at the receiver [8.59].[34]

---

[32] We are using here the traditional measure of SNR, which is the ratio of peak-to-peak allowable signal to rms noise, not the ratio of actual signal spectral density to actual noise spectral density.

[33] Grey code is reflected binary code. The principle is readily generalized.

[34] This technique is essentially the same as the hybrid transmission method previously described.

Rather than combining two subbands in this way, the quantized data-under signal can be unrelated digital information of any kind, such as audio, data, synchronization signals, adaptation information, coding data, etc. For a given channel CNR, the SNR of the analog component can be adjusted by choosing the number of bits/sample, not necessarily an integer, used in the related digital signal.

In this method, the capacity of the analog-over channel is that portion of the capacity of the entire channel not used for digital transmission – the portion that is discarded in conventional digital transmission. The channel capacity thus made available to the analog signal is normally quite small. As a practical matter, this method is useful only if the analog signal is an adaptively modulated dc-free subband. The small capacity is sufficient for typical subbands because, due to adaptive modulation, the transmitted signal amplitude in the blank areas is comparable to what it would have been had normal transmission been used. For example, with a maximum adaptation factor of 32, and with 4 bits/sample in the digital signal, the analog component is reduced in amplitude to fit between the quantization steps by about the same factor by which it is increased by adaptation in the blank areas.

8.13 Digital Terrestrial Broadcasting

In a remarkable development, digital terrestrial transmission of HDTV is now being seriously considered in the US. Digital transmission is already widely used in wired systems as well as in satellite transmission, where signal quality is rather good, channel characteristics are fixed or slowly changing, and the signal level is uniform at all receivers. The proposal to use it in terrestrial broadcasting, which has none of these characteristics, is truly revolutionary. Although channel coding is not a main subject of this book, some additional treatment at this point is essential for a more complete understanding of the new systems.

8.13.1 The General Instrument Proposal

Current interest in digital television broadcasting in the US began in June 1990, with a proposal from the General Instrument Corporation, apparently based on their earlier work in satellite transmission. The proposed system used motion-compensated prediction, DCT coding of the prediction error, and 16-QAM digital channel coding with error protection. The total transmission rate was about 20 Megabits/s with a net rate of 12.6 Megabits/s for video. The same data rate was used at all receivers and the threshold SNR was said to be 19 dB. Picture quality was very good in the simulation. A similar method was demonstrated that permitted the transmission of four NTSC signals in a 30-Megabits/s satellite channel.

Reaction to the proposal was very strong. Although digital transmission had been discussed earlier, most commentators thought of it as a prospect for the distant future. (The earlier European work in DAB, discussed in Sect. 8.9.5 and 8.9.6, was largely unknown in the US.) With the GI announcement, there was almost instantaneous acceptance of the approach. Within a year, there were three more proposals for similar systems. Whatever the eventual outcome of the GI proposal, the company deserves a great deal of credit for permanently changing the prognosis for new television systems. Current developments are discussed in the Appendix.

8.13.2 Claims for Digital Broadcasting

The enthusiasm for digital broadcasting was fueled by a number of claims made by its supporters. These included higher compression ratio, better noise performance, suppression of ghosts and interference, and easier interoperability with nonbroadcasting applications. Sometimes claims of higher transmission efficiency were made. What was rarely mentioned was the subject of spectrum efficiency, discussed in Sect. 8.8.1 and 8.9.1.

It is the opinion of the author that all these claims are erroneous. They seem to arise from the widespread notion that "digital" means modern and efficient and "analog" means old and inefficient, including everything that is wrong with PAL and NTSC.

Noise and Interference Rejection. Quantization in digital systems suppresses a certain amount of additive noise and interference. The quantizing noise so introduced is always larger than the unwanted signals that can be rejected. Quantization is useful where the quantized signal must be recovered exactly. (The original signal, of course, is not exactly recoverable on account of the quantization noise.) However, exact recovery is not required in TV transmission unless the transmitted data has been entirely stripped of redundancy and thus bears no simple relationship to the original. In that case, any error is catastrophic. For most coding systems, some of the coded data does require error-free reception, but other data (e.g., the amplitude of DCT coefficients) does not.

Multipath Rejection. Digital transmission does not reject multipath. On the contrary, multipath and other linear channel impairments, such as frequency distortion, must first be eliminated by other means in order to permit digital transmission at useful rates.

Transmission Efficiency. Since digital transmission necessarily collapses a range of output states of the channel into a single state, it has a lower, not a higher, efficiency than analog transmission. Shannon showed that this reduction in capacity can be made as small as desired by sufficiently good (and necessarily long) codes. In practice, it is a rare digital system that achieves even 50% of the Shannon rate. Of course, digital transmission gives more freedom to the source coder, whose performance is equally important for the overall system efficiency. It is therefore necessary to examine the effect of the channel coder on the compression ratio that can be achieved by the source coder.

Interoperability. This is the most surprising claim of all, the least justified, and yet the one most often made. In fact, source coders with a high level of compression inevitably produce signals that have a very complex relationship to the original, and that are therefore more difficult, not less difficult, to transcode. In most cases, complete decoding is required before transformation into any other format. Since the need for interoperability will rarely be from the coded broadcast signal, but rather from the production or international exchange format, complex coding for broadcasting is not necessarily a drawback. But in that case it is obvious that digital *broadcasting* has no relationship whatever to interoperability.

Due to the natural evolution of technology, television postproduction is already implemented largely by digital processing. Digital video recorders are rapidly replacing analog recorders. HDTV receivers will surely be almost entirely digital. Whatever contribution digital processing can make to interoperability has been and will be made in these domains and not by digital broadcasting. On the other hand, there are many aspects of system design that do affect transcoding, such as appropriate choices of resolution, frame rate, aspect ratio, and the use of progressive, rather than interlaced, scanning. There is certainly a need to develop coding systems that permit decoding in a variety of formats from a single coded representation, and that do not thereby place any limitations on compression ratio.

8.13.3 Interaction Between Source Coding and Channel Coding

To understand the extent to which the use of digital transmission permits the use of source-coding methods that have higher compression ratios than otherwise, we must examine the best source codes used at present to see whether equal compression would be possible if all or part of the transmitted data were in analog form. MPEG, discussed in Sect. 8.14.6, is a good example. The information transmitted consists of velocity vectors as well as the amplitude and identity of the adaptively selected DCT coefficients. The last two items are jointly coded, although the correlation between them is probably not large. In fact, some of the HDTV systems discussed in the Appendix depart from the MPEG standard in that a vector code is used to identify the adaptive selection data and the amplitudes of the selected coefficients are transmitted separately. In that case, analog transmission of the coefficients could be used, since noise on the coefficients has a similar effect to noise added to an analog video signal.

MPEG transmits differential information, so that each frame is formed by adding new information to an image that is already stored at the receiver. This is often thought to require digital transmission to prevent the accumulation of errors in what is, in effect, an integrator loop. This problem can easily be solved by transmitting the dc component separately and using differential data only to recover the high frequencies, as discussed in Sect. 5.4.2. Since some nondifferential information is already transmitted in MPEG to deal with scene changes and channel switching, this procedure would not materially detract from the effectiveness of the code.

There are some kinds of source codes that do require error-free transmission. For example, the ultimate code for still pictures is a vector code in which each possible picture is represented by a unique code word whose length depends on the picture probability. (Such codes are quite practical for restricted classes of images such as typographical characters.) However, these codes are not in use at present, nor, for very good reasons, are they likely to be used in the forseeable future. The inescapable conclusion is that the best present-day codes can use hybrid analog/digital transmission with little, if any, loss of efficiency.

8.13.4 Digital Transmission with Soft Thresholds

In normal terrestrial broadcasting, there is a large variation of SNR, and therefore channel capacity, among receivers in the reception area. Systems that transmit at equal data rates to all receivers necessarily waste channel capacity in the central cities, where the excess capacity is the highest and where the demand for spectrum for mobile services is growing rapidly. They are spectrum-efficient only in the fringe areas.

The achievement of maximum spectrum efficiency requires that each receiver recover a substantial proportion of the local Shannon capacity and make use of it in obtaining the highest possible image quality. Such a scheme, which is, effectively, what we have now with NTSC, gives higher average picture quality with less bandwidth than that of a system that gives only fringe-area quality everywhere. This implies a multiresolution source coder as well as a channel coding system with a soft threshold. Soft thresholds were shown to be theoretically possible by Cover [8.60].

No methods of achieving continuous thresholds with digital transmission have been discovered, but several methods of getting a stepwise increase in data rate with SNR have been found. A rather good method, whose origin seems to be unknown, is the use of nonuniform-level QAM [8.61].

In this method, several binary data streams are added together with appropriate weightings for each of the in-phase and out-of-phase components, producing a constellation as shown in Fig. 8.16.

At very low SNR, only the two most significant digits are recovered. At higher and higher SNR, more and more digits are recovered, until finally the transmission appears to be 256-QAM, with 8 bits/symbol. With a source coder that can take advantage of the additional data recoverable at higher SNR to make better pictures, a system that has good spectrum efficiency over a wide range of SNR can be devised [8.62]. Other methods for getting a step-wise threshold in digital transmission are discussed in [8.63].

8.13.5 Hybrid Transmission

Another way to get a soft threshold while transmitting at least some digital data is simply to add an analog data stream to a binary signal, producing a constellation of the type shown in Fig. 8.17.

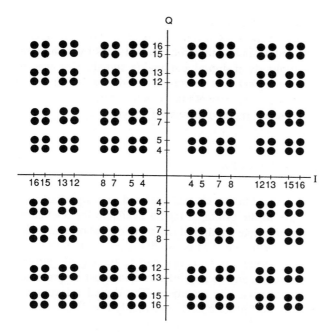

Fig. 8.16. Nonuniform QAM with 256 Levels. The level spacings are 1, 2, 4, and 8, corresponding to thresholds separated by 6 dB. Errors of the least-significant digit result from an instantaneous noise level of 4, while errors of the most-significant digital result from a noise level of 0.5. Such a modulation method is intended to be used with a progressive source-coding scheme in which each additional recovered bit augments the resolution.

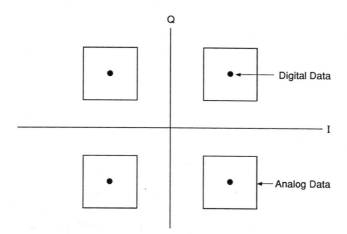

Fig. 8.17. A Hybrid Transmission Constellation. Analog information may be superimposed on multilevel digital signals as long as it is small enough not to cause excessive digital errors. Here analog data is superimposed on a conventional 4-QAM constellation. For a Nyquist bandwidth of 5 MHz, this arrangement permits the transmission of 10 Megasamples plus 10 Megabits/s

Such a constellation comprises two bits plus two analog samples/symbol. At very low SNR, only the digital data is recovered. At higher SNR, the analog data is recovered. Its SNR, of course, is less than the channel SNR by an amount that depends on the relative amplitude of digital and analog signals. One way to use such a scheme is to use the digital data to convey a low-resolution image and the analog data to convey high-frequency enhancement information. This is obviously related to the "data-under" method discussed above.

8.14 Some Sample System Designs

In this section, the methods discussed in this chapter are used to design transmission systems appropriate for use in the principal media available for television. Exclusively digital coding is appropriate for digital fiber and digital optical disks. AM can be used on cable and over the air. FM is usually used in satellite transmission and magnetic recording service. In all cases, part of the signal, particularly the dc components, lows, audio, and coding information, must be protected carefully, while the highs do not need this treatment. The digital information, and in some cases the highs, can be further coded statistically if desired. All the examples are for illustrative purposes only and are not meant to be optimized designs. They are meant to illustrate the possibilities of these methods. Further detail is given in a recent paper [8.40].

As pointed out previously, one important consideration in transmission systems is the ease of transcoding between formats optimized for different media. In these examples, we have taken two steps to simplify transcoding. One is to use the same spatiotemporal bandwidth for all subbands, which therefore are represented by equal numbers of samples per second. The other is to group the data into packages nominally 1/12 second long, a period chosen to be the least common integral multiple of the field duration of all commonly used systems – 24, 50, 59.94, and 60 fps. Encoding and decoding the data from and to conventional video signals requires temporal filtering with down- or upconversion ratios of 2, 4, or 5.

If the resolution of all of the subbands is the same, transcoding between transmission systems with the same field rate requires only rearrangement of data within each 1/12-second frame, and no filtering of any kind, either spatial or temporal. For field rates having 12 fps as a common submultiple, such as 24 and 60, filtering is likewise not needed. For more complicated relationships, such as between 50 and 60, transcoding is possible by adjusting the program duration slightly as is done in 50-Hz countries at present when utilizing 24-fps film.[35] Naturally, this cannot be done in real time, which still would require temporal interpolation by nonintegral factors. However, nonreal-time transcoding between transmission systems optimized for different media, which is often done, requires only repacking the components. It does not require any filtering at all, and thus

---

[35] The audio pitch can be restored, if desired, by a separate operation.

can be implemented with a frame store plus some circuitry for arithmetic and logic operations. Naturally, when transcoding between systems with different quality, the resulting quality is that of the inferior system, but in the desired format.

8.14.1 A 90-Mb/s Digital System

It is convenient to use subbands $144 \times 256 \times 12$ fps,[36] each needing 36,864 samples per 1/12-second frame or 442,000 samples/s. If we reserve 10 Megabits/s out of a data rate of 90 Megabits/s for audio and data, we can have 166 components at 1 bits/sample or 83 at 2 bits/sample, etc. Assigning the number of bits/sample and choosing components so as to achieve an overall diamond-shaped response as shown in Fig. 8.18, we have 43 components ranging from 2 to 8 bits/sample, with an average of 3.86 bits/sample. The SNR of the components is high at low frequencies and low at high frequencies, as known to be generally correct. The spatial resolution ranges from 720×1280 at 12 frames/s to 288×512 at 60 fps. Chrominance is transmitted at 12 fps, which has been shown to be adequate due to the lower temporal frequency response of the HVS for color than for luminance [8.64]. Scrambling is not required in digital transmission systems, although Roberts' randomized quantization may be used, if desired.

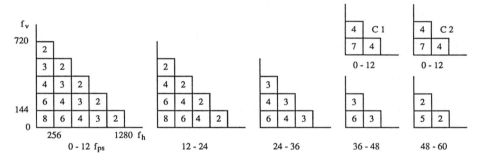

Fig. 8.18. A 90-Mb/s Digital System. This diagram shows the spatiotemporal frequency response achievable by the methods discussed in the text. The general idea is to select the subbands to be used and to assign bits/pel, indicated in each box, according to the desired CNR for that component. With adequate allowance for digital audio and data, the resolution ranges from 720×1280 at 12 frames/s to 288×512 at 60 frames/s

Even if experiment shows that some adjustment is required in the bit assignments, it is clear that substantial improvement in quality per unit channel capacity, as compared with straightforward systems, is made possible by adaptive modulation of subband components. All of this improvement is ascribable solely to the use of a more efficient signal description based on the needs of the HVS. Further compression is available by statistical coding of the digital data

---

[36] The choice of 144×256 offers some simplification of circuitry and also provides so-called square pixels, a feature of great interest in the computer-graphics community.

stream (e.g., adaptive selection of signal samples) that emerges from the coding process so far described, although the redundancy can be expected to be smaller than that found in normal digital video signals, due to the adaptive modulation.

If subbands are to be chosen adaptively, the only change needed in the system is the incorporation of a selection algorithm at the encoder and the transmission, with each block, of information as to which components have been used and their location within the data stream. The synthesis process at the decoder would be nonadaptive, with only the received components being input to the interpolator, its other inputs being set to zero.

8.14.2 A System Using Amplitude Modulation

We now perform a similar exercise for an analog channel such as coaxial cable, where the CNR at each terminal can be guaranteed to exceed some minimum value, which we choose to be 36 dB for the purposes of this example. The lowest-frequency components can be transmitted digitally, and since there may be insufficient capacity for PCM with a high enough number of bits/sample, some elementary coding system such as DPCM may be used.

In Fig. 8.19, the choice of components is shown. The resolution is the same as that of the 90-Megabits/s digital system at all temporal frequencies except the highest, where it is slightly lower. A total of 41 components is used, 6 of which are digital. The digital image components require a total data rate of $36{,}864 \times 12 \times (1 \times 4 + 5 \times 3) = 8.48$ Megabits/s. Additional data is required for audio and adaptation information. To provide a channel for the required digital information, the data-under method is used to transmit data along with 15 of the analog components. With 9 components at 3 bits/sample, 5 at 2, and 1 at 1, the total digital data rate is $36{,}864 \times 12 \times (9 \times 3 + 5 \times 2 + 1 \times 1) = 16.81$ Megabits/s, which is twice the required rate. This allows a more conservative design to be used or a slightly higher resolution and/or SNR to be achieved.

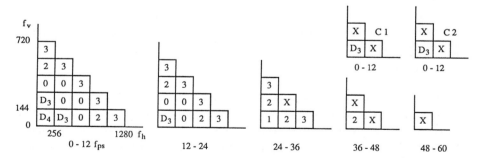

Fig. 8.19. A System Using Amplitude Modulation. In this arrangement, designed for a 6-MHz channel with a CNR of at least 36 dB, a digital transmission rate of about 18 Mb/s is assumed using the data-under method. A number in a box indicates the number of bits/pel hidden under the analog signal. D_n means digital transmission (DPCM) at n bits/pel. The character "x" indicates that a component is hidden under another marked "0." There are 35 analog and 6 digital components

Of the 35 analog components, 20 are combined in 10 pairs to distribute the CNR between the components in a near-optimum way, leaving 25 to be transmitted. Each of these requires $144 \times 256 \times 12$ samples/s, for a total of 11.5 Megasamples/s, which is appropriate to the 6-MHz channel. In real channels, in an environment where conventional systems and systems such as described here were intermixed, this rate would have to be reduced perhaps as much as 10% to allow for adequate spacing between channels. Were this system to be used on all channels, the superior interference performance would permit the use of much less sharp-cutting filters to define the band edges. Some band overlap would be permissible. In that case, a full 12 Megasamples/s would be allowed in accordance with the sampling theorem.

With systems such as this, which combine digital and analog transmission, if the CNR falls below the design level, in this case 24 dB, the analog components suffer reduced SNR and picture quality is reduced. There are several ways this situation may be handled. One is to measure the CNR using some calibration signal, and to discard components when their use would not be beneficial. Even when no analog components are used, a low-resolution color image is still attained. If further investigation shows that this method is useful, then the CNR of the analog components can be graded in such a way as to achieve a graceful degradation at lower and lower CNR.

For actual transmission of such a signal in an rf channel, the method of choice is double-sideband quadrature amplitude modulation of a single carrier, centered in the channel, by two signals, each of 3-MHz bandwidth. To minimize the effect of errors in carrier phase recovery at the receiver, it is desirable that the two signals be derived from vertically adjacent lines in the image [8.59]. In this way, phase errors cause only a small loss of vertical resolution rather than highly visible crosstalk between unrelated image elements. Fig. 8.20 shows the format of the two 3-MHz (6 Megasamples/s) modulating baseband signals, using the parameters from an earlier example.

8.14.3 A System Using Frequency Modulation

Most TV programs, at some point, are transmitted in a satellite transponder channel using frequency modulation. In relay service, from point of origin to local television station or cable head end, large receiving antennas are used and quality is very high. In direct broadcasting service (DBS) the emphasis is on small receiving antennas. Because of the rather sharp FM threshold, system parameters must be chosen to guarantee adequate CNR at the receiver. Since all receivers are at essentially the same distance from geostationary satellites, this is primarily a matter of deciding on the severity of the weather conditions under which reception is to be maintained.

FM has a triangular noise spectrum, so that the noise power at the receiver rises with frequency. This is quite desirable for the luminance signal, but not for the color components in a composite system such as NTSC in which the color information is impressed on a high-frequency subcarrier. In any event,

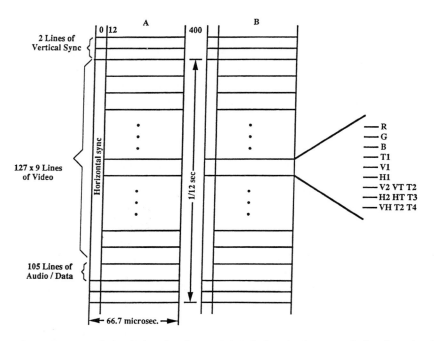

Fig. 8.20. Transmission Using Quadrature Modulation. Analog transmission through a 6-MHz channel for an earlier system than that of Fig. 8.17 is shown here. The components are transmitted sequentially at a data rate corresponding to 3-MHz bandwidth for each of two baseband signals, A and B. The two signals quadrature-modulate a single carrier in the middle of the band. At each instant, A and B represent vertically adjacent picture elements so as to reduce the effect of demodulation errors

since it is a one-dimensional signal that is used for modulation in analog FM systems, the desirable frequency distribution of noise is obtained only in the horizontal direction in frequency space. In the vertical and temporal directions, it is uniform. The use of subband coding makes it possible to achieve a perceptually more appropriate overall distribution, just as in AM.

When a signal is divided into n subbands that are then transmitted sequentially, each is time-compressed by the factor n so as to maintain the original bandwidth. All have the same SNR. In order to achieve different SNR for different components in FM, the relative time compression and therefore the relative modulation index of the components can be selected as desired. There is no need to pair components using the data-under method as in the AM case.

In Fig. 8.21, a selection of 35 luminance and 4 chrominance components is shown. Using the same $144 \times 256 \times 12$ components as in the previous examples, each of which comprises 442,000 samples/s, plus an additional 2.88 Megasamples/s for audio and data, the total rate is 20.1 Megasamples/s. We choose an rf bandwidth of 27 MHz and a CNR of 15 dB. The relative time-compression factors shown are chosen to achieve about a 6 dB difference in SNR between adjacent components using modulation indices chosen according to Carson's rule

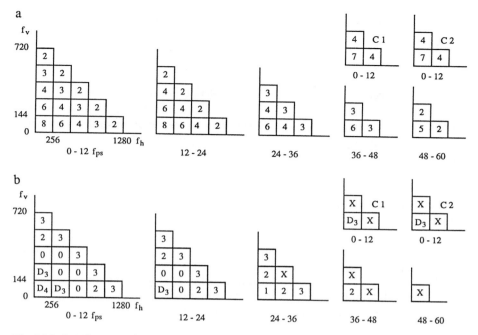

Fig. 8.21a,b. A System Using Frequency Modulation. In order to achieve the appropriate CNR for each component in the presence of triangular noise, different degrees of time compression are used. The CNRs shown in (a) are those achieved before adaptive modulation. When the latter technique is also employed, the SNR is substantially increased. A resolution somewhat higher than the AM case is achieved in a standard transponder channel. A similar format would be used for analog magnetic tape recording The labels in (b) are as in Fig. 8.19

[8.65]. Experience shows that better results are obtained by using a higher index and permitting slight distortion due to spectrum truncation, since this gives a higher SNR.[37]

Further large improvements in perceived SNR are achieved by adaptive calculation of the modulation index in accordance with local image activity in a manner quite similar to the adaptive modulation used in the previous examples. In FM, because of the complicated relationship between the signal, the modulation index, the CNR, and the distortion, calculating the optimum index is more difficult than in the AM case, and is best done by using a recursive procedure [8.66].

Although not further discussed here, a format for magnetic tape recording would be similar to that for FM transmission by satellite. The two cases are somewhat different. In satellite transmission, which is power-limited, the CNR is low but the bandwidth is high. In tape recording, the bandwidth is lower and the CNR is much higher, but the signal level is subject to large variations that can be ignored by using FM.

---

[37] These calculations were done by Julien Piot while at MIT.

8.14.4 Systems Using Adaptive Selection of Subbands

Preliminary work has been done on schemes that select subbands for transmission, not on the basis of average spatiotemporal frequency response, but on the basis of energy actually present in the spectrum for each signal package, in our case 1/12 second. If x, y, t space is divided into blocks small enough so that the motion within most blocks is nearly uniform, it is found that many of the subbands are empty in each block. This is easy to see, since, in stationary blocks, all the energy is on the zero temporal-frequency plane. With motion, this plane tilts. Therefore many of the subbands are near zero and may be omitted with negligible loss in quality.

In Fig. 8.22, we show an example of the potential of this method. With an average of only 10 out of 80 subbands transmitted, the resolution of moving objects is very near to that of the original image. Of course, the subband selection information must now be transmitted accurately. This information is highly correlated, so that the required channel capacity is very low. The division of the transmitted information into some that must be transmitted digitally without error, and the rest that can be transmitted using analog amplitude modulation, lends the system to use in over-the-air transmission, where the CNR, and hence the Shannon channel capacity, varies substantially from point to point.

8.14.5 JPEG

JPEG is a protocol for still-picture data compression developed by the Joint Photographic Experts Group under the auspices of the CCITT and the International Standards Organization (ISO) [8.67].

It is intended to encourage the production of low-cost integrated circuits that would facilitate the use of standardized compression systems in a variety of applications, all of which assume digital storage or transmission. (Channel coding is not a part of the JPEG protocol.) We need not concern ourselves here with

---

→

Fig. 8.22a–d. Adaptive Selection of Subband Components. Shown here are single frames from a sequence with quite rapid motion. The three sections of the image are taken from a TV film, a computer-generated test pattern featuring a rotating test target, and an artificial sequence made by computer zoom and pan through a high-resolution fixed image. The original, having 80 subbands (4 × 4 spatial decomposition and 5:1 temporal decomposition) is shown at (**a**), while a four-times compressed version (fixed selection of 20 out of 80 subbands, optimized for this sequence) is shown at (**b**). Note that there is a loss of resolution in the moving areas as well as some uncancelled aliasing. In (**c**), an adaptive selection of 20 subbands is made using 2 × 2 blocks (8 × 8 in the final image) and .11 bits/pel of side information is needed for the selection data. A big improvement is achieved, although there are still some artifacts. In (**d**), we use 1 × 1 blocks (4 × 4 in the final image), the selection data requiring .2 bits/pel. However, quality very near the original is achieved using only 10 out of 80 subbands. It is believed that the lack of visible aliasing is due to the fact that the omitted components have very low energy. Note that all three of the systems shown in the previous figures would have had higher effective resolution with adaptive selection of components as used in this case. The example shown here would have had higher resolution if nonseparable filters had been used in the analysis and synthesis processes

a

b

Fig. 8.22a–b. For caption see opposite page

269

270

c

d

Fig. 8.22c–d. For caption see page 268

those aspects of JPEG that deal with standards questions or with accommodation of the system to a wide variety of specific coding methods. Rather, we shall use JPEG as an example of typical current good practice in still image compression. Since it involves many tradeoffs, such as between efficiency and complexity, it does not achieve the highest possible efficiency in all cases. For the main intended purpose, which is to code color images of moderate complexity in the resolution range typical of computer displays, it claims to achieve the following performance:

- 0.25–0.5 bits/pel: moderate to good quality, sufficient for some applications
- 0.5–0.75 bits/pel: good to very good quality, sufficient for many applications
- 0.75–1.5 bits/pel: excellent quality, sufficient for most applications
- 1.5–2.0 bits/pel: usually indistinguishable from the original, sufficient for the most demanding applications

As any reader of this book understands, these are extremely imprecise measures of image quality. Nevertheless, typical JPEG algorithms work acceptably, and have already been widely adopted. This is particularly true in the graphic arts, where the much higher resolutions lead to substantially higher compression ratios.

Since JPEG embraces many different specific schemes, we describe here just one of many possible algorithms. Image data is first put into luminance/chrominance form with the linear resolution of the latter set at one-half that of the former. Each of the three components is coded separately. The basic compression method is based on the DCT applied to 8×8 blocks of the luminance and subsampled chrominance images. The dc coefficients of successive blocks are transmitted differentially. The ac coefficients are quantized with a coarseness that depends on spatial frequency. This is accomplished by first dividing the 63 coefficients by a matrix of factors provided as a table and then applying a uniform quantizer. (The tradeoff between compression ratio and quality is made by altering the table.) The quantization operation effectively discards a large percentage of low-amplitude coefficients. The coarse quantization of the higher-frequency coefficients provides further compression. The final (lossless) compression step uses Huffman or arithmetic coding to transmit the identity and amplitudes of the selected coefficients, using a second table, which is also provided to the decoder.

The selected coefficients in each 8×8 block are placed in a string by zig-zag scanning through the frequency plane. Each location is specified as a run length up to 15 along the zig-zag path. Two code words are transmitted for each coefficient. The first gives the run length and the number of bits that are used in the second word; the second word denotes the amplitude; a unique word identifies the end of the block. The first code word is taken from a table that can be changed for each picture, if desired. The second code word is taken from a fixed table. Run lengths greater than 15 require additional code words.

The weakest part of the JPEG scheme is probably the zig-zag scanning in the frequency plane as a means of denoting the retained coefficients, and the joint coding of their position and amplitude. As discussed in the Appendix, a number of the proposed HDTV systems instead specify the selected coefficients by using a unique vector code for each possible pattern. In addition, higher efficiency can probably be achieved by locating the coefficients in the space plane rather than the 8×8 frequency plane, since the former is generally much larger than 8×8.

8.14.6 MPEG

MPEG is a protocol for coding of moving pictures being developed under the auspices of ISO and the International Electrotechnical Commission (IEC) [8.68].

Like JPEG, it is not a single algorithm, but a class of algorithms intended to stimulate the development of integrated circuits that would facilitate the general use of compression methods for moving pictures in a variety of applications, particularly in computer workstations. Its original target was "video cassette recorder quality" of NTSC or PAL imagery at a total data rate of no more than 1.5 Megabits/s, primarily for compact-disk storage. Eventually, full video quality (CCIR Recommendation 601) is to be achieved at 5-10 Megabits/s and HDTV at 20-40 Megabits/sec. The algorithms are strictly source-coding methods intended for digital storage media and not for transmission through analog channels. We need not concern ourselves here with many aspects of MPEG dealing with standards and with special properties (such as playing in reverse), but instead will treat it as representative of good modern practice in motion video coding.

MPEG leans heavily on JPEG, but obtains additional compression by exploiting temporal redundancy, which must be high to produce the illusion of continuous motion. A common opinion is that this additional compression factor is about 3, on average, although there may be no gain at all for certain kinds of material. An obvious such example is a sequence of unrelated images, perfectly legitimate material for systems that do not use temporal processing. A rapid series of scene changes or extremely complex motion are less drastic examples of cases with reduced temporal correlation.

The basic principle of MPEG is to use the DCT for removal of spatial redundancy (as in JPEG, but different quantization tables are used) and motion-compensated forward and bidirectional "prediction" (interpolation) for removal of temporal redundancy. This is implemented by calculating each frame from one or two neighboring frames using a local motion estimate, and subjecting the error, sometimes called the "residual," to the DCT. Since it is necessary to accommodate scene changes and channel switching on receivers, some nondifferential information must also be transmitted. Therefore, there is provision for transmitting entire pictures on an intraframe basis, as often as desired. Typically, this amounts to about three frames/sec. The rate of transmission of such frames is a result of a tradeoff between coding efficiency (it takes much more data to transmit on an intraframe basis) and the time required to recover from scene changes and from channel errors.

The choice of the coding mode – intra, predicted, or interpolated – can also be done on a block-by-block basis to minimize errors. In this way, for example, scene changes can be detected and the first new frame coded for the most part using the intraframe method.

Motion estimation is usually done by block-matching using 16×16 blocks. (The decoder need not know how the motion vectors are computed, so any method can be used.) Motion vectors are transmitted on a block-to-block differential basis. The vectors are used both for prediction and for bidirectional interpolation. Only the forward-predicted frames and the intraframe coded frames are used for prediction or interpolation. The latter is useful, even though the interpolated frames are not used for prediction, since newly uncovered as well as newly covered areas are then available for motion estimation and interpolation, thus giving higher compression.

In many cases, the coded data rate must be equal to or less than some particular value, such as 1.5 Megabits/s for compact disks. To maintain a constant rate in the face of input material of varying complexity, a channel buffer must be used. The quantization step size is changed as a function of buffer occupancy. It is also possible to change the step size on a block-by-block basis using a perceptual criterion so as to improve the picture quality for a given data rate.

8.15 Lessons for the System Designer

The fundamental principles governing the performance of all imaging systems, as developed in the earlier chapters, are fully applicable to important current problems, such as the design of improved television broadcasting systems. Efficient systems must deal with the properties of images to be transmitted, the visual performance of the human viewer, and the physical characteristics of the channels to be used. In this short review, we have found that existing TV systems, based on the transmission of video signals derived from raster scanning, are very inefficient in their use of channel capacity and extremely vulnerable to analog channel impairments such as echoes, noise, interference, and frequency distortion. They ignore viewers' varying sensitivity to noise as a function of frequency and as a function of the image on which the noise is superimposed. They also do not take advantage of the revolutionary improvement in signal-processing capability made available by modern, inexpensive, integrated circuits.

Analysis shows that the highly nonuniform spectrum of raster-scanned video signals is an important element in both spectrum inefficiency and poor interference performance. This can be remedied by adaptive modulation of subband components, which makes the spectrum more uniform and which increases noise immunity. In effect, channel capacity can be assigned to the components in accordance with their visual importance. Interference performance is greatly enhanced by scrambling. The combination of these techniques has an unexpectedly favorable result, in that unwanted signals that are randomized by scrambling are reduced in amplitude by adaptive demodulation in the relatively blank image areas where they would otherwise be most noticeable. The resulting signal looks like white noise and is totally encrypted. The transmission format can easily be optimized for the physical properties of different channels, and the different formats can readily be transcoded one to the other.

The introduction of the idea of digital broadcasting and the clamor from nonbroadcasting interests for systems with a great degree of interoperabilty have introduced new elements into TV system design. The MPEG and JPEG developments promise substantially higher compression than previously thought possible, but it is too early to tell whether the resultant quality will be seen to be good enough when applied to the full range of images that must be processed. Digital broadcasting presents a host of new problems never before dealt with, so that the prospects for success are far from clear. Current attempts to standardize compression and broadcasting systems that have not been adequately tested are most regrettable.

Nevertheless, new television systems based on the principles discussed in this chapter will eventually provide better image quality under typical transmission conditions in many different media, including the terrestrial broadcast channel. At the same time, less spectrum will be needed to give each viewer a choice of at least as many services as exist today.

Acknowledgements. Much of the work reported in Chap. 8 was performed at MIT under the Advanced Television Research Program, a research project sponsored by the Center for Advanced Television Studies. The members of CATS have, at various times, included ABC, CBS, NBC, PBS, HBO, Ampex, General Instrument, Harris, Kodak, 3M, Motorola, RCA, and Zenith. A number of colleagues and students contributed to the developments. Most of the images used in Chap. 8 were made by students, including Warren Chou, Ashok Popat, Paul Shen, Adam Tom, and Kambiz Zangi. The help of all these people and organizations is gratefully acknowledged.

Appendix
Development of High-Definition Television
in Japan, Europe, and the United States

Plans for the practical implementation of improved television systems in the three principal TV regions of the world have been influenced at least as much by considerations of economic gain and national or company pride as by the scientific and technological principles discussed in Chap. 8. For this reason, it is necessary to deal with many issues that are matters of opinion rather than fact. The opinions expressed here are those of the author alone. Another view is found in a report to the French parliament [A.1]. The final outcome of what has turned out to be a vigorous contest is still far from clear at the time of writing (August 1992). No doubt, many of the contentious arguments now being conducted will eventually disappear into well earned oblivion. Nevertheless, enough new knowledge of the purely technical aspects of providing much better pictures and sound for the world's living rooms and classrooms has emerged to make it worthwhile to give an up-to-date account.

A1. Introduction

France had an 819-line monochrome system in commercial operation from shortly after WWII until 1985, but it shifted its principal transmissions to 625-line SECAM in 1965 when all of Europe adopted color. However, the recent history of HDTV started in Japan in 1970, when the Japan Broadcasting Company (NHK) set about to design the next generation of television systems. The main systems work was done by NHK, but the development of particular equipment, such as cameras, recorders, and displays, was delegated to the large Japanese electronics companies. The first demonstrations were held in the late seventies. Japanese attempts to have their system accepted as an international standard for production and international exchange of programs stimulated the Europeans to pool their efforts in Project Eureka 95 in 1985. A preliminary demonstration with most of the hardware, but not the final system design, was held in 1988 in Brighton, England. A more advanced version was demonstrated in 1989 in Berlin. Extensive coverage was provided of the Albertville and Barcelona Olympics in 1992. In the US, there has been no agreement among the interested parties. The system to be used will be selected in a contest run by the television industry under supervision of the Federal Communications Commission (FCC). In June 1990, the nature of the contest in the US was changed dramatically by the proposal from the General Instrument Corporation for an all-digital system using source-coding technology closely related to MPEG. (See Sect. 8.14.6)

A1.1 Compatibility

Until the advent of the all-digital proposals, no aspect of new TV systems had excited such intense discussion as the question as to whether the new signals should be viewable on old receivers. For the sake of the then-existing 10 million receivers (there are more than 180 million NTSC receivers and 60 million NTSC VCRs in the US, and more than 700 million TV sets in the world at present) the US added color in 1953 in a completely compatible manner. Old receivers could use the new signals and new color receivers could use the old signals. This was considered a technological triumph at the time, although it caused problems that have only recently been properly addressed. On the other hand, when Europe adopted 625-line color in 1965, although it was compatible in some countries, the new signals were not compatible with old receivers in France and Great Britain. A decision was made to serve existing sets in these countries by separate transmissions for a period of 20 years. This procedure, called simulcasting, had generally been viewed with disfavor by commercial broadcasters because of the extra expense.

Although, to many in the industry in the US and Europe in the 1980s, it seemed self-evident that HDTV *had* to be compatible to be successful, compatibility removes one of the principal incentives to buy new receivers – the prospect of seeing attractive programs that are not otherwise available. Viewed in this light, the near-failure of NTSC compatible color (it took 10 years to reach the 1% penetration point) may perhaps be attributed to the fact that viewers were able to see all the new programs on their existing receivers, albeit in black and white.

Aside from providing more incentive to buy new receivers, a system free of the constraint of compatibility can surely be a better system, particularly with respect to the efficiency of spectrum utilization, as has been shown in Chap. 8. Simulcasting deals with the "chicken-and-egg" problem, in which viewers fail to buy receivers because there are few new programs and producers are reluctant to make programs since there are so few receivers. When one adds to that the difficulty of conforming different aspect ratios in compatible systems as well as the fact that simulcasting need not be as expensive as once thought, the argument for compatibility seems not so overwhelming.

A1.2 Spectrum Considerations

With about twice the horizontal and vertical resolution as NTSC and with a wider aspect ratio, HDTV transmitted with the same efficiency as NTSC or PAL would require about five times the bandwidth. This seems impractical in all existing transmission media, and would probably not be feasible even in fiber-optic cable. An important goal, therefore, is to compress the required bandwidth to no more than is used at present. While bandwidth-reduction schemes have never been popular with broadcasters since they did not believe it was possible to maintain quality while reducing bandwidth, the exigencies of the HDTV situation and the demonstration of some effective compression methods have changed attitudes considerably.

Aside from reducing the bandwidth required for single HDTV signals, the other important spectrum consideration relates to the number of stations that can be accommodated within a given overall allocation. Here the question of interference, both within a single channel and between adjacent channels, is of paramount importance. It has now been accepted that much better interference performance is possible with systems that abandon current transmission formats. (The highest possible spectrum efficiency requires the use of single-frequency networks, as shown in Chap. 8.) The pressure for using some spectrum now allocated to television for other services, such as mobile radio and cellular telephones, is very strong, particularly in the United States. To accommodate these new services without reducing the current level of TV service is only possible with entirely new systems.

A1.3 Alternative Transmission Channels

As shown in Fig. A.1, the overall television system, from the scene in front of the camera to the image on the home receiver, is very complicated. Any new system design must take account of this. It must be adaptable to every transmission medium and, if different formats are required in the different media, provide for adequate transcoding among them.

Even though more than half of American TV homes subscribe to cable, the over-the-air (terrestrial) channel is likely to remain the primary distribution means in the US for some time. It is the terrestrial broadcasters who provide most of the programming, even on cable. For another thing, it is by far the cheapest way to distribute broadband signals to the home, particularly as the system is already installed and paid for. Finally, it is free in the US and cheap to viewers in most countries. All other media involve higher cost to the consumer, so there is an understandable reluctance to eliminate terrestrial transmission entirely. As discussed in Chap. 8, delivering really high-quality signals in this medium is not easy.

→

Fig. A.1. Universal Production and Distribution System. In the future, TV may be distributed to the home by four different kinds of channels in addition to tape or disk recorders. Production of programs is a complicated business involving multiple sources and a good deal of communication. The finished programs must then be transmitted to the TV stations, cable head-ends, satellite up-links, etc., for international exchange and/or retransmission to the viewers. If, as anticipated, different transmission formats are used for each section of the pathway, transcoding is required at many points

278

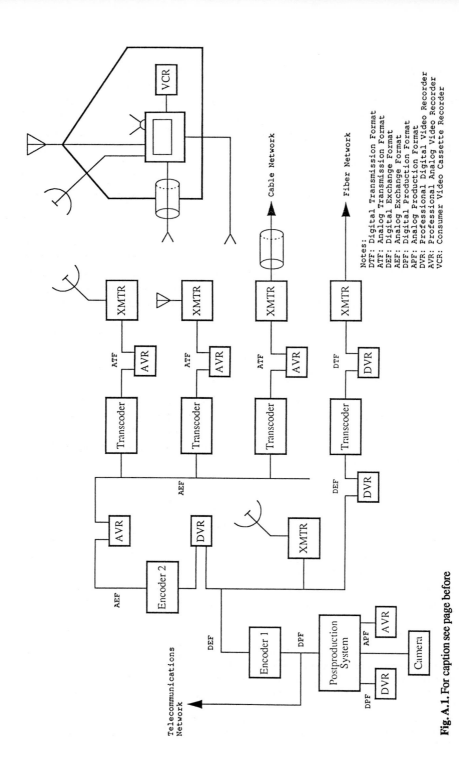

Fig. A.1. For caption see page before

Notes:
DTF: Digital Transmission Format
ATF: Analog Transmission Format
DEF: Digital Exchange Format
AEF: Analog Exchange Format
DPF: Digital Production Format
APF: Analog Production Format
DVR: Professional Digital Video Recorder
AVR: Professional Analog Video Recorder
VCR: Consumer Video Cassette Recorder

Cable, in principal, should give better signal quality. It should also have fewer spectrum constraints. In practice, cable systems suffer from transmission distortions of various types that significantly reduce image quality in most homes. However, it is very important to be able to distribute, by cable, the same programs that are transmitted over the air. Therefore, in spite of the differences, the two media have used exactly the same transmission format. Such proposals as have been made to make cable the primary distributor for HDTV are based on commercial rather than technological considerations.

Direct-broadcast satellites (DBS) are already in use. As compared to terrestrial transmission, the signal quality is normally much better, with substantial freedom from ghosts and interference, and with an adequate SNR. In addition, in Europe and Japan, which have far fewer terrestrial channels available than the US, DBS is a way of providing additional channels at acceptable distribution costs. The consumer must buy a DBS antenna and receiver, which may cost as little as $300 at today's prices. Since it has been repeatedly demonstrated that consumers will pay for program choice, such prices would not be an insurmountable barrier in countries where there is not a well established cable network already serving this purpose.

Fiber-optic cable is already in use for telephone trunk lines, and there is a move to put it into every home in order to serve all telecommunication purposes. However, at present, fiber installations are more expensive than cable. In order for fiber to the home to become economically viable, viewers must be willing to pay substantially more for a wider variety of services than they now get from cable. There is no market evidence that they want these additional services, so fiber remains an intriguing possibility rather than a certainty.[1] If fiber does materialize, it will have a large effect, since it can provide a high-capacity high-quality digital channel. This is especially true if the telephone companies, which are committed to all-digital transmission, become the main providers of this service.

Note that all these transmission media themselves are physically analog. At present, terrestrial, cable, and DBS use analog transmission exclusively for television, while both analog and digital transmission are used on fiber. In the future, digital transmission is likely to be the dominant mode on fiber and DBS. This is also possible, though less likely, on cable. The biggest uncertainty is terrestrial transmission. Until 1990, there was little discussion of using digital transmission for this medium. There is now a lot of activity in this area, but it is too early to predict the outcome.

One medium that will certainly be of great importance in HDTV, as it is in today's television distribution system, is the video recorder, both erasable, as the VCR, and not, as the optical disk. It was once thought (feared, actually) by American broadcasters that enough viewers might watch recorded HDTV to cause them to lose significant audience share. However, since sports are sure to

---

[1] Ingenious entrepreneurs are trying to find a way around this problem. The most recent idea is to make current telephone subscribers pay for fiber by permitting accelerated depreciation of the existing copper telephone plant. See the NY Times, 9 June 1992, p 1.

be among the most popular HDTV subjects, there certainly will be a role for live high-definition broadcasting. To the extent that magnetic video recorders remain analog, they can use technology quite similar to that of DBS. If sufficiently inexpensive digital magnetic recorders are developed for home use, digital coding schemes such as those suitable for optical disks or optical cable can be used.

If we were to wire the country "from scratch," then fiber in the cities and satellite service in the countryside would be very attractive. However, the country is already fully wired. Therefore conversion to different media must meet different, and far more difficult, economic tests.

A1.4 Transition Scenarios

A final element that influences HDTV strategy is the means by which a new TV system can be introduced. This, in turn, depends a great deal on whether a compatible or noncompatible (to some, evolutionary or revolutionary) approach is taken, and on whether today's TV is to be abandoned or is to exist side-by-side with HDTV.

For compatible systems, broadcasters must take the first step of transmitting HDTV programs while there are few HDTV receivers. Much of the cost would be borne by advertising directed primarily at the installed base.[2] The incentive to bear the increased cost of HDTV programming would presumably be competition from HDTV delivered in alternative media. The presumption is that viewers would start to buy HDTV receivers to see the same programs in higher quality. As this happened, production values could gradually be changed to take advantage of the higher definition, with the result that the visual impact on the old receivers would decline. Presumably, this would provide added incentive to buy new sets. Eventually, all broadcasting might be in HDTV, but viewers would never be forced to buy new sets, a situation identical to that of adding color to NTSC. In the latter case, color receivers have become so cheap, and color is so obvious an improvement over monochrome, that the market for black-and-white receivers is quite small, and virtually all broadcasting is in color. Since high definition gives a much smaller increment in visual impact than color, we would expect the transition to be much longer, and the market for old-style sets to remain substantially larger. These trends would be intensified if the HDTV sets were to remain considerably more expensive than today's or if the picture quality on HDTV receivers did not live up to expectations.

In the compatible scenario, it has also been suggested that the ultimate picture quality might be reached in a series of steps to be taken both by broadcasters and viewers. The former would start with low-cost steps such as first increasing the aspect ratio without changing the resolution, thus not requiring much new

---

[2] In the case of NTSC compatible color, RCA, the largest receiver manufacturer at that time, also owned NBC, a prominent broadcaster. By producing programs as well as manufacturing receivers, it simultaneously promoted both the "chicken" and the "egg." Nevertheless, it took ten years to reach the 1% penetration level. An entirely different situation prevails today, making the prospects for HDTV less certain.

studio equipment. Additional channel capacity and new cameras might be added at a later date to provide higher resolution. Viewers could start by buying new receivers to match, which presumably would be less expensive than those for full HDTV. Of course, this requires that each step be compatible with the previous step, particularly in the case of receivers. We would also expect slower penetration with this scenario, since the increment of picture quality at each step would be smaller.

In the case of noncompatible systems that are to supplement the existing service, rather than to replace it, the scenario is simpler, being comparable to the coexistence of AM and FM. HDTV broadcasters will start up with almost no audience. The higher cost of production and broadcasting in HDTV will be borne in the hope that an audience will develop because of the possibility of seeing unique programs that are also of higher quality. Simulcasting will be avoided in most cases so as to provide the greatest incentive to buy new receivers. This scenario clearly works much better where an insufficient number of channels is currently available, as is the case in Europe and Japan. The additional programs available, whether in HDTV or not, would most likely be the main reason for buying the receivers. Some have suggested that this scenario lends itself to the introduction of HDTV via cable or other pay services, rather than as a free service.

For noncompatible systems intended ultimately to replace existing services, simulcasting will probably be used as a matter of equity to viewers who might otherwise be deprived of service unless they were willing to buy a new set immediately as well as to provide an audience for the programmers and advertisers while the number of new sets remained small. The incentive to buy new sets is initially the better quality, and, eventually, the demise of the old broadcasts. If announced far enough in advance (10 years or more), this scenario meets the test of fairness. It also leaves open the option of retaining the old broadcasts for a longer period, depending on the acceptance of high definition in the market place. The FCC has now ruled that NTSC is to go off the air 15 years after HDTV transmissions begin. This will require the purchase of new receivers or of set-top converters, should they become available. The price of set-top converters and of low-performance receivers for the HDTV signal will have a good deal of influence on the transition scenario. Unfortunately, this has been inadequately recognized.

It is difficult to know the extent to which authorities in the various countries thought through the implications of the fundamental choices that they made in their quite distinct approaches. The desire to capture market share in receivers and other HDTV equipment, as well as the equal desire not to be captured, have no doubt figured prominently in the various strategies.

A2. Japan

To many observers, HDTV is simply the new television system developed by the Japanese. Since they started so early and spent so much more money than anyone else, those outside the field felt, until recently, that there was no point in anyone trying to catch up, in particular the Americans. As we shall see, starting early has its disadvantages as well as its advantages, both for the Japanese and the Americans.

A2.1 History of the Japanese Developments

With typical foresight and willingness to invest money many years in the future in order to secure markets, the Japan Broadcasting Corporation (NHK) began development of HDTV in 1970. The program was orchestrated by NHK, which did the system development and preliminary development of equipment, while the commercial equipments – cameras, displays, recorders, and ancillary products – were developed by the major domestic electronics corporations, including Hitachi, Matsushita, Sony, NEC, Toshiba, Ikegami, etc. From the first, the plan was to implement HDTV in Japan as an entirely new, noncompatible, service, delivered to viewers by DBS, and intended to supplement, rather than to replace, the existing over-the-air (terrestrial) system, which would continue to employ NTSC.[3]

Scanning standards of 1125 lines, 60 fields/s, 2:1 interlace, and 5:3 aspect ratio (later changed to 16:9) were chosen with the intention of making the picture quality comparable to that of 35-mm motion pictures.[4] These numbers may be contrasted with NTSC's 525 lines, 59.94 fps, 2:1 interlace, and 4:3 aspect ratio, and the 625/50/2:1/4:3 PAL and SECAM systems used in the 50-Hz countries. The 1125/60 system, perhaps intentionally, is unrelated to, and rather difficult to convert to, either NTSC or PAL.

Demonstrations outside of Japan with this system, now called the "studio system," began in 1981. Experiments were carried out to show terrestrial transmission at 38 GHz and analog transmission in optical fiber. Successful satellite transmissions were carried out in 1978. Because the system had a basebandwidth about five times that of NTSC, standard satellite transponder channels, which have 24 or 27 MHz bandwidth,[5] were not adequate, so that a special transponder of more than 100 MHz bandwidth was required.

---

[3] The color TV system originally proposed in 1953 by the National Television System Committee is used in Japan and in most 60-Hz countries, including the United States. (Most 50-Hz countries use PAL or SECAM.) Most of the satellite capacity would be used by NHK, but some would be made available to independent broadcasters, who presumably would use NTSC or some enhanced version of NTSC. It was not originally intended to transmit HDTV via terrestrial broadcasting.

[4] This is not a very well defined quality, since movie quality is highly variable and steadily improving. It should be noted that 35-mm amateur slides have more than twice the area of movie film. The 1125-line system does not have nearly the resolution of amateur slides.

[5] FM is used today for satellite TV transmission. The rf bandwidth of the satellite channel is usually four to five times that of the basebandwidth of the TV signal. It is quite likely that digital transmission will be used for DBS in the future.

Evidently, it was deemed impractical to use such wide transponder channels. Therefore, a transmission system, MUSE [A.2], was developed so that a compressed version of the signal could be transmitted in a single normal satellite channel. This system was announced in 1984. MUSE has been demonstrated many times by DBS, and was also demonstrated in terrestrial UHF service in Washington in January 1987, using two adjacent channels, for a total of 12 MHz rf bandwidth. A MUSE DBS system is operating eight hours per day in Japan at the present time. The cheapest receivers are about $8,000, so that there are few sets in the hands of the public.

During the period when most Americans thought that compatibility was essential, the National Association of Broadcasters (NAB) and the Association of Maximum Service Telecasters (MST) urged NHK to develop systems that were compatible with (i.e., could be received by) NTSC receivers and that would be suitable for the 6-MHz channel-allocation scheme used in the US. Several such systems, called the "MUSE Family," were demonstrated by computer simulation at the NAB convention in April 1988 and in hardware at NAB in April 1989.

System and equipment developments were paralleled by efforts to have the "studio" system adopted as an international standard for program production and international exchange, but this effort failed. There was intense opposition in Europe, largely on protectionist grounds, and there was enough objection in the US to halt efforts here as well. With the advent of the proposed digital systems, the standardization effort, for the most part, has been abandoned.

A2.2 The "Studio" System

A definitive description of the original 1125-line system was given by NHK in a detailed technical report issued in 1982 [A.3]. It was stated that the system parameters were chosen to give a more psychologically satisfying viewing experience in which the viewing angle would be large enough to give some sense of depth (30°) and the resolution high enough so that at this viewing distance (about three times the picture height or 3H) the image would not be noticeably blurry. A wider screen (5:3 rather than NTSC's 4:3) was felt to be very important for an improved sense of realism.[6]

The validity of these experiments is open to question. The relationship of the desired aspect ratio to the picture material was not studied at all. The experiments were carried out with synchronized interlaced scanning of camera and display and with no storage in the system, an arrangement most unlikely to be used in any modern television system. Up to and including the date of the 1982 report, the concept of a studio standard, as distinct from a transmission or display standard, was never mentioned.

---

[6] When widescreen movies were introduced in the 1950s, 2.35:1 was the favored aspect ratio, not only to enhance the viewing experience, but also to be as different from television as possible. At the present time, 1.85:1 is the most common film aspect ratio, partly because aspect ratios much wider make cropping for 4:3 television much more damaging to the artistic effect. Note that actual practice varies a great deal. Typical TV viewing angles are much smaller than typical movie viewing angles.

Actually, as in any system design, the final parameters were the result of compromises between ideal and affordable values. In this case, widening the screen and increasing the viewing angle as well as the resolution all lead to wider video bandwidths. When the bandwidth is increased, camera sensitivity and signal-to-noise ratio (SNR) go down, costs go up, and fewer programs can be accommodated within an overall spectrum allocation.

Generally speaking, the NHK choices were reasonable for the intended purpose. There were two questionable decisions, however. One was to use interlace and the other was to use vestigial-sideband transmission (VSB) for AM transmission. The use of interlace is justified only on the basis that HDTV is to be a straight-through system without any storage. As a result, the cameras, channel signal, and display are all synchronous, just as in the NTSC monochrome standard adopted in 1941. Even in 1970, many would have argued that progressive scan would give much better results and would simplify transcoding between various standards. Twenty years later, it appears that all new TV systems will use frame stores, so that the argument for interlace is now even weaker.[7]

Since the "studio" system was originally designed for FM satellite transmission, it is possible that the decision to use VSB for AM applications simply did not receive enough attention at the time. At present, it appears to be a simple error. Double-sideband quadrature modulation, as is used for the color information in NTSC, is much more effective. For example, MUSE would require an rf bandwidth of just over 8 MHz with DSBQM, whereas with VSB, there is considerable doubt as to whether the claim that it will fit within 9 MHz is actually justified. The latter point became moot when the FCC ruled against 9-MHz systems.

A2.3 MUSE

The "studio" system required a bandwidth of 32.5 MHz – 20 for luminance and 7 and 5.5 MHz for the two chrominance signals. To fit into one normal FM transponder channel, a bandwidth reduction of at least 4:1 is needed. MUSE accomplishes this by two methods. Halving the diagonal resolution by means of transmitting only every other signal sample, interleaved line to line so as to give a 45° sampling pattern, provides a factor of 2. Another factor of two is gained by reducing the frame rate to 15 fps by sending alternate samples on alternate frames.[8]

At the receiver, a clever motion-adaptive interpolation method completely eliminates the sampling structure due to the effective 4:1 interlace and also eliminates much of the blurring of moving objects that would be expected with nonadaptive interpolation. The basic scheme is to interpolate temporally in the

---

[7] Some workers still insist that a progressive-scan HDTV camera is beyond the state of the art, although the Europeans have successfully built such cameras using quincunx (45° offset) sampling.

[8] MUSE receivers must use frame stores to derive a 60-field display from the transmitted 15-frame signal. This removes the principal obstacle to using progressive scan in the studio system, making the use of interlace even more questionable.

stationary image areas and spatially in the moving areas. This technique reduces the resolution of moving objects by a factor of two. The effect would be most noticeable when the cameras were panned. The original system used one motion vector for panning, the receiver performing *motion-compensated* interpolation. MUSE pictures are better than might be expected with this scheme, and it is possible that more motion vectors are now being used. It is noteworthy that in the Sony commercial 1125-to-PAL transcoder, an independent vector is used for each 48 × 32-pel block [A.4].

MUSE pictures look very good. Although there is some loss of sharpness with motion, the reduction in picture quality is much less than might be expected with a 4:1 reduction in the transmission rate of image samples. It may well be that limited camera-tube resolution is masking some of the loss in resolution. The lack of interline flicker due to interlace (interline flicker is often noticeable in NTSC) gives strong evidence for this speculation. If this assumption is true, then the difference between the "studio" system and MUSE will eventually become more evident with the development of better camera tubes. Figure A.2 shows the resolution of both the "studio" system and MUSE.

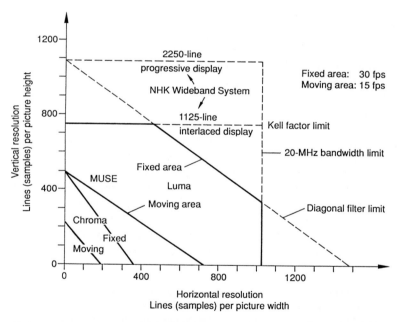

Horizontal resolution
Lines (samples) per picture width

Fig. A.2. Resolution of the NHK Wideband System and MUSE. Both systems have about 1070 active lines, so that the resolution on an interlaced display is about 750 lines/picture height. The exact horizontal resolution of the wideband system is governed by camera resolution and bandwidth assigned to each component. At bandwidths of 20, 7, and 5 MHz, this comes out to 1007, 353, and 252 for L, $C1$, and $C2$ respectively. The use of a diagonal filter and $4\times$ subsampling for bandwidth compression in MUSE lowers the diagonal resolution as shown. The moving-area horizontal resolution is also halved to give the results indicated in the diagram. The static luminance resolution is further limited by the camera and display, but the other resolutions are set by the system parameters

285

A2.4 Systems Specifically Designed for US Broadcast Use

Since the original development work was intended for DBS use, and not to replace NTSC and PAL in terrestrial broadcasting, no attention was given in Japan to conforming to terrestrial channel widths used in the US. Likewise, the overwhelming preference of US broadcasters for a compatible system seems not to have been made evident to NHK until the specific request of NAB and MST to develop such systems.[9] For this and other reasons, the Japanese were never really ahead of the rest of the world in system design, although they still are in the design of HDTV equipment. They never seem to have appreciated the difficulties of transmitting in the terrestrial channel, and they were apparently taken entirely off guard by the sudden emergence of all-digital systems.

A2.4.1 Narrow MUSE

This is a version of MUSE in which the transmitted signal is 750 lines, 2:1 interlace, 30 fps, with a bandwidth of 6 MHz. The main performance difference from the original version of MUSE (now called MUSE-E) should be a reduction in diagonal resolution and some loss of vertical resolution. As actually exhibited, the motion compensation did not seem to be as effective, but no doubt this deficiency has been removed. The 750-line signal is derived from an 1125-line source and reconverted to 1125 lines in the receiver for display.[10] A receiver capable of handling NTSC and Narrow MUSE as well as MUSE-E would cost little more than the normal MUSE receiver.

A2.4.2 NTSC "MUSE"-6

Compatible systems must accommodate the difference in aspect ratio between NTSC and HDTV, and this may well be one of the most difficult problems. Certainly it has been the subject of a great deal of shallow thinking. NHK showed two different methods of accomplishing this. In their "top-and-bottom-mask" method, 25% of the theoretical picture height is left empty of picture material on the NTSC receiver, so that the remaining image area is 16:9. Thus the image content and shape shown on the old and new receivers are the same. The bar areas can be used for enhancement information. This is the method once proposed for

---

[9] In my opinion, the insistence on receiver compatibility by American and European broadcasters, although conforming to their short-term interests, was never in their long-term interests. Compatible systems cannot possibly provide the high channel efficiency or resistance to analog channel impairments that are now available in completely new systems, since the deficiencies of NTSC and PAL in this regard are inherent in their design. If broadcast TV cannot compete in picture quality provided by alternative media, then as these media become able to provide programming of equal attractiveness as that of the broadcasters, the latter will lose audience share and eventually go out of business. With various rulings of the FCC, and now with the proposal for digital transmission, broadcasters seem to have reluctantly accepted simulcasting. The moment of truth, however, will not come until the date arrives for the final abandonment of NTSC.

[10] This results in a higher Kell factor than if 750 lines had been used in camera and display. NHK claims that, as a result, the vertical resolution is almost as high as that of MUSE-E.

the MIT-RC system and is much like the "letter-box" format used in Europe for wide-screen movies shown on TV. It is also used in PALplus. (See Sect. 8.6.4)

The "side-panel" method, like that of the Sarnoff system (ACTV), transmits information for the extra picture area by hiding it within the NTSC signal. The extra area encompassed within the side panels is not seen on the NTSC receiver. It therefore must be devoid of significant picture information, greatly reducing the visual impact of the wider screen. NHK advocates the bar method rather than the side-panel method, partly for this reason and partly because of their experience that the seams between the side panels and central section were often made visible by channel impairments.

The receiver-compatible NHK systems are not MUSE systems at all, since they do not use subsampling. Several methods are used to hide information, in addition to placing it in the top and bottom bars. Some is hidden in the vertical and horizontal blanking intervals. High-frequency enhancement information is transmitted at a lower frame rate, only in the stationary areas of the image, by multiplexing it with the upper half of the NTSC luminance and chrominance signals. This is an indirect way of using the "Fukinuki hole," [A.5] the portion of 3-dimensional frequency space diametrically opposite to that occupied by the color subcarrier and its side bands.

A2.4.3 NTSC "MUSE"-9

This system is much like the previous one, but in addition uses another 3-MHz channel for augmentation information. In the version shown, this extra capacity was used for digital audio and for increasing the vertical and horizontal resolution in moving areas, which are of quite low resolution in the 6-MHz version.

An interesting point is that NHK itself places, in rising order of quality, NTSC MUSE-6, NTSC MUSE-9, Narrow MUSE, and MUSE-E. Thus Narrow MUSE, with a 6-MHz bandwidth, is better than NTSC MUSE-9, which requires 9 MHz. This shows quite clearly the penalty that is paid for building NTSC compatibility into any new television system.

As exhibited at the 1989 NAB meeting, Narrow MUSE was distinctly inferior to MUSE-E in resolution, and the moving-area interpolation did not seem to be working properly. Both NTSC-compatible systems showed such serious defects in moving areas that it is hard to understand why they were exposed to public view. Table A.1 and Figs. A.3 and A.4 show some aspects of the MUSE Family [A.6].

A2.5 Conclusions: Japan

The 1125/60 "studio" system makes rather good pictures, but the use of interlace is a major drawback in other than the originally intended applications. MUSE, while quite sophisticated, is closely tied to the original system. Although clearly inferior in quality, it makes very good pictures by DBS under conditions of adequate CNR. Since it requires at least 9 MHz bandwidth, it is probably permanently

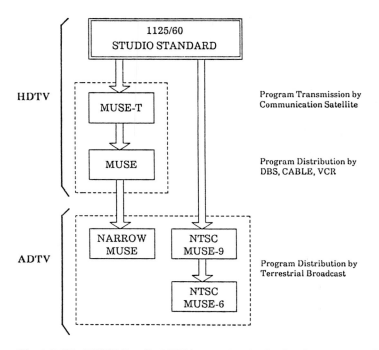

Fig. A.3. The MUSE Family. NHK has proposed a family of systems that fit within the US 6-MHz channel and/or are compatible with NTSC receivers. All these systems are derived from the wideband system by filtering and subsampling. MUSE-T is a higher-resolution version of standard MUSE, now sometimes called MUSE-E. Narrow MUSE is a 6-MHz version that achieves its compression by reducing the number of scan lines, and thus, the vertical resolution. It is otherwise quite similar to MUSE and can be processed almost entirely in standard MUSE coders and decoders. The NTSC-compatible systems are not MUSE at all, but utilize a variety of methods to add information to a standard NTSC signal. NTSC MUSE-6 uses a single channel, while NTSC MUSE-9 uses an additional 3-MHz augmentation channel. NHK rates the quality of NTSC MUSE-9 as slightly below Narrow MUSE

excluded from consideration for over-the-air use or for cable. Narrow MUSE has been tested under the FCC program, but it is highly unlikely that it can meet all the interference requirements. The 6-MHz compatible versions developed specifically for US use have all been abandoned due to subsequent FCC rulings against such systems. (See below.)

Although the Japanese are persisting, for the time being, with their very elaborate plans to encourage the use of MUSE in Japan, evidence has been accumulating, such as the high prices and very small sales of receivers, that it will be abandoned eventually. While many observers are ascribing this to the rising interest in digital TV in the US and Europe, it is more probable that the fundamental reason is that those two regions have decisively rejected the Japanese systems. It would be most unusual for modern Japan to engage in large-scale manufacture of anything that could not also be sold abroad with minimal modification. It has been recent Japanese manufacturing strategy to

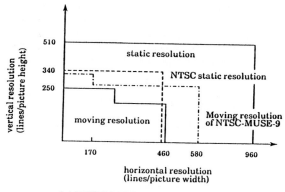

(a) NTSC-MUSE-6

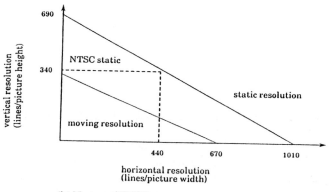

(b) Narrow-MUSE

Fig. A.4a,b. Resolution of the MUSE Family. These NHK figures show their assessment of the moving and static resolution of the MUSE Family. Note that Narrow MUSE is much like MUSE-E except for the lower vertical resolution. The odd shapes of the resolution outlines for the NTSC-compatible systems is due to hiding various pieces of enhancement information in various places in the NTSC signal

achieve economies of scale overseas, even if at small profit, and to make most profits at home in what amounts to a captive market. On that basis, Japan is most likely to adopt the American system, if one is selected that uses good technology and seems to have a chance of success in the marketplace.

A3. The United States

The most striking feature of the HDTV scene in the US is that the domestic consumer-electronics industry, whose research and product-development expenditures would normally be expected to produce systems particularly adapted to

Table A.1. The MUSE Family. There are many variants here because of the necessity of conforming aspect ratio. NHK, showed two different methods. The "top-bottom mask" method is generally called "letterbox" format in this country. Blank bars are left at top and bottom so that the same image material is shown on the two kinds of receivers. In the "side-picture" method, generally called "side-panel" here, the sides of the wide-screen (16:9) HDTV picture are cut off for display on the NTSC (4:3) screen

	System	Bandwidth	Compatibility	Aspect ratio expansion	Resolution (lines/picture width)	HDTV/*2 ADTV aspect ratio	NTSC*3 aspect ratio
HDTV	MUSE	9MHz	no*1	-	1020	16:9	16:9/4:3
ADTV	NARROW MUSE	6MHz	no*1	-	1010	16:9	16:9/4:3
	NTSC-MUSE-9	9MHz	yes	top-bottom mask	960	16:9	16:9
					900		
				side picture	900		4:3
	NTSC-MUSE-6	6MHz	yes	top-bottom mask	960	16:9	16:9
					680		
				side picture	900		4:3

*1 An experimental low-cost downconverter has already been realized.
*2 Aspect ratio displayed on HDTV and ADTV receivers.
*3 Aspect ratio displayed on NTSC receivers.

the many unique features of American television broadcasting, has been decimated by foreign competition.[11]

The result was that no work on HDTV was being done in the US before the middle eighties, leaving the field open to the Japanese. There was no denying the high image quality produced both by the 1125/60 "studio" system as well as MUSE, although some did voice objections to motion blurring in the latter. Consequently, in the absence of American alternatives, there was a substantial amount of support in the US for adopting 1125/60 as a production and international exchange standard. The question of transmission in the American environment was put aside. However, when it appeared that HDTV might first be introduced by alternative media and that current over-the-air broadcasters might need extra spectrum to compete, the industry petitioned the FCC for a formal inquiry. It began in September 1987 with the appointment of an Advisory Committee made up primarily of TV industry executives. Congressional hearings were also held, at which the trade implications of adopting a foreign standard surfaced. When transmission began to be considered, the broadcasters' preference for compatible systems resulted in a turning away from the Japanese approach. Ironically, it was not the absence of compatibility that eventually defeated the Japanese proposals.

---

[11] Some of this has been "fair," in that it represents the results of superior product design, but the most important factor was systematic and illegal dumping [A.7]. The departure of a number of American companies from this low-profit field for more lucrative areas such as military and medical electronics is another. In any event, at this date, only one large American-owned company, Zenith, still makes TV receivers, and virtually all of those are assembled in Mexico.

290

A3.1 The FCC Inquiry

Although originally intended simply to assess the probable impact of HDTV on TV broadcasting in the US, the Inquiry has developed into a selection process for an American system for terrestrial transmission service. Under the general supervision of the FCC, the Advisory Committee has set up a testing procedure using the facilities of the Advanced Television Test Center, which in turn was set up and financed by the TV industry. Systems were to be tested, in full hardware implementations only, during a period of about one year beginning in 1990, with the FCC due to make a selection in 1992. This schedule slipped many times. It is now hoped that laboratory testing will be completed by fall of 1992. A field test of the most successful system is planned for the end of 1992 and a decision by the FCC is hoped for by the end of 1993. This seems overly optimistic, particularly in view of the digital proposals, which seem to require much more development.

The decisions on and arrangements for testing, in the author's view, have had very unfortunate consequences. The levy of a substantial testing fee cut down the field of proponents drastically.[12] The requirement of hardware testing ensured that building the hardware (which turned out to be very complex) would be the principal preoccupation of the participants. In view of their very poor financial condition, they were precluded from spending much effort examining alternatives. This has had its most important effects on the digital proponents, who have apparently put very little emphasis on channel coding.

In September 1988, the FCC issued a Tentative Decision in which it was stated that the government would set standards for terrestrial broadcasting only, the other media being free to adopt whatever standards they pleased. The most important part of the ruling stated that, at least for an initial period, all programs broadcast in HDTV must be made available to NTSC receivers and that no more than 12 MHz could be used to serve both kinds of receivers. The three methods available to do this are single-channel compatible systems such as ACTV [A.8], augmentation systems such as that of Philips [A.9], and single-channel noncompatible systems that would utilize simulcasting [A.10, 11].

This decision was widely misreported as indicating that the FCC had ruled in favor of compatible systems. In fact, nearly the reverse was true. During the Inquiry, it had become evident that, in order to find enough channel capacity to accommodate both NTSC and HDTV during a transition period, the HDTV and/or augmentation signals would have to operate at much lower D/U ratios than NTSC, which means that they could not use the NTSC signal format. The decision did eliminate standard MUSE, which would have required at least 15 MHz for both services.

---

[12] It is true that a number of bizarre proposals were made at first, and some weeding out was required. However, the impact of the testing decisions was mainly on the academic laboratories, which had generated many fruitful ideas.

A3.2 Originally Proposed Systems

At a time when no American systems had been proposed, the Association of Maximum Service Telecasters (MST) and the National Association of Broadcasters (NAB) requested the Japanese to produce versions of MUSE that were compatible with NTSC receivers and/or were compatible with the American spectrum allocation of 6 MHz for each channel.[13] As a result, the MUSE family, discussed in Sect. A2.4, was developed. NHK had planned to present two of these systems for testing, a 6-MHz version of standard MUSE, called Narrow MUSE, and a compatible system as well. The latter was later dropped.

Aside from the Japanese entries, the largest amount of funding had gone to ACTV, the single-channel compatible system developed at the Sarnoff Laboratories. (See Sect. 8.6.3.) This system was eventually implemented in hardware and tested early in 1992, before being withdrawn. ACTV-II, like the first version but with a second channel for enhancement information, was proposed at one time, but was never seriously pursued and was quickly dropped. The compatible augmentation system of North American Philips was also a contender. Faroudja Laboratories had proposed an enhanced NTSC system [A.12]. Both of these systems were later withdrawn from consideration. In addition to NHK's Narrow MUSE, single-channel noncompatible systems using hybrid analog/digital transmission were offered by Zenith and MIT.

Although noncompatible systems that would require simulcasting were originally derided by the entire American TV industry except Zenith, the effect of the Inquiry, perhaps unintended, was to change many opinions.[14] It is now generally accepted that true HDTV is possible within 6 MHz if an entirely new system is designed, and that such new systems could use much lower transmitter power than now needed for NTSC, making simulcasting less expensive. The importance of achieving much better interference performance for the purpose of higher spectrum efficiency is now widely understood, especially within the FCC itself. There seems to be a consensus that this cannot be done with NTSC. As a result, in a most remarkable move, North American Philips and Thomson Consumer Electronics (formerly the GE and RCA consumer-electronics divisions) agreed to cooperate with the National Broadcasting Corporation (NBC) and the Sarnoff Laboratories to develop a simulcast system. The ultimate effect of this decision was to replace the Philips augmentation system with an all-digital system based on MPEG.

In March 1990, the FCC issued another ruling that affected the competition for the US transmission standard. It stated that augmentation systems would

---

[13] Most of the effort subsequently expended by NHK was wasted. The FCC eventually decided that NTSC was to be abandoned after a transition period of 15 years, eliminating entirely the role of compatible and augmentation systems.

[14] Zenith deserves a lot of credit for its simulcasting proposal, radical as it appeared at the time to most TV industry executives. Progress towards a high-quality system would have been greatly delayed without this initiative.

no longer be considered, as they were not sufficiently bandwidth-efficient.[15] It also said that it would issue a standard for EDTV, *if ever*, only after an HDTV standard was selected.[16] Since all the compatible systems then proposed were EDTV and not HDTV,[17] that meant that the FCC had concluded that the transmission standard to be selected would be noncompatible with NTSC, although this was not stated explicitly. A corollary is that simulcasting would be used to service existing NTSC receivers. Of course, such a decision depends on the successful operation of the simulcast systems, which had yet to be conclusively demonstrated.

As a result of these rulings, Faroudja dropped out of the competition, the NHK compatible system was withdrawn, and all the augmentation systems were abandoned. Surprisingly, ACTV remained alive, although its only hope appears to rest on the possibility that none of the simulcast systems will work.

A3.3 The General Instrument Proposal

The GI proposal, referred to previously in Sect. 8.13.1, was made at the end of June 1990, on the last day for accepting new proposals. Regardless of how it eventually fares in the competition, it permanently changed the character of the search for an advanced television system for the US, probably for the better. It has undoubtedly also affected the outcome in Japan and Europe. As remarkable as the all-digital technology seemed,[18] the most surprising aspect of the proposal was that it was almost instantaneously widely accepted in the industry, even by many who earlier had supported compatibility or had been advocating using the NHK system.

What was revolutionary about the proposal was digital transmission, not digital source coding.[19] The basic idea in the source coding – application of the DCT to the motion-compensated frame-to-frame prediction error as in MPEG – had been worked on in many laboratories. The general principles of digital communications had been known about 50 years, but it had almost never been used for broadcasting.[20] No new developments in digital transmission were reported

---

[15] This should not have taken an FCC ruling. Augmentation systems, which transmit NTSC in one channel and enhancement information in a second channel, combine all the worst features of new and old systems. In addition, they preserve the deficiencies of NTSC forever. It is a mystery to the author how this simple conclusion was not evident to everyone at an early date.

[16] EDTV is generally defined as vertical resolution between 500 and 1000 lines, while HDTV usually means vertical resolution of at least 1000 lines.

[17] It is the author's opinion that it is not possible to design a true HDTV system that is also compatible.

[18] In the press releases, GI referred to its system as "all-digital and all-American." It is worth noting that both of these descriptors are slogans. Commendable though these properties may be on other grounds, they do not guarantee technical merit.

[19] Hardly any journalists understand the distinction. As a result, there has been a large amount of erroneous reporting. Even worse, many in the industry also seem unaware of the difference, which the author fears will lead to faulty decision-making.

[20] One such system is the Joint Tactical Information Distribution System (JTIDS) in use by the American military for communicating among ships, planes, and ground stations. Although JTIDS

as the basis of the decision to use this method, nor was any reference made to the European DAB work. In fact, the channel coding method was essentially the same as used in telephone or satellite service, where the channels have much more benign characteristics.

The basic parameters of the original proposal are shown in Table A.2. [A.13] The use of 16-QAM leads to a gross data rate of about 20 Megabits/s, and, with the channel coding then proposed, a threshold of 19 dB. It is clear that, because of the rather low data rate possible in the terrestrial channel, a very high compression ratio is needed. After allotting data for audio, error correction, and other purposes, less than 14 Megabits/s is available for video. This represents a compression ratio of about 50, something never before attempted. Even allowing for some as yet undisclosed advances in MPEG-type source coding, such high ratios are bound to produce a good deal of scene dependence. Whether the quality will remain adequate (it certainly will be good enough if there is little motion) to meet the demands of broadcasters remains to be demonstrated.

Another aspect of such systems is the abrupt threshold (now often called the "cliff effect") which is likely to lead to erratic reception in the fringe area. This is in contrast to analog systems, where picture quality degrades slowly with poorer reception conditions.

Another aspect of digital transmission to be taken into account is that, even to achieve the quite modest rate of 20 Megabits/sec, nearly perfect channel equalization, including the elimination of echoes, is required. Whether such equalizers can work properly under practical conditions of large and rapidly changing multipath remains to be seen, as is their effect on the SNR. In addition, since equalizers cannot cope with unlimited echoes, rather good antennas are needed.

Finally, systems that deliver the same data rate to all receivers in the viewing area are spectrum-efficient only in the fringe area. They necessarily waste channel capacity in the central cities, where it is in the shortest supply. To deal with this problem requires multiresolution source coding as well as channel coding that permits each receiver to recover an amount of data related to its local SNR. (See Sect. 8.13.4.) Of course, if it were possible to transmit HDTV in, say, 1 Megabit/s, this argument would have little force. No one has yet claimed that, however.

It is quite interesting to see the changes made by GI 14 months later. These are shown in Table A.3. [A.13] It appears that one motivation for these changes is that much better pictures can be had if the video data rate is raised somewhat. In addition, a threshold of 12 dB, not 19 dB, is needed to meet the FCC interference requirements. These are conflicting requirements, and were met by GI by proposing two versions of the system. The 16-QAM version uses more error-correction data and a more sophisticated code (trellis coding plus Reed-Solomon

---

is not classified, there seems to be no regularly published information available. Much of the data to be transmitted is numerical, so digital transmission is natural. Voice service is also provided. Spread spectrum is used for protection against multipath and for encryption. The data rate as compared to the bandwidth is much too low for HDTV transmission in 6 MHz.

Table A.2. General Instrument Proposal, 8 June 1990. As originally proposed, the GI scheme used 16-QAM with a total data rate of 19.43 Megabits/s, a video data rate of 13.83 Megabits/s, and a threshold (not listed) of 19 dB. Error correction consumed about 18.5% of the available communications capacity. The threshold was too high to meet the FCC requirement to find an extra channel for all license holders

Parameters	Value
VIDEO	
Aspect Ratio	16:9
Raster Format	1050/2:1 Interlaced
Frame Rate	29.97 Hz
Bandwidth	
Luminance	22 MHz
Chrominance	5.5 MHz
Horizontal Resolution	
Static	660 Lines per Picture Height
Dynamic	660 Lines per Picture Height
Horizontal Line Time	
Active	27.18 μsec
Blanking	4.63 μsec
Sampling Frequency	51.8 MHz
Active Pixels	
Luminance	960(V) × 1408(H)
Chrominance	480(V) × 352(H)
AUDIO	
Bandwidth	15 kHz
Sampling Frequency	44.05 kHz
Dynamic Range	85 dB
DATA	
Video Data	13.83 Mbps
Audio Data	1.76 Mbps
Async Data and Text	126 Kbps
Control Data	126 Kbps
Total Data Rate	15.84 Mbps
TRANSMISSION	
FEC Rate	130/154
Data Transmission Rate	19.43 Mbps
16-QAM Symbol Rate	4.86 MHz

error correction) to reduce the threshold to 12 dB.[21] To raise the video data rate to 17.5 Megabits/s, a 32-QAM version was introduced having a gross data rate of 24.4 Megabits/s and a threshold of 16.5 dB. Obviously, these two versions cannot be used at the same time; it is impossible to use 17.5 Megabits/s for video and at the same time have a 12 dB threshold. Of course, it is true that it is only in a few very crowded big cities that 12 dB is really needed.

---

[21] One result of such coding is to produce an extremely sharp threshold. The corrected bit error rate changes by a factor of 10,000 with a .1 dB change in SNR!

Table A.3. General Instrument Proposal, 22 August 1991. In the revised proposal, the threhold was lowered to 12.5 dB by using 31.6% of the gross capacity for error correction. In spite of a better audio compression system, this reduced the video data rate to 12.59 Megabits/s, which is evidently marginal. An alternative configuration was added using 32-QAM. This yielded 17.47 Megabits/s for video, but raised the threshold to 16.5 dB. It also reduced the NTSC protection by 5 dB

Operating Mode	16-QAM	32-QAM
VIDEO		
Raster Format	1050/2:1 Interlaced	1050/2:1 Interlaced
Aspect Ratio	16:9	16:9
Frame Rate	29.97 Hz	29.97 Hz
Bandwidth		
Luminance	21.5 MHz	21.5 MHz
Chrominance	5.4 MHz	5.4 MHz
Active Pixels		
Luminance	960(V) × 1408(H)	960(V) × 1408(H)
Chrominance	480(V) × 352(H)	480(V) × 352(H)
Horizontal Resolution		
Static	660 Lines per Picture Height	660 Lines per Picture Height
Dynamic	660 Lines per Picture Height	660 Lines per Picture Height
Sampling Frequency	53.65 MHz	53.65 MHz
Colorimetry	SMPTE 240M	SMPTE 240M
Horizontal Line Time		
Active	26.24 μsec	26.24 μsec
Blanking	5.54 μsec	5.54 μsec
AUDIO		
Number of Channels	4	4
Bandwidth	20 kHz	20 kHz
Sampling Frequency	47.2 kHz	47.2 kHz
Dynamic Range	90 dB	90 dB
DATA		
Video Data	12.59 Mbps	17.47 Mbps
Audio Data	503 kbps	503 kbps
Async Data and Text	126 kbps	126 kbps
Control Channel Data	126 kbps	126 kbps
Total Data Rate	13.34 Mbps	18.22 Mbps
TRANSMISSION		
FEC Rate	6.17 Mbps	6.17 Mbps
Data Transmission Rate	19.51 Mbps	24.39 Mbps
QAM Symbol Rate	4.88 MHz	4.88 MHz
Adaptive Equalizer Range	−2 to 24 μsec	−2 to 24 μsec
SYSTEM THRESHOLD		
Noise (C/N)	12.5 dB	16.5 dB
ATV Interference (C/I)	12.0 dB	16.0 dB
NTSC Interference (C/I)	0.0 dB	5.0 dB

A3.4 Other Digital Proposals

Following in GI's footsteps, three additional digital proposals emerged. One, from the Advanced Television Research Consortium (North American Philips, Thomson Consumer Electronics (the combined former consumer-electronics divisions of RCA and GE) NBC, Sarnoff, and Compression Labs) is called MPEG++. MIT dropped its hybrid system, made an agreement with GI, and proposed a progressively scanned digital system. Zenith dropped its hybrid system, joined with AT&T, and also proposed an all-digital system. All four of these systems will have been tested by the fall of 1992 [A.14].

The similarities of these systems, which are listed in Table A.4, are so great that a single overall block diagram would serve for all four. The main differences are shown in Table A.5. Two systems use progressive scanning and two use interlace, but this is not a vital difference, as all systems could be configured to use either.[22]

Table A.4. Common Features of the Proposed Digital Systems. The four all-digital systems proposed for use in the US have many features in common, in addition to using the same overall compression method. This and the following table, which were distributed to the systems proponents for their comments, are based on information made public as of May 1992. Zenith* appears not to use trellis coding. Resynchronization speed** varies somewhat, system to system

---

COMMON FEATURES of the PROPOSED DIGITAL SYSTEMS

- 16×9, all-digital, SMPTE 240M color
- Motion estimation by block matching, motion-compensated prediction, discrete cosine transform (DCT) applied to prediction error
- Transmission of motion vectors
- Adaptive selection and quantization of DCT coefficients
- Scrambling (interleaving)
- Reed-Solomon forward error correction; trellis coding*
- Channel buffer control
- Error concealment
- Use of data cells; rapid resynchronization**
- Use of automatic channel equalization
- New kinds of impairments
- Extremely sharp threshold(s)
- 24 f/s film mode
- Claims for picture quality, coverage area, reliability, flexibility, usefulness in alternative media

---

All systems have two modes of operation that give either lower quality and lower threshold or higher quality and higher threshold, but the GI and MIT systems use manual selection.[23] The other two divide the data into two streams (high and low priority), so that a lower-priority ("viewable") image is available when only the lower-threshold data is recovered. This somewhat extends the range at which some service is provided. Although this is sometimes said to eliminate the cliff effect, it really provides two very sharp thresholds instead of one. ATRC makes good use of this scheme by using two different carriers, so that the high-priority data is in a band below the picture carrier in a cochannel

---

[22] Of course, it is very important which is selected. The computer and electronic imaging industries, which have been advocating interoperability, will be very disappointed if interlace is used.

[23] These systems appear to use identical channel coders. The hardware for both is being built by GI.

Table A.5. **Differences Among the Proposed Digital Systems.** The main system differences are shown here. Most of these features could be used in the other systems. E/C: error correction; C/I: carrier-to-interfenceratio; FIR: finite impulse response; GCR: ghost-correctionreference; IIR: infinite impulse response; VC: vector coding; VSB: vestigial sideband. The transmitter power* figures, which represent effective radiated power, are only approximate, since all proponents varied them to achieve the desired coverage area.

DIFFERENCES AMONG THE PROPOSED DIGITAL SYSTEMS

System	GI	AT&T/Zenith	ATRC	MIT
Modulation Method	16-QAM 32-QAM	2-VSB 4-VSB	16-QAM 32-QAM	like GI
Raster	960 × 1408	720 × 1280	960 × 1440	720 × 1280
Frame Rate, f/s	30, interlaced	60, prog	30, interlaced/prog	60, prog
Pixel Rate, Msmpls/s	40.6	55.3	41.5	55.3
Raw Data Rate, Mb/s	19.5 24.4	10.8 21.0	19.2 24.0	21.0 26.4
E/C Video Rate, Mb/s	12.6 17.5	8.5 16.9	12.5 17.7	13.6 18.8
Adaptive Selection Coding Method	MPEG style	VC	MPEG	VC
Threshold Extension	none	2/4 VSB	2 carriers	like GI
Threshold CNR, dB	12.5 16.5	10 16	8/13 11.1/16.1	like GI
Transmitter Power* Relative to NTSC, dB	−14.5	−14.5	−12	like GI
NTSC Protection	none	comb filter	2 carriers	like GI
NTSC sync	+	++	+	like GI
NTSC C/I, dB	0 5	−6 0	−8/−4 −7/−2	like GI
Auto Equalization	FIR, blind	FIR/IIR, GCR	FIR, blind	like GI
Audio	Dolby AC2	Dolby AC2	Musicam	MIT-AC

NTSC station and the low-priority data is in a band above the NTSC carrier. The high-priority carrier is transmitted using 5 dB more power density than that of the low-priority carrier. This greatly reduces the susceptibility to NTSC interference and somewhat extends the range for low-priority reception. It is a very worthwhile feature. (Of course, it could be used in the other systems.[24]) AT&T/Zenith use a comb filter for interference protection, but it reduces the usable SNR by 3 dB. The filter can be removed if not needed; the receiver responds automatically, thus avoiding the SNR loss if NTSC interference is not present. The MIT and GI systems have no special NTSC interference protection.

Two of the proponents have abandoned MPEG diagonal scanning to code the adaptive-selection data, instead using a unique vector code for each possible pattern of selected coefficients. The amplitudes of the coefficients are transmitted separately.[25] All of the systems use some kind of scrambling, so that NTSC interference appears as random noise. Since the amount of noise depends on the

---

[24] Narrow MUSE, as submitted for testing, also has a split spectrum that avoids the NTSC picture carrier, greatly improving its cochannel interference performance. In addition, the lower-frequency components can be separately decoded for use in low-cost low-performance receivers.

[25] This means that hybrid modulation, with its higher channel efficiency, could be used rather than all-digital modulation.

picture content, a situation may arise in which reception comes and goes on an HDTV receiver along with the changing video content of an NTSC cochannel interferer.

In spite of the obvious similarities of the systems, there may well be differences in the source-coding effectiveness. For this reason, since picture quality is likely to be markedly dependent on data rate, and since the data rate is certainly marginal, a rather small difference in compression ratio may make a noticeable difference in picture quality for certain critical scenes. If the systems worked properly in all other respects, this might be crucial to the selection process.

A3.5 Problems of All-Digital Systems

The most obvious deficiency of the digital proposals is that the channel-coding scheme is vastly inferior to OFDM, as developed in Europe. The performance of automatic equalizers, no matter how well designed, is bound to be inferior to that of a system that is inherently immune to multipath, even with omnidirectional antennas, up to a certain time spread. Even with expensive highly directional antennas, it is likely that reception will be erratic in the fringe area and under any condition in which there are large rapidly moving echoes.

The tradeoff between picture quality and cochannel interference in all the systems is quite marginal, even if they work precisely as predicted. The clear solution to this is to use a soft threshold and multiresolution source coding. Then, at least in much of the reception area, a high-enough data rate would be achieved to get very good pictures. For the sake of coverage, quality would have to be compromised only in the fringe areas or where NTSC interference was severe.

Related to this tradeoff is spectrum efficiency. The dual-threshold schemes of ATRC and AT&T/Zenith are better than nothing, but they do not improve picture quality at all in areas where (because of the higher SNR) the channel capacity is as much as five times that at the fringe. Systems that do take advantage of this higher capacity could provide better average picture quality with no more than half the bandwidth

The most recent description of the MIT system [A.14] proposes a form of single-frequency network for threshold extension. However, the performance of the channel equalizer does not appear to be adequate for such service unless highly directional antennas are used. In the cellular arrangement of low-power transmitters proposed in Europe, echoes can be as large as the main signal. Even if an equalizer could be made that could cope with 0-dB ghosts (Specifications of the MIT and GI equalizers have been given; they cannot deal with such ghosts.) it would greatly reduce the SNR, whereas OFDM constructively adds the echoes.

Aside from this mention in the MIT proposal, there has been little talk of single-frequency networks in the US. It remains true, however, that this is the only way known at present to get maximum spectrum efficiency. The second best way is to use a soft threshold. Faced with these two possibilities, it is hard to see how the regulatory authorities could select any of these systems, which use neither.

A3.6 Conclusions: United States

The most prominent feature of the American scene is the moribund consumer-electronics industry. As a result, an inadequate effort has been put into research and development that might support the design of a new television system intended to last for decades. This is particularly true for transmission technology. Nevertheless, a contest is being held to select a system with little regard for these conditions. In the opinion of the author, none of the proposed systems, or any combination of the proposed systems, will meet the real needs of broadcasters, manufacturers, viewers, and the regulatory authorities.

Of course, the proposed systems are a great improvement over the compatible systems that were offered earlier in accordance with broadcasters' strongly held (and erroneous) views. Zenith and General Instrument deserve much credit for helping to bring about the new climate of opinion that made it possible to give consideration to totally new systems. The Federal Communications Commission has, if anything, been in the lead in this matter. However, the mistaken idea that digital transmission would automatically solve every problem has distorted the search, and made it impossible to select a suitable system on the wished-for time scale. At the very least, European transmission technology must be given careful consideration.

Some hope for a way out of this situation is held out by a recent agreement among AT&T, Zenith, GI, and MIT, to share royalties, and possibly technology, in the event that one of them wins. ATRC, the one proponent not included in the agreement, is probably the one with the most knowledge of OFDM.

During the period that the trade implications of HDTV were in the news, many hoped that an American HDTV system might spark a revival of the domestic consumer-electronics industry. This now seems less and less likely, as the last sections of the industry have departed. It is almost certain that the system eventually used in the US will be designed in the US, but it is highly probable that the proportion of receivers manufactured in the US will be no greater than is now the case with NTSC. Evidence for this is provided by the infusion of funds into Zenith by Goldstar and the licensing agreement between GI and MIT on the one hand and Toshiba on the other. The remaining system proponent, ATRC, is primarily European-owned.

A4. Europe

Although there was originally some indication of acceptance of 1125/60 for studio use by the European Broadcasting Union, the potential economic effects eventually caused Europe to reject that system emphatically in favor of one more in harmony with its own view of the future of television. The major broadcasting and manufacturing interests agreed on a system in 1986, although all details had not been worked out by then. The principal decisions were that HDTV was to be delivered exclusively by DBS and that it was to be 'compatible' with

PAL and SECAM. [26] Thus, many viewers are likely to be watching the new broadcasts on their old receivers for a long time to come. Work on the system is being carried out under project Eureka 95, which involves both industrial and government investment. The US proposals for digital broadcasting have sparked a great deal of interest in Europe, but there has been, as yet, no official change in the originally adopted Eureka plan.

A4.1 Multiplexed Analog Components

Independently of HDTV, the Europeans had developed, primarily in connection with DBS, a number of multiplexed analog component (MAC) systems in which the composite video formats of PAL and SECAM were abandoned in favor of transmitting luminance and two color components in sequence. The time bases of these components are changed so that they can fit within the line duration in proportion to their relative bandwidth. MAC systems eliminate luminance/chrominance cross-talk and are also better suited to the triangular noise spectrum characteristic of FM. One such system, D2-MAC [A.15], also has provision for four independent channels of digital audio. Conversion to PAL or SECAM is not very difficult since the scanning standards are identical. It would be implemented by adding some circuitry to the satellite receiver required, in any case, to utilize DBS. MAC systems are also potentially suitable for digital transmission or storage in accordance with CCIR Recommendation 601, which calls for a total data rate of 216 Megabits/s.

A4.2 Overall European Strategy

A fundamental aspect of European strategy has been to prevent Japanese domination of the European market, and, preferably, of the American market as well. The decision to maintain at least the appearance of compatibility with the 625-line, 50 Hz environment, may well have been as much related to this point as to concern over the usefulness of the existing large installed base of receivers. In any event, the European decisions were as different from Japanese strategy as possible; the HDTV system was to be compatible and it was to be implemented in an evolutionary manner. In view of the implication that existing systems are not to be dropped for many years and since there are very few spare terrestrial channels available, a decision that is hard to criticize was to transmit high-definition signals solely by DBS.[27] Thus, 1250-line HD-MAC is really

---

[26] It is not really compatible, since the signal format is quite different, but the cost of transcoding is potentially quite low, so that it can be hidden within the cost of the satellite receiving equipment.

[27] It would have been possible to retain 'compatibility' with PAL receivers by using a much more sophisticated satellite channel coding system – perhaps even digital – and placing a transcoder in the satellite ground station. While this would have somewhat increased the cost of the ground station, it would have made for a much better HDTV system by removing important design constraints. In particular, there would have been no need to trade off the picture quality on the old and new receivers, which is a major limitation of HD-MAC.

compatible with 625-line MAC, and not directly with PAL or SECAM.[28] The latter are accommodated in the same way as when any MAC format is used. It is a format-conversion process in which everything about the signal is changed except the line and frame rate.

A4.3 Technological Approach

The more obvious methods by which a high-definition signal can be made compatible with existing receivers are discussed in Sect. 8.6. One is to hide augmentation information within the existing signal and the other is to utilize an additional half or full channel. A third method, proposed for the US by *Iredale* [A.16] but since abandoned, involved straightforward subsampling of an HDTV signal to obtain the normal format. For example, if an HDTV signal has exactly twice the number of lines/frame and twice the number of pels/line and the same frame rate as NTSC, taking half the pels on half the lines would give a signal with the same line and frame rate as NTSC. (Subsampling is complicated somewhat by interlace.) The principal problem with this simple-minded approach is that the subsampled signal exhibits aliasing, as the higher vertical- and horizontal-frequency components are folded back into the baseband spectrum by sampling at less than twice the bandwidth. The form of the aliasing depends on the sub-sampling grid, but it cannot be avoided entirely. In a grid such as in MUSE, four fields are used, so that 1/4 of all the pels in a frame are sampled from each field. As a result, 15-Hz flicker (12.5 Hz for PAL) appears around sharp edges. The effect is totally unacceptable, even in stationary areas [A.17].

To reduce the aliasing on old receivers, the image can be blurred by low-pass filtering, which, of course, also reduces sharpness on new receivers. The sharpness can be partly restored by high-pass filtering at the expense of an increase in the noise level. The question is whether there is a compromise that does not reduce the compatible quality excessively while at the same time provides worthwhile improvements on new receivers. The answer seems to be "no." The reason is simple and is independent of the cleverness of the system designer. The augmentation information required to make the 'HDTV' picture better than the 'PAL' picture comprises precisely the signal components that cause the aliasing on the old receivers.

A possible way out of this dilemma is to use motion-adaptive prefiltering in combination with motion-compensated interpolation in the new receivers. This is only a possible solution, as can be seen by considering the zero-motion case. If the four-field sampling pattern is a 2×2 square, the four pels of the high-resolution picture will appear in the same location on the low-resolution raster. If these pels are not identical, as required to retain the higher resolution at the HDTV receiver, flicker *must* result on the standard receiver. The motion-adaptive

---

[28] From the beginning, HD-MAC was planned as a 16:9 aspect-ratio system, but with the capability of dealing with 4:3 images by using blank side panels. Likewise, 16:9 images can be displayed on 4:3 receivers by using blank top and bottom panels, much as in the letterbox format now used for wide-screen movies.

filtering and interpolation in this case can only be parallel to the time axis and thus can only remove the flickering on old receivers at the price of reduced sharpness and resolution on HDTV receivers.

These considerations indicate a fundamental superiority of MUSE to HD-MAC. The requirement of compatibility in the latter means that the HDTV image quality must be reduced in order to minimize artifacts on the old receivers. It may well be true that the more sophisticated motion compensation in HD-MAC makes it inherently superior to MUSE, absent the compatibility requirement. Should that be the case, MUSE could be improved using similar techniques, so that MUSE, not having the constraint of compatibility, would still be superior.

A4.4 Three-Branch Motion Compensation

The algorithm now used in Eureka classifies each 16×16 block of each frame as having rapid (more than 12 pels/frame), slow (0.5 to 12 pels/frame), or zero (less than 0.5 pels/frame) motion. The slow blocks are reconstructed using motion-compensated interpolation. The velocity vectors are determined by a phase-matching technique and are transmitted in a side channel with a capacity of about 1 Megabit/s. Different prefilters, postfilters, and sampling patterns are used for the three branches. The image is reconstructed at the coder independently from all three branch signals, the one producing the smallest error being chosen in each block. The frequency response that results is shown in Fig. A.5 [A.18]. Obviously, there are many parameters that can be used for optimizing this technique, such as the filters, the accuracy of velocity information, the block size, the subsampling pattern, etc., so that the quality that may ultimately be obtained is not definite. The fact that the motion-compensation algorithm has not been settled definitely after more than four years of experience seems to indicate that no completely satisfactory scheme exists. Further evidence of a fundamental problem with the system is that the HD-MAC proponents have been extremely reluctant to give side-by-side demonstrations of the performance on 625- and 1250-line receivers.

A basic assumption that underlies the use of multiple branches depending on the amount of motion is that the spatial acuity of the human visual system depends on velocity. While this must be true to some extent (convincing data on this point is lacking), in the case of tracking of objects moving at constant velocity within the image, 200 ms or so after the beginning of movement, the required resolution is not much less than needed for stationary objects. In the original BBC demonstrations of digitally assisted TV (DATV, the technique now used in HD-MAC), the halving of resolution of objects in motion when the motion compensation was turned off was clearly visible and most annoying [A.19].

A4.5 Implementation of D2-MAC and HD-MAC

The 1986 European plan called for initiating the new services with D2-MAC high-power satellite and cable broadcasting; in accordance with this approach, a

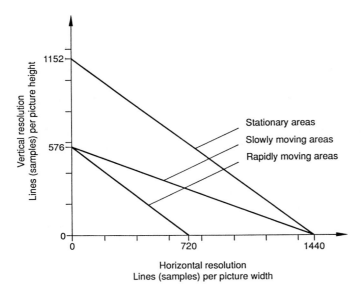

Horizontal resolution
Lines (samples) per picture width

Fig. A.5. Luminance Resolution of HD-MAC. The original image of 1152 active lines and 1440 resolvable elements per line is sampled in an offset (quincunx) pattern. In the stationary areas, all samples are transmitted in 4 frames, in the slowly moving areas, half the samples are transmitted in 2 frames, and in the rapidly moving areas, one-fourth of the samples are transmitted every frame. This results in the theoretical spatial-frequency responses shown here. For comparison, see Fig. A.2

DBS service went on the air in the UK in 1990. A competing service using PAL had been initiated in 1989. The MAC venture failed and was combined with the PAL scheme. As a result, when the MAC service is finally terminated, about 120 thousand MAC receivers will be exchanged for free PAL receivers or a year's paid-up subscription to the PAL service, making this a very costly experiment. Since the evolutionary plan for HD-MAC called for widespread use of D2-MAC as a starting point, this represents a serious setback for the basic plan.

A fundamental problem with D2-MAC, which is a DBS-only service, is that it requires the purchase of new receivers, and the perceived picture quality is not much better than that of PAL delivered by DBS. Lay viewers are not so much bothered by cross-color and cross-luminance as are the professionals.[29] In addition, even these benefits of component transmission are not achieved unless the programs are produced in MAC, and not PAL. To do so requires a complete and very expensive refitting of studios with very little direct benefit to broadcasters.

In spite of this setback, the initial plan for advanced television services in Europe still has a good deal of political support. The Council of Ministers of the European Communities (EC) therefore decided in May 1992 to *require* the use

---

[29] It must not be forgotten that people buy television sets primarily to watch programs. For any new service to be successful, it must provide adequate programming. In a new format, such as MAC, this is very expensive.

of D2-MAC and HD-MAC for all nondigital transmissions by satellite as well as the incorporation of a D2-MAC decoder in all wide-screen receivers. These regulations are to be made acceptable by providing subsidies for both programming and equipment. [A.20] Several more steps are required to put this new plan in place and to find the required financing. If the plan is fully implemented,[30] it may make some difference to the so-far slow progress in getting started with D2-MAC. In spite of this new initiative, many observers believe that D2-MAC and HD-MAC will never be widely used, partly because of the difficulty of getting D2-MAC started, but especially in view of developments in digital transmission.

Meanwhile, progress in developing the cameras, displays, recorders, and other equipment required to realize HD-MAC has been good. Extensive coverage was provided for the recent Albertville Winter and Barcelona Summer Olympics. The motion compensation was still said to be unsatisfactory, and it apparently was still necessary to reduce the resolution on 1250-line receivers to get acceptable pictures on 625-line receivers.

A4.6 PALplus

Although there is surface unity in Europe on the subject of D2-MAC and HD-MAC, there are still many who feel that improvements to PAL are more likely to be of immediate benefit, as well as being much less expensive to implement. There is thus a good deal of support for PALplus, a wide-screen format that can be received on unaltered PAL receivers, but with blank bars at top and bottom. On wide-screen receivers, already on the market, the bars are not visible. There is a potential for inserting digital audio and enhancement information into these areas, as well.

It should be noted that D2-MAC uses a different method of conforming 16:9 transmissions to the screen of 4:3 receivers. Side panels can be clipped off so as to fill the screen, using a variant of the pan-and-scan technique in which the area to be used is controlled by data transmitted along with the signal. Some observers believe that this is so much to be preferred to the letterbox scheme of PALplus that there will be a significant preference for D2-MAC widescreen service. However, there seems to be little data to support this view. Since the preference for one or the other method is so dependent on picture content, it is inherently difficult to prove this point.

For a period of time, it appeared that PALplus had a good chance to be implemented. It provides very attractive pictures at relatively low cost to broadcasters and it can be received (albeit with bars) on PAL receivers. (The full benefit requires new wide-screen 625-line receivers.) It does not have to be thought of as an alternative to HD-MAC, but simply as a means of improving PAL while HD-

---

[30] When the plan was finally enacted, it was watered down by exempting existing services and by being made to expire in 1998.

MAC is developing. Recent interest in digital transmission may prove an obstacle however, since the attention of many in the industry has been diverted.[31]

A4.7 Digital Television in Europe

Work on digital terrestrial broadcasting began in Europe in the middle eighties. The initial goal was digital audio broadcasting (DAB) with the intention of replacing FM with an audio service comparable in quality with that of compact discs. Since many radio receivers are in moving automobiles, the technical problems are formidable. As in television, both source coding and channel coding are needed. MUSICAM [A.21] was selected for the former and OFDM [A.22] for the latter. The source coder reduces the data rate per audio channel to 128 Kbits/sec; the channel coder transmits 33 monophonic audio signals, for a total data rate of 4.224 Megabits/s in one 8-MHz video channel. *The system is absolutely immune to echoes of any amplitude with a time-spread of no more than 16 microsec.* The transmission rate is 1 bit/s per Hertz, but the prototype equipment did not utilize all of the available capacity. The same simple antennas now used for FM are adequate.

It is not clear when the idea emerged of using OFDM for television, but by 1990 some work was being done, based on the earlier work in audio [A.23].

It may well be that the work was sparked in part by US proposals for digital broadcasting. One problem for the European researchers was how to maintain the appearance of support for HD-MAC while developing digital techniques that are completely antithetical. There is some evidence that the earliest digital experiments were not publicized for just this reason.

There are now a number of projects underway and more proposals. All are based on OFDM and a number also involve single-frequency networks, as discussed in Chap. 8. Thus the Europeans, unlike the Americans, are building on a firm basis of experimentation with a system shown by both theory and practice to be well suited for terrestrial broadcasting. Some of the projects are as follows:

SPECTRE. This comprises the work being done at the National Transcommunication Laboratories and the Independent Television Commission (descendants of the old Independent Broadcasting Authority) in Winchester, UK. In the spring of 1992, successful field tests were conducted of a system designed to deliver 12 Megabits/s for CCIR 601 quality in an 8-MHz channel. Later in 1992, field tests including an MPEG-type source coder are planned.

dTTb. This project, Digital Television Terrestrial Broadcasting, is a consortium of the major electronics and broadcasting establishments in Europe, its members coming from France, UK, Italy, Germany, and the Netherlands. It plans to build on the work done in SPECTRE so as to develop a complete system, including HDTV. It has not yet been fully funded by the EC [A.24].

---

[31] The author cannot understand why improvements to PAL and NTSC must be seen as roadblocks to HDTV. Current broadcasting formats will be important for the next 15 to 20 years. It would seem that low-cost improvements would have important benefits.

HDTV-T. This German consortium, comprising Bosch, DAB-Plattf., DLR/NT, DTB, FI/FTZ, Grundig, HHI, IRT, and ITT, plans to carry out a full program of research and development, leading to HDTV systems for DBS, cable, and terrestrial broadcasting. It is being led by the Heinrich-Hertz-Institut in Berlin.

HD-DIVINE. This is a Scandinavian consortium, including Swedish Telecom, Telecom Denmark, Norwegian Telecom and the Swedish Broadcasting Corporation. The intial work of the group led to a partially successful demonstration at the International Broadcasting Convention (IBC) meeting in Amsterdam in July 1992 [A.25].

Thomson CSF. Some work is being done by Thomson CSF. It was first publicized at IBC 92, although a successful field test was held in the US in December 1990. It appears that this test was not disclosed for fear of undermining support for HD-MAC [A.26].

STERNE. This is the work going on at CCETT in Rennes. CCETT played an important role in the earlier DAB work, and is the source of the most significant theoretical study of OFDM [A.27].

Other papers describing this work were presented at the 1992 National Association of Broadcasters (NAB) meeting in Las Vegas, Nevada, and at IBC 92 in Amsterdam.

With so much work going on and with a consensus about the technical approach, it is highly likely that a number of different practical systems will result.[32] However, it is not decided at the present time that single-frequency networks will be widely used in Europe. If not, then systems with a soft threshold will become quite important. Since OFDM, as implemented to date, does not have a soft threshold, some additional work must be done. Any of the methods discussed in [A.28] or [A.29] can be used for this purpose.

A4.8 Conclusions: Europe

The spirit of cooperation shown in Europe in agreeing on a common approach in 1986, and in coordinating the work of many laboratories in the various countries to put together a workable system in a very short period of time, is most remarkable. What appears to have been one of the primary objectives of the work – to keep the NHK system out of Europe, has been entirely successful. However, HD-MAC has two serious problems. The first is that the quality of the 1250-line picture must be compromised to get acceptable results on 625-line receivers so as to retain compatibility. The decision to use a compatible system was not technological, but was based on an assessment of the nature of the dynamics of the TV market and on the likely behavior of consumers. Clearly, non-Europeans are not in as good a position to judge these matters as the Europeans themselves.

---

[32] This is in contrast the excellent coordination shown in the development of HD-MAC. It is not clear why the various countries are not cooperating more closely.

However, depending entirely on improved picture quality to motivate prospective purchasers would seem to be a risky approach anywhere. The success of HD-MAC has been further endangered by the difficulties of D2-MAC.

The second problem is that rapid progress in digital broadcasting may well make HD-MAC technologically obsolete before it gets started. Particularly to the degree that DBS may be an important factor in HDTV in Europe, there is no serious technical objection to digital transmission. Conversion to PAL or MAC in the satellite ground station should be quite practical, so even if 'compatibility' is to be retained, the satellite signal itself need not be like PAL.

The extent to which these problems will determine the fate of HD-MAC is not clear at present. It would seem likely that if convincing demonstrations can be made of digital broadcasting – terrestrial, cable, and satellite – within the next 2 or 3 years, that Europe will prefer to go digital, at least for satellite transmission. With the head start Europe has in OFDM, it is even possible that digital terrestrial broadcasting will become feasible.

The fate of PALplus is also unclear. Even if digital broadcasting were adopted, it would seem that a modest investment in PAL improvements would be worthwhile. With adequate forethought, the wide-screen PALplus receivers could be made digital-ready so that they could receive the digital HDTV broadcasts, albeit at lower resolution.

A5. Overall Conclusions

The recent history of HDTV in all three areas of the television world – Japan, the US, and Europe – underscores certain principles of successful research and development that have been known and appreciated for a long time, but that are evidently quite hard to apply in practice.

- The objectives and requirements for a system should be carefully established before the design is started. For systems that entail large investments on the part of the public, economic and political factors may be as important as the technical factors.
- The technical feasibility of each part of a complex system should be established before it is incorporated into system design.
- Competition is not sufficient to guarantee good products. Enough resources must be brought to bear on complex design problems to explore all their aspects in sufficient depth.

All these rules were violated in the three regions. The Japanese did not appreciate the special position of terrestrial broadcasting in the US, and therefore tried to have their system, designed for satellite delivery, accepted as a standard. They also did not appreciate either the fierce determination in Europe to protect its consumer electronics industry or its ability to join forces in a common goal when threatened. The Europeans, in their resolve to keep out the Japanese, put their system together in great haste, failing to make use of already existing knowledge of digital audio broadcasting and digital satellite transmission technology. Like many Americans, they took a very narrow view of compatibility, mistakenly thinking that the HD-MAC signal itself had to be much like a 625-line PAL signal. They also failed to perform a careful cost/benefit analysis of MAC. In the US, we mistakenly believed that American ingenuity and unfettered free enterprise could make up for the loss of the consumer electronics industry, and that a revolution in television broadcasting could be wrought by a small group of very bright people in a very short period of time without spending much money. In the US and Europe, the stampede to digital transmission without adequate consideration of its advantages and disadvantages, and in spite of the higher transmission efficiency of analog and hybrid systems, may well delay an optimum solution. In all three regions, the technicians seem to have forgotten that television receivers are purchased and used primarily to watch interesting programs, not for the appreciation of small differences in technical image quality.

The growth of HDTV will be slow, whether as a new and separate service in Japan, or as integrated into existing broadcasting systems as planned in Europe and the United States. It may well get started in nonbroadcasting applications or in some kind of pay service directed at specific audiences, such as hotel guests or school children. Industrial and military applications may be among its early uses. If inexpensive set-top converters can be developed, then NTSC and PAL broadcasting can be abandoned at an early date, freeing up spectrum for cellular telephony and other mobile applications. The advent of ubiquitous fiber may also speed up HDTV, but significant economic and policy questions must be dealt with first.

The technological problems of terrestrial HDTV remain the most difficult. Their solution would bring very large economic benefits, since it would then be possible to secure the future of over-the-air broadcasting and at the same time release spectrum for rapidly growing mobile services. An OFDM/SFN system with a gross data rate of 30 to 40 Megabits/s per channel, using some kind of progressive source coding, might permit very high picture quality using intraframe coding for the high-frequency components and confining motion processing to the low-frequency components only. Inexpensive receivers could be built for applications not requiring high resolution. The quality and reliability of service within precisely defined service areas could be greatly improved, and interoperability with nonbroadcasting applications could be facilitated.

To bring this happy result into being, cooperation between researchers in Europe and the US is required. While this is probably possible in the European Community, it will require a substantial change in habits in the United States.

References

Chapter 1

1.1   *Image Processing at the Jet Propulsion Lab*. See J. Geophys. Res. **82** (28), 3951–4684 (1977) (entire issue); see also Proceedings of the CALTECH/JPL Conference on Image Processing, JPL SP 43-30, California Institute of Technology, Pasadena, California, November 3–5 (1976);
W.B. Green: "Computer Image Processing, the Viking Experience", IEEE Trans. **CE-23** (3), 281–299 (1977);
J.M. Soha et al.: "IPL Processing of the Mariner 10 Images of Mercury," J. Geophys. Res. **80** (17), 2394–2414 (1975)

1.2   J. Allnatt: *Transmitted-picture Assessment* (Wiley, New York 1983)

1.3   J.M. Barstow, H.N. Christopher: "The Measurement of Random Video Interference to Monochrome and Color Television Pictures," Trans. AIEE (Comm. and Electron.) **81**, Part I, 313–320 (1962)

1.4   F.T. Percy, T. Gentry Veal: "Subject Lighting Contrast for Color Photographic Films in Color Television," J. SMPTE **63**, 90–94 (1954)

1.5   P. Mertz, F. Grey: "A Theory of Scanning," Bell System Tech. J. **13**, 464–515 (1934)

1.6   D.G. Fink: *Principles of TV Engineering* (McGraw-Hill, New York 1940)

1.7   W.M. Goodall: "Television by PCM", Bell System Tech. J. **30** (1), 33–49 (1951)

1.8   W.F. Schreiber: "Psychophysics and the Improvement of Television Image Quality," J. SMPTE **93** (2), 717–725 (1984)

1.9   D.H. Kelly: "Theory of Flicker and Transient Responses, I: Uniform Fields," J. Opt. Soc. Am. **61** (4), 537–546 (1971); "II. Counterphase Gratings," J. Opt. Soc. Am. **61** (5), 632–640 (1971); Ph.D. Thesis, University of California, Los Angeles (1960); "Adaptation Effects on Spatio-temporal Sine-wave Thresholds," Vision Res. **12**, 89–101 (1972)

1.10 M.W. Baldwin, Jr.: "Demonstration of Some Visual Effects of Using Frame Storage in Television Transmission," IRE Conv. Rec., p. 107 (1958) (abstract only)

1.11 J.L. Cunningham: "Temporal Filtering of Motion Pictures," Sc.D. Thesis, Massachusetts Institute of Technology, Electrical Engineering Department (1963)

1.12 J.R. Ratzel: "Discrete Representation of Spatially Continuous Images," Sc.D. Thesis, Massachusetts Institute of Technology, Electrical Engineering and Computer Science Department (1980)

1.13 R.D. Kell, A.V. Bedford, M.A. Trainer: "An Experimental Television System," Proc. IRE **22** (11), 1246–1265 (1934)

1.14 S.C. Hsu: "The Kell Factor: Past and Present," J. SMPTE **95** (2), 206–214 (1986)

1.15 N.T.S.C.: *Television Standards and Practice* (McGraw-Hill, New York 1943)

1.16 B.M. Oliver, J.R. Pierce, C.E. Shannon: "The Philosophy of PCM," Proc. IRE **36**, 1324–1331 (1948)

Chapter 2

2.1 A. Chappel (ed.): *Optoelectronics, Theory and Practice*, Texas Instruments Electronics Series (McGraw-Hill, New York 1978);
D.E. Gray (ed.): *American Institute of Physics Handbook* (McGraw-Hill, New York 1957)
2.2 A.C. Hardy: *Handbook of Colorimetry* (MIT Press, Cambridge, MA 1936)
2.3 W.G. Driscoll (ed.): *Handbook of Optics*, Optical Society of America (McGraw-Hill, New York 1978)
2.4 M. Born, E. Wolf: *Principles of Optics*, 6th ed. (Pergamon, New York 1980)
2.5 A. Papoulis: *Systems and Transforms with Applications in Optics* (McGraw-Hill, New York 1968)
2.6 R.E. Hopkins, M.J. Buzawa: "Optics for Laser Scanning," Opt. Eng. **15** (2), 90–94 (1976)
2.7 J.W. Goodman: *Introduction to Fourier Optics* (McGraw-Hill, New York 1968)
2.8 J.B.J. Fourier: *Theorie analytique de la chaleur* (Didot, Paris 1822)
2.9 A. Papoulis: In [2.5], p. 145
2.10 E.R. Kretzner: "Statistics of TV Signals," Bell System Tech. J. **31**, 763 (1952)
2.11 E. Dubois, M.S. Sabri, J.Y. Ouellet: "Three-Dimensional Spectrum and Processing of Digital NTSC Color Signals," J. SMPTE **91**, 372–378 (1982)
2.12 D.G. Fink: "The Future of High Definition Television," J. SMPTE **89**, 89–94 (1980)
2.13 M. Ritterman: "An Application of Autocorrelation Theory to TV," Sylvania Technologist 70–75 (1952)

Chapter 3

3.1 S.S. Stevens: *Handbook of Experimental Psychology* (Wiley, New York 1951)
3.2 T.N. Cornsweet: *Visual Perception* (Academic, New York 1970)
3.3 S. Hecht: "Quantum Relations of Vision," J. Opt. Soc. Am. **32**, 42 (1942)
3.4 A. Rose: *Vision: Human and Electronic* (Plenum, New York 1973)
3.5 M.E. Chevreul: *Laws of Color Contrast* (Routledge, London, 1868; Imprimerie Nationale, Paris, 1889; Reinhold, New York, 1967) (first published 1839). This excellent book is an example of the extent to which detailed knowledge about human vision was available long before the advent of modern technology.
3.6 Koenig and Brodhun data (1884), quoted by S. Hecht: J. Gen. Physiol. **7**, 421 (1924)
3.7 R.M. Evans: *The Perception of Color* (Wiley, New York 1975); "Fluorescence and Gray Content of Surface Colors," J. Opt. Soc. Am. **49**, 1049–1059 (1959)
3.8 S.M. Newhall: "Preliminary Report of the O.S.A. Subcommittee on the Spacing of the Munsell Colors", J. Opt. Soc. Am. **30**, 617 (1940)
3.9 R.M. Evans: *An Introduction to Color* (Wiley, New York 1948);
G. Wyszecki, W.S. Stiles: *Color Science* (Wiley, New York 1967) pp. 451–453;
B.C.J. Bartleson, E.J. Breneman: "Brightness Perception in Complex Fields," J. Opt. Soc. Am. **57**, 953–957 (1967)
3.10 B. Hashizume: "Companding in Image Processing," B.S. Thesis, Massachusetts Institute of Technology, Electrical Engineering and Computer Science Department (1973)
U. Malone: "New Data on Noise Visibility," M.S. Thesis, Massachusetts Institute of Technology, Electrical Engineering and Computer Science Department (1977)
3.11 W.F. Schreiber: "Image Processing for Quality Improvement," Proc. IEEE **66** (12), 1640–1651 (1978)
3.12 D.H. Kelly: "Theory of Flicker and Transient Responses, I: Uniform Fields," J. Opt. Soc. Am. **61** (4), 537–546 (1971); "Adaptation Effects on Spatio-temporal Sine-wave Thresholds," Vision Res. **12**, 89–101 (1972)
3.13 E.M. Lowry, J.J. DePalma: "Sine Wave Response of the Visual System: I. The Mach Phenomenon," J. Opt. Soc. Am. **51** (7), 740–746 (1961); "Sine Wave Response of the Visual Sys-

tem: II. Sine Wave and Square Wave Contrast Sensitivity," J. Opt. Soc. Am. **52** (3), 328–335 (1962)

3.14   D.H. Kelly: "Theory of Flicker and Transient Response, II: Counterphase Gratings," J. Opt. Soc. Am. **61** (5), 632–640 (1971)

3.15   O.R. Mitchell: "The Effect of Spatial Frequency on the Visibility of Unstructured Spatial Patterns," Ph.D. Thesis, Massachusetts Institute of Technology, Electrical Engineering and Computer Science Department (1972)

3.16   A. Rosenfeld, A.C. Kak: *Digital Picture Processing* (Academic, New York 1976), Chaps. 6 and 7

3.17   D.H. Kelly: "Image-Processing Experiments," J. Opt. Soc. Am. **61** (10), 1095–1101 (1971)

3.18   D.E. Troxel, W. Schreiber, C. Seitz: "Wirephoto Standards Converter," IEEE Trans. **COM-17** (5), 544–553 (1969)

3.19   W.F. Schreiber: "Wirephoto Quality Improvement by Unsharp Masking," Pattern Recognition **2**, 117–121 (1970)

3.20   J.A.C. Yule: *Principles of Color Reproduction* (Wiley, New York 1967) p. 74

3.21   R.M. Evans: "Sharpness and Contrast in Projected Pictures," presented at the 1956 SMPTE Convention, Los Angeles

3.22   D.H. Kelly: Private communication

3.23   G. Sperling: "Temporal and Spatial Visual Masking," J. Opt. Soc. Am. **55**, 541–559 (1965)

3.24   O. Braddick et al.: "Channels in Vision: Basic Aspects," in *Handbook of Sensory Physiology*, Vol. 8, ed. by R. Held, H.W. Leibowitz, H.L. Teubner (Springer, Berlin, Heidelberg 1978)

3.25   A.N. Netravali, B. Prasada: "Adaptive Quantization of Picture Signals Using Spatial Masking," Proc. IEEE **65** (4), 536–548 (1977)

3.26   W.F. Schreiber, R.R. Buckley: "A Two-Channel Picture Coding System: II-Adaptive Companding and Color Coding," IEEE Trans. **COM-29** (12), 1849–1858 (1981)

3.27   A.J. Seyler, Z. Boudrikis: "Detail Perception After Scene Changes in TV," IEEE Trans. **IT-11**, 31–43 (1965)

3.28   W.E. Glenn: "Compatible Transmission of HDTV Using Bandwidth Reduction," videotape demonstration (Natl. Assn. of Broadcasters, Las Vegas, April 12, 1983)

3.29   L.A. Riggs, F. Ratliff, J.C. Cornsweet, T.N. Cornsweet: "The Disappearance of Steadily Fixated Test Objects," J. Opt. Soc. Am. **43**, 495–501 (1953)

3.30   R. Fielding (ed.): *A Technological History of Motion Pictures and Television* (University of California, Berkeley 1967, 1983)

3.31   R.A. Kinchla, L.G. Allan: "A Theory of Visual Movement Perception," Psychol. Rev. **76**, 537–558 (1969)

3.32   J. Korein, N. Badler: "Temporal Anti-Aliasing in Computer Generated Animation," Comput. Graph. **17** (3), 377–388 (1983)

3.33   Association for Computing Machinery, MOTION: Representation and Perception, SIGGRAPH/SIGART Interdisciplinary Workshop, Toronto, April 4–6, 1983. See especially A.B. Watson, A.J. Ahumada, "A Look at Motion in the Frequency Domain," pp. 1–10

Chapter 4

4.1   S.J. Mason, H.J. Zimmermann: *Electronic Circuits, Signals, and Systems* (Wiley, New York 1960) p. 281;
A. Papoulis: *Signal Analysis* (McGraw-Hill, New York 1977)

4.2   B.M. Oliver, J.R. Pierce, C.E. Shannon: "The Philosophy of PCM," Proc. IRE **36** (11), 1324–1331 (1948)

4.3   J.N. Ratzel: "The Discrete Representation of Spatially Continuous Images," Sc.D. Thesis, Massachusetts Institute of Technology, Electrical Engineering and Computer Science Department (1980);
W.F. Schreiber, D.E. Troxel: "Transformation Between Continuous and Discrete Representations of Images: A Perceptual Approach," IEEE Trans. **PAMI-7** (2), 178–186 (1985)

4.4 H.S. Hou, H.C. Andrews: "Cubic Splines for Image Interpolation and Digital Filtering," IEEE Trans. **ASSP-26** (6), 508–517 (1978)

4.5 F.C. Harris: "On the Use of Windows for Harmonic Analysis with the Discrete Fourier Transform," Proc. IEEE **66** (1), 51–83 (1978)

4.6 R.W. Grass: "An Image Compression/Enhancement System," M.S. Thesis, Massachusetts Institute of Technology, Department of Electrical Engineering and Computer Science (1978)

4.7 C.U. Lee: "Image Rotation by 1-D Filtering," M.S. Thesis, Massachusetts Institute of Technology, Department of Electrical Engineering and Computer Science (1985)

4.8 P. Roetling: "Halftone Method with Enhancement and Moiré Suppression," J. Opt. Soc. Am. **66** (10), 985–989 (1976)

4.9 R. Mersereau: "Hexagonally Sampled 2-D Signals," Proc. IEEE **67** (6), 930–949 (1979)

4.10 L.G. Roberts: "Picture Coding Using Pseudorandom Noise," IRE Trans. **IT-8** (2), 145–154 (1962)

4.11 N. Ziesler: "Several Binary Sequence Generators," Massachusetts Institute of Technology Lincoln Laboratory Tech. Rep. 95 (9) (1955) "Linear Sequences," J. SIAM **7**, 31–48 (1959); S.W. Golomb: *Shift Register Sequences* (Aegean Park, Laguna Hills, CA 1967)

4.12 D.N. Graham: "Two-Dimensional Filtering to Reduce the Effect of Quantizing Noise in Television," M.S. Thesis, Massachusetts Institute of Technology, Electrical Engineering Department (1962);
R.A. Bruce: "Optimum Pre-emphasis and De-emphasis Networks for Transmission of Television by PCM," IEEE Trans. **CS-12** (9), 91–96 (1964);
E.G. Kimme, F.F. Kuo: "Synthesis of Optimal Filters for a Feedback Quantization System," IEEE Trans. **CT-10** (9), 405–413 (1963);
A.E. Post: B.S. Thesis, Massachusetts Institute of Technology, Department of Electrical Engineering (1966)

4.13 K.P. Wacks: "Design of a Real Time Facsimile Transmission System," Ph.D. Thesis, Massachusetts Institute of Technology, Electrical Engineering and Computer Science Department (1973)

Chapter 5

5.1 P.A. Wintz: "Transform Picture Coding," Proc. IEEE **60** (7), 809–820 (1972); see also A.G. Tescher's article on transform coding in W.K. Pratt (ed.): *Image Transmission Techniques* (McGraw-Hill, New York 1979)

5.2 J.O. Limb, C.R. Rubinstein, J.E. Thompson: "Digital Coding of Color Video Signals – A Review," IEEE Trans. **COM-25** (11), 1349–1385 (1977);
R.R. Buckley: "Digital Color Image Coding and the Geometry of Color Space," Ph.D. Thesis, Massachusetts Institute of Technology, Electrical Engineering and Computer Science Department (1982)

5.3 A.N. Netravali, J.D. Robbins: "Motion-Compensated TV Coding", Bell System Tech. J. **58** (3), 631–670 (1979)

5.4 W.F. Schreiber: "Color Reproduction System," US Patent No. 4,500,919 (1985)

5.5 W.-H. Chen, W.K. Pratt: "Scene Adaptive Coder," IEEE Trans. **COM-32** (3), 225–232 (1984)

5.6 C.E. Shannon, W. Weaver: *The Mathematical Theory of Communication* (University of Illinois Press, Urbana 1949). Everyone interested in statistical coding should read Shannon's landmark paper.

5.7 D.A. Huffman: "A Method for the Construction of Minimum Redundancy Codes," Proc. IRE **40** (9), 1098-1101 (1952)

5.8 W.F. Schreiber: "The Measurement of Third Order Probability Distribution of TV Signals," IRE Trans. **IT-2** (9), 94–105 (1956)

5.9 S. Ericsson, E. Dubois: "Digital Coding of High Quality TV," Paper 3.5, High Definition Television Colloquium, Canadian Dept. of Communications, Ottawa (1985)

5.10 D. Bodson, S. Urban, A. Deutermann, C. Clarke: "Measurement of Data Compression in Advanced Group IV Facsimile Systems," Proc. IEEE **73** (4), 731–739 (1985)

5.11 W.F. Schreiber: "Reproduction of Graphical Data by Facsimile," Quarterly Progress Report, Research Laboratory of Electronics, Massachusetts Institute of Technology, No. 84 (1967); R.B. Arps, R.L. Erdmann, A.S. Neal, C.E. Schlaepfer: "Character Legibility Versus Resolution in Image Processing of Printed Matter," IEEE Trans. **MMS-10** (3), 66–71 (1969)

5.12 R.N. Ascher, G. Nagy: "A Means for Achieving a High Degree of Compaction on Scan Digitized Printed Test," IEEE Trans. **C-23** (11), 1174–1179 (1974); W.K. Pratt, P. Capitant, W. Chen, E. Hamilton, R. Wallis: "Combined Symbol Matching Facsimile Data Compression System," Proc. IEEE **68** (7), 786–795 (1980)

5.13 D.R. Knudson: US Patent No. 4,281,312 (1981)

5.14 S.W. Golomb: "Run Length Encodings," IEEE Trans. **IT-15** (7), 399–400 (1969); J. Capon: "A Probabilistic Model for Run-Length Coding of Pictures," IRE Trans. **IT-5** (12), 157–163 (1959); C.G. Beaudette: "An Efficient Facsimile System for Weather Graphics," in Proceedings of Symposium on Picture Bandwidth Compression, MIT, ed. by T.S. Huang, O.J. Tretiak (MIT Press, Cambridge, MA 1969) pp. 217–229; H.G. Mussman, D. Preuss: "Comparison of Redundancy Reducing Codes for Facsimile Transmission of Documents," IEEE Trans. **COM-25** (11), 1425–1433 (1977); T.S. Huang, A.B.S. Hussain: "Facsimile Coding by Skipping White," IEEE Trans. **COM-23** (12), 1452–1460 (1975); T.S. Huang: "Bounds on the Bit Rate of Linear Run-Length Codes," IEEE Trans. **IT-21** (11), 707–708 (1975); R.B. Arps: "An Introduction and Digital Facsimile Review," in Proc. Int. Conf. on Communications Vol. 1 (San Francisco 1975) p. 7. A number of other interesting papers are included in this issue; Kalle Infotec U.K. Ltd.: "Redundancy Reduction Technique for Fast Black and White Facsimile Apparatus," CCITT Study Group XIV, Temp. Document No. 15 (1975); R.B. Arps: "Bibliography of Digital Graphic Image Compression and Quality," IEEE Trans. **IT-20** (1), 120-122 (1974)

5.15 M.W. Baldwin: "The Subjective Sharpness of Simulated TV Pictures," Proc. IRE **28** (10), 458–468 (1940)

5.16 T.S. Huang: "Run Length Coding and its Extensions," Proceedings of Symposium on Picture Bandwidth Compression, MIT, ed. by T.S. Huang, O.J. Tretiak (MIT Press, Cambridge, MA 1969) pp. 231–264

5.17 F. DeCoulon, M. Kunt: "An Alternative to Run Length Coding for Black and White Fascimile," in Proceedings of IEEE International Zurich Seminar on Digital Communication, Paper C-4 (1974); F. DeCoulon, O. Johnsen: "Adaptive Block Scheme for Source Coding of Black and White Facsimile," Electron. Lett. **12** (3), 61 (1976)

5.18 R.M. Gray: "Vector Quantization," IEEE ASSP Mag. 4–29 (April 1984)

5.19 W.F. Schreiber, T.S. Huang, O.J. Tretiak: "Contour Coding of Images," Wescon Technical Papers, paper 8/3, IEEE (1968); also in *Picture Bandwidth Compression* (Gordon Breach, New York 1972); and in N.S. Jayant (ed.): *Waveform Quantization and Coding* (IEEE, New York 1976); D. Gabor, P.C.J. Hill: "Television Band Compression by Contour Interpolation," Proc. IEE (England) **108**, Part B, 303–315 (1961); T.H. Morrin: "Recursive Contour Coding of Nested Objects in Black/White Images," in Proc. Int. Conf. on Communications, Vol. 1 (San Francisco, 1975) pp. 7-17

5.20 P.M.J. Coueignoux: "Compression of Type Faces by Contour Coding," M.S. Thesis, Massachusetts Institute of Technology, Electrical Engineering and Computer Science Department (1973)

5.21 D.N. Graham: "Image Transmission by Two-Dimensional Contour Coding," Ph.D. Thesis, Massachusetts Institute of Technology, Electrical Engineering Department (1966); also Proc. IEEE **55** (3), 336–346 (1967)

5.22 J.W. Pan: "Reduction of Information Redundancy," Sc.D. Thesis, Massachusetts Institute of Technology, Electrical Engineering Department (1963)

5.23 H.S. Hou, H.C. Andrews: "Cubic Splines for Image Interpolation and Digital Filtering," IEEE Trans. **ASSP-26** (4), 508 (1978)

5.24 R. Hunter, A.H. Robinson: "International Digital Facsimile Coding Standards," Proc. IEEE **68** (7), 854–867 (1980)

5.25 J.R. Ellis: Ph.D. Thesis, Massachusetts Institute of Technology, Electrical Engineering and Computer Science Department (1977);
D. Spencer: M.S. Thesis, Massachusetts Institute of Technology, Electrical Engineering and Computer Science Department (1968)

5.26 H. Freeman: "On the Encoding of Arbitrary Geometrical Figures," IRE Trans. **EC-10**, 260–268 (1961);
I.T. Young: "TV Bandwidth Compression Using Area Properties," MS Thesis, Massachusetts Institute of Technology, Electrical Engineering Department (1966)

5.27 P. Elias: "Predictive Coding," IRE Trans. **IT-1** (3), 16–32 (1955);
R.E. Graham: "Predictive Quantization of TV Signals," Wescon Conv. Record **2** (4), 147–157, IRE (1958);
W. Zschunke: "DPCM Picture Coding with Adaptive Prediction," IEEE Trans. **COM-25** (11), 1295–1302 (1977);
C.M. Harrison: "Experiments with Linear Prediction in TV," Bell System Tech. J. **31**, 764–783 (1952)

5.28 L.D. Davisson: "Data Compression Using Straight-Line Interpolation," IEEE Trans. **IT-14** (3), 390–394 (1968);
A.N. Netravali: "Interpolative Picture Coding Using a Subjective Criterion," IEEE Trans. **COM-25** (5), 503–508 (1977)

5.29 C.C. Cutler: US Patent No. 2,605,361 (1952);
J.B. O'Neal: "Predictive Quantizing System (DPCM) for the Transmission of TV Signals," Bell System Tech. J. **45**, 689–721 (1966)

5.30 J. de Jaeger: "Deltamodulation, a Method of PCM Transmission Using a 1-bit Code," Philips Res. Rep. **7** (6), 442–466 (1952)

5.31 J. Max: "Quantizing for Minimum Distortion," IRE Trans. **IT-6** (3), 7–12 (1960);
A.N. Netravali: "On Quantizers for DPCM Coding of TV Signals," IEEE Trans. **IT-23** (3), 360–370 (1977);
T. Berger: "Optimum Quantizers and Permutation Codes," IEEE Trans. **IT-18** (11), 759–776 (1972)

5.32 H.G. Mussman: "Predictive Coding of TV Signals," in *Image Transmission Techniques*, ed. by W. Pratt (Academic, New York 1979)

5.33 D.K. Sharma, A.N. Netravali: "Design of Quantizers for DPCM Coding of Picture Signals," IEEE Trans. **COM-25** (11), 1267–1274 (1978)

5.34 B. Prasada: Private communication

5.35 A. Habibi: "Comparison of nth Order DPCM Encoder with Linear Transformations and Block Quantization Techniques," IEEE Trans. **COM-19** (6), 948–956 (1971);
D.J. Connor, R.F.W. Pease, W.G. Scholes: "TV Coding Using Two-Dimensional Spatial Prediction," Bell System Tech. J. **50**, 1049–1061 (1971)

5.36 P.A. Ratliff: "Digital Coding of the Composite PAL Colour Television System for Transmission at 34 Mbits/sec," BBC Eng. No. 115, 24–35 (1980)

5.37 A. Habibi: "Survey of Adaptive Image Coding Techniques," IEEE Trans. **COM-25** (11), 1257–1284 (1977);
B. Prasada, A. Netravali: "Adaptive Companding of Picture Signals in a Predictive Coder," IEEE Trans. **COM-26** (1), 161–164 (1978)

5.38   N.S. Jayant: "Adaptive Quantization with a One-Word Memory," Bell System Tech. J. **52**, 1119–1143 (1973)

5.39   A.P. Zarembowitch: "Forward Estimation Adaptive DPCM for Image Data Compression," M.S. Thesis, Massachusetts Institute of Technology, Electrical Engineering and Computer Science Department (1981)

5.40   C.C. Cutler: "Delayed Encoding: Stabilizer for Adaptive Coders," IEEE Trans. **COM-19** (6), 898–907 (1971)

5.41   A.N. Netravali, B. Prasada: "Adaptive Quantization of Picture Signals Using Spatial Masking," Proc. IEEE **65** (4), 536–548 (1977)

5.42   S.K. Goyal, J.B. O'Neal: "Entropy Coded DPCM systems for TV," IEEE Trans. **COM-23** (6), 660–666 (1975)

5.43   E.G. Kimme, F.F. Kuo: "Synthesis of Optimum Filters for a Feedback Quantization Scheme," IEEE Trans. **CT-10** (9), 405–413 (1963)

5.44   W.F. Schreiber, D.E. Troxel: US Patent No. 4,268,861 (1981)

5.45   D.E. Troxel: "Application of Pseudorandom Noise to DPCM," IEEE Trans. **COM-29** (12), 1763–1167 (1981)

5.46   E.R. Kretzmer: "Reduced Alphabet Representation of Television Signals," IRE Nat. Conv. Rec. **4** (4), 140–153 (1956)

5.47   W.F. Schreiber, C.F. Knapp, N.D. Kay: "Synthetic Highs, An Experimental TV Bandwidth Reduction System," J. Soc. Motion Picture Television Eng. **68** (8), 525–537 (1959);
D.N. Graham, "Image Transmission by Two Dimensional Contour Coding," Proc. IEEE **55** (3), 336–346 (1967)

5.48   D.E. Troxel, W. Schreiber, P. Curlander, A. Gilkes, R. Grass, G. Hoover: "Image Enhancement/Coding Systems Using Pseudorandom Noise Processing," Proc. IEEE **67**, 972–973 (1979);
D.E. Troxel, W.F. Schreiber, R. Grass, G. Hoover, R. Sharpe: "Bandwidth Compression of High Quality Images," International Conference on Communications, 31.9.1–5, Seattle (1980); also "A Two-Channel Picture Coding System: I – Real-Time Implementation," IEEE Trans. **COM-29** (12), 1841–1848 (1981)

5.49   W.F. Schreiber, R.R. Buckley: "A Two-Channel Picture Coding System: II – Adaptive Companding and Color Coding," IEEE Trans. **COM-29** (12), 1849–1858 (1981)

5.50   R.B. Sharpe: "Statistical Coding of the Highs Channel of a Two-Channel Facsimile System," M.S. Thesis, Massachusetts Institute of Technology, Electrical Engineering and Computer Science Department (1979)

5.51   M. Kocher, M. Kunt: "A Contour-Texture Approach to Picture Coding," in Proc. ICASSP-2, Paris (IEEE, New York 1982) pp. 436–440

5.52   P.J. Burt, E.H. Adelson: "The Laplacian Pyramid as a Compact Image Code," IEEE Trans. **COM-31** (4), 532–540 (1983)

5.53   J. Alnatt: *Transmitted-picture Assessment* (Wiley, New York 1983)

Chapter 6

6.1   H.J.P. Arnold: *William Henry Fox Talbot* (Hutchinson Benham, London 1977)

6.2   R. Higonnet, L. Moyroud: US Patent No. 3,188,929 (1965)

6.3   C. Bigelow: "Technology and the Aesthetics of Type," Seybold Report on Publishing Systems **10** (24) (1981); "The Principles of Digital Type," ibid. **11** (11) (1982)

6.4   D.E. Troxel, W. Schreiber, S. Goldwasser, M. Khan, L. Picard, M. Ide, C. Turcio: "Automated Engraving of Gravure Cylinders," IEEE Trans. **SMC-11** (9), 585–596 (1981)

6.5   "Lasergravure" (product literature of Crosfield Electronics)

6.6   J.A.C. Yule, D.S. Howe, J.H. Altman: "The Effect of the Spread-Function of Paper on Halftone Reproduction," TAPPI **50** (7), 337–344 (1967)

6.7   W.F. Schreiber: "An Electronic Process Camera," in Proceedings, Tech. Assn. of the Graphic Arts (1983) pp. 101–127

6.8   Yao-ming Chao: "An Investigation into the Coding of Halftone Pictures," Ph.D. Thesis, Massachusetts Institute of Technology, Electrical Engineering and Computer Science Department (1982)

6.9   C.N. Nelson: "Tone Reproduction", in *Theory of the Photographic Process*, ed. by T.H. Jones (Macmillan, New York 1977) pp. 536–560

6.10   J. Sturge (ed.): *Neblette's Handbook of Photography and Reprography* (Van Nostrand Reinhold, New York 1977). (This is an excellent reference for many of the subjects of this chapter. It contains a number of review articles with good bibliographies.)

6.11   B.E. Bayer: "An Optimum Method for Two-Level Rendition of Continuous-Tone Pictures," ICC 73, IEEE **I**, 26.11–26.15

6.12   J.F. Jarvis: "A Survey of Techniques for the Display of Continuous Tone Pictures by Bilevel Displays," Comput. Graph. Image Proc. **5**, 13–40 (1976)

6.13   R.W. Floyd, L. Steinberg: "An Adaptive Algorithm for Spatial Grayscale," Proc. Soc. for Information Display **17** (2), 75–77 (1976)

6.14   B. Woo: "A Survey of Halftoning Algorithms and Investigation of the Error Diffusion Technique," B.S. Thesis, Massachusetts Institute of Technology, Electrical Engineering and Computer Science Department (1984)

Chapter 7

7.1   R.M. Evans: *The Perception of Color* (Wiley, New York 1975)

7.2   W.T. Wintringham: "Color TV and Colorimetry," Proc. IRE **39** (10), 1135 (1951)

7.3   D.L. MacAdam: "Visual Sensitivities to Color Differences in Daylight," J. Opt. Soc. Am. **32** (5), 247–274 (1942)

7.4   W.F. Schreiber: "A Color Prepress System Using Appearance Variables," J. Imag. Tech. **12** (4), 200–210 (1986)

7.5   M. Baldwin: "Subjective Sharpness of Additive Color Pictures," Proc. IRE **39** (10), 1173–1176 (1951)

7.6   U.F. Gronemann: "Coding Color Pictures," Ph.D. Thesis, Massachusetts Institute of Technology, Electrical Engineering Department (1964)

7.7   A.V. Bedford: "Mixed Highs in Color Television," Proc. IRE **38** (9), 1003–1009 (1950)

7.8   R.R. Buckley: "Digital Color Image Coding and the Geometry of Color space," Ph.D. Thesis, Massachusetts Institute of Technology, Electrical Engineering and Computer Science Department (1981)

7.9   Second Color Television Issue, Proc. IRE **42** (1) (1954) (entire issue)

7.10   E. Dubois, W.F. Schreiber: "Improvements to NTSC by Multidimensional Filtering," J. SMPTE **97**, 446–463 (1988)

7.11   N.M. Nasrabadi, R.A. King: "Image Coding Using Vector Coding: A Review," IEEE Trans. **COM-36** (8), 957–971 (1988)

7.12   W. Bender: "Adaptive Color Coding Based on Spatial/Temporal Features," SPSE Electronic Imaging Devices and Systems Symposium, Los Angeles (1988)

Chapter 8

8.1   W.F. Schreiber: "Psychophysics and the Improvement of Television Image Quality," J. SMPTE **93** (2), 717–725 (1984)

8.2   E.F. Brown: "Low-Resolution TV: Subjective Comparison of Interlaced and Non-Interlaced Pictures," Bell System Tech. J. **46** (1), 199–232 (1967);
       T. Fujio (ed.): "HDTV," NHK Tech. Monograph **32** (1982)

8.3   B. Wendland: "Extended Definition TV with High Picture Quality," J. SMPTE **92** (10), 1028–1035 (1983)

8.4   T.G. Schut: "Resolution Measurements in Camera Tubes," J. SMPTE **92** (12), 1270–1293 (1983)

8.5 S.C. Hsu: "The Kell Factor: Past and Present," J. SMPTE **95** (2), 206–214 (1986)

8.6 D.H. Kelly: "Visual Response to Time-Dependent Stimuli," J. Opt. Soc. Am. **51** (4), 422–420 (1961)

8.7 T.N. Cornsweet: *Visual Perception* (Academic, New York 1970)

8.8 E.M. Lowry, J.J. DePalma: "Sine Wave Response of the Visual System," J. Opt. Soc. Am. **51** (10), 474 (1961)

8.9 A.N. Netravali, B. Prasada: "Adaptive Quantization of Picture Signals Using Spatial Masking," Proc. IEEE **65** (4), 536–548 (1977)

8.10 W.F. Schreiber, R.R. Buckley: "A Two-Channel Picture Coding System: II – Adaptive Companding and Color Coding," IEEE Trans. **COM-29**, 12 (1981)

8.11 L.G. Roberts: "Picture Coding Using Pseudo-Random Noise," IRE Trans. **IT-8** (2), 145–154 (1962)

8.12 D.G. Fink: *TV Standards and Practice* (McGraw-Hill, New York 1943)

8.13 Proc. IRE **39** (10) (1951) and **42** (1) (1954) (entire issues)

8.14 U.F. Gronemann: Ph.D. Thesis, Massachusetts Institute of Technology, Electrical Engineering Department (1964)

8.15 A.V. Bedford: "Mixed Highs in Color Television," Proc. IRE **38** (9), 1003 (1950)

8.16 F. Gray: "Electro-Optical Transmission System," US Patent No. 1,769,920 (1929)

8.17 E. Dubois, W.F. Schreiber: "Improvements to NTSC by Multidimensional Filtering," J. SMPTE **97** (6), 446–463 (1988)

8.18 M.W. Baldwin: "The Subjective Sharpness of Simulated Television Pictures," Proc. IRE **18** (10), 458–468 (1940)

8.19 Y. Faroudja, J. Roizen: "Improving NTSC to Get Near-RGB Performance," J. SMPTE **96** (8), 750–761 (1987)

8.20 "Hierarchical High Definition Television System Compatible with the NTSC Environment," North American Philips Corp., Briarcliff Manor, NY (1987)

8.21 W.E. Glenn, K.G. Glenn: "HDTV Compatible Transmission System," J. SMPTE **96** (3), 242–246 (1987)

8.22 W.R. Neuman: MIT Audience Research Facility, private communication

8.23 J.S. Wang: "Motion-Compensated NTSC Demodulation," Ph.D. Thesis, Massachusetts Institute of Technology, Electrical Engineering and Computer Science Department (1989)

8.24 Y. Ninomiya, Y. Ohtsuka: "A Single Channel HDTV Broadcast System, – The MUSE," NHK Lab. Note 304, 1–12 (1984)

8.25 W.R. Neuman: MIT Audience Research Facility, private communication

8.26 T. Fukinuki, H. Hirano: "Extended Definition TV Fully Compatible with Existing Standards," IEEE Trans. **COM-32** (8), 948–953 (1984)

8.27 M. Isnardi, J. Fuhrer, T. Smith, J. Koslov, B. Roeder, W. Wedam: "A Single-Channel Compatible Widescreen EDTV System," 3rd HDTV Colloquium, (Canadian Dept. of Communication, Ottawa 1987)

8.28 Y. Yasumoto, S. Kageyama, S. Inouye, H. Uwabata, Y. Abe: "An Extended Definition Television System Using Quadrature Modulation of the Picture Carrier with Inverse Nyquist Filter," Consumer Electronics Conference, Chicago (1987)

8.29 J.M. Barstow, H.N. Christopher: "The Measurement of Random Video Interference to Monochrome and Color Television Pictures," Trans. AIEE (Communication and Electronics) **81** (1), 313–320 (1962)

8.30 W.F. Schreiber et al.: "A Compatible High-Definition Television System Using the Noise-Margin Method of Hiding Enhancement Information," J. SMPTE **98** (12), 873–879 (1989)

8.31 W.J. Butera: "Multiscale Coding of Images," B.S. Thesis, Massachusetts Institute of Technology, Media Laboratory (1988)

8.32 W.F. Schreiber: "Improved Television Systems: NTSC and Beyond," J. SMPTE **86** (8), 734–744 (1987)

8.33 W.F. Schreiber, H. Lippman, A. Netravali, E. Adelson, D. Staelin: "Channel-Compatible 6-MHz HDTV Distribution Systems," J. SMPTE **98** (1), 5–13 (1989)

8.34 A. Toth, J. Donahue: "ATV Multiport Receiver – Preliminary Analysis," EIA Multiport Receiver Subcommittee (1989)

8.35 A. Hirota, S. Hirano, H. Kitamura, T. Tsushima: "Noise Reducing System for Video Signal," US Patents No. 4, 607, 285 (1986);
A. Hirota, T. Tsushima: "Noise Reducing System for Video Signal," US Patent No. 4, 618, 893 (1986);
L. Pham Van Cang: "Circuit for Processing a Color Television Signal," US Patent No. 4, 007, 483 (1977)

8.36 P.P. Vaidyanathan: "Quadrature Mirror Filter Banks, M-Band Extensions, and Perfect-Reconstruction Techniques," IEEE ASSP Mag. 4–20 (July 1987)

8.37 E.R. Kretzmer: "Reduced-Alphabet Representation of Video Signals," IRE Convention Record, Part 4, 140–146 (1956)

8.38 W.F. Schreiber, A.B. Lippman: "Reliable EDTV/HDTV Transmission in Low-Quality Analog Channels," J. SMPTE 98 (7), 496–503 (1989)

8.39 W.F. Schreiber: "A Friendly Family of Transmission Standards for All Media and All Frame Rates," Proceedings NAB Engineering Conference pp. 417-423 (1989)

8.40 W.F. Schreiber: "The Role of Technology in the Future of Television," Telecommun. J. 57 (11), 763–774 (1990)

8.41 CCETT, Groupement D'Interêt Economique, "Description of the COFDM System," Cesson Sévigné, France (1990)
M. Alard, R. Lasalle: "Principles of Modulation and Channel Coding for Digital Broadcasting to Mobile Receivers," EBU Review Technical 8 168–190 (1987)

8.42 "Advanced Digital Techniques for UHF Satellite Sound Broadcasting - Collected Papers," EBU (8) (1988)

8.43 F. Conway, S. Edwards, D. Tyrie, R. Voyer: "Initial Experimentation with DAB in Canada," Proceedings, NAB Engineering Conference, Las Vegas, pp. 281–290 (1991)

8.44 A.G. Mason, N.K. Lodge: "Digital Terrestrial Television Development in the SPECTRE Project," National Transcommunications Ltd, and Independent Television Commission, UK, IBC, Amsterdam (1992)

8.45 S.B. Weinstein, P.M. Ebert: "Data Transmission by Frequency Division Multiplexing using the Discrete Fourier Transform," IEEE Trans on Communication Tech., COM-19 (15) (1971)
B. Hirosaki: "An Orthogonally Multiplexed QAM System using the Discrete Fourier Transform," IEEE Trans. on Communication Tech., COM-29 (7) (1981)

8.46 D. Pommier, P.A. Ratliff, E. Meier-Engelen: "A Hybrid Satellite/Terrestrial Approach for Digital Audio Broadcasting with Mobile and Portable Receivers," Proceedings, NAB Engineering Conference, pp. 304–312 (1990)

8.47 Spectrum Compatible HDTV System (Zenith Electronics Corp., Glenview, IL 1988)

8.48 W. Luplow: Zenith Electronics Corp., private communication

8.49 B.G. Schunck: "The Image Flow Constraint Equation," Comp. Vision, Graph. Image Proc. 35, 20–46 (1986)

8.50 E. Dubois, S. Sabri: "Noise Reduction in Image Sequences Using Motion-Compensated Temporal Filtering," IEEE Trans. COM-73 (7), 502-522 (1984);
D. Martinez, J.S. Lim: "Implicit Motion-Compensated Noise Reduction of Motion Picture Scenes," Proc. Intl. Conf. on Acoustics, Speech, and Signal Processing (1986) pp. 375–378

8.51 E.A. Krause: "Motion Estimation for Frame-Rate Conversion," Ph.D. Thesis, Massachusetts Institute of Technology, Electrical Engineering and Computer Science Department (1987);
D.M. Martinez: "Model-Based Motion Estimation and its Application to Restoration and Interpolation of Motion Pictures," Ph.D. Thesis, Massachusetts Institute of Technology, Electrical Engineering and Computer Science Department (1986)

8.52 M.A. Krasner: "The Critical Band Coder – Digital Encoding of Speech Signals Based on the Perceptual Requirements of the Auditory System," ICASSP (IEEE, New York 1980) pp. 327–331

8.53 R.E. Crochiere et al.: "Digital Coding of Speech in Subbands," Bell System Tech. J. **55** (10), 1069–1085 (1976)

8.54 D. Baylon: "Adaptive Amplitude Modulation for Transform/Subband Coefficients," M.S. Thesis, Massachusetts Institute of Technology, Electrical Engineering and Computer Science Department (1990)

8.55 N. Ziesler: "Several Binary Sequence Generators," Massachusetts Institute of Technology Lincoln Laboratory Tech. Rep. 95 (9) (1955); "Linear Sequences," J. SIAM **7**, 31–48 (1959); S.W. Golomb: *Shift Register Sequences* (Aegean Park, Laguna Hills, CA 1967)

8.56 A.B. Watson: "The Window of Visibility: A Psychological Theory of Fidelity in Time-Sampled Visual Motion Displays," NASA Tech. Paper 2211 (1983); "A Look at Motion in the Frequency Domain," *MOTION: Representation and Perception*, SIGGRAPH/SIGGART Workshop, Association for Computing Machinery, Toronto 1–10 (1983)

8.57 E. Chalom: "Video Data Compression Using Quadrature Mirror Filters and Adaptive Coding," M.S. Thesis, Massachusetts Institute of Technology, Electrical Engineering and Computer Science Department (1989); W.F. Schreiber et al.: "Robust Bandwidth-Efficient HDTV Transmission Formats," J. SMPTE Conference, Los Angeles (1989)

8.58 C.E. Shannon, W. Weaver: *The Mathematical Theory of Communication* (University of Illinois Press, Urbana 1949)

8.59 W.F. Schreiber: "6-MHz Single-Channel HDTV Systems," HDTV Colloquium, Canadian Dept. of Communications, Ottawa (1987)

8.60 T.M. Cover: "Broadcast Channels," IEEE Trans. on Information Theory, **IT-8** (1) pp. 2014 (1972)

8.61 W.F. Schreiber: "All-Digital HDTV Terrestrial Broadcasting in the US: Some Problems and Possible Solutions," International Symposium Europe-USA, Paris, (1991)

8.62 D. Anastassiou, M. Vetterli: "All Digital Multiresolution Coding of HDTV," Proceedings NAB Engineering Conference (1991)

8.63 P.G. deBot: "Multiresolution Transmission over the AWGN Channel," Nederlandse Philips Bedrijven B.V., (1992) (in press)

8.64 W.J. Butera: "Multiscale Coding of Images," M.S. Thesis, Massachusetts Institute of Technology, Media Laboratory (1988)

8.65 C.L. Ruthroff: "Computation of FM Distortion in Linear Networks for Bandlimited Periodic Signals," Bell Syst. Tech. J. **47** (6), 1043–1063 (1968)

8.66 W.F. Schreiber, J. Piot: "Video Transmission by Adaptive Frequency Modulation," IEEE Commun. Mag. **MCOM 26** (11), 68–76 (1988)

8.67 G. K. Wallace: "The JPEG Still Picture Compression Standard," Communications of the ACM **34** (4) 31–44 (1991)

8.68 D.J. LeGall: "The MPEG Video Compression Algorithm," Signal Processing: Image Communications **4** (2) 129–140 (1992); L. Chiariglione: "The International Approach to Audio-Visual Coding," CSELT, Via Reiss Romoli, Torino, Italy **274** 1–1048 (1992)

Appendix

A.1 R. Forni, M. Pelchat: Deputies, "La Télévision à Haute Définition," Rapport de l'Office parlementaire d'évaluation des choix scientifiques et technologiques, Assemblée Nationale-Senat (Ed. Economica, Paris 1989)

A.2 Y. Ninomiya, Y. Ohtsuka: "A Single Channel HDTV broadcast System, – The MUSE," NHK Lab. Note 304, 1–12 (1984)

A.3 T. Fujio et al.: "High-Definition Television," NKH Technical Monograph 32 (1982)

A.4 J.W. Richards, et al.: "Experience with a Prototype Motion Compensated Standards Converter for Down-Conversion of 1125/60/2:1 SMPTE-240 M High Definition to 625/50/2:1 Video," Proceedings IBC, Amsterdam, pp. 56-61 (1992)

A.5   T. Fukinuki, H. Hirano: "Extended Definition TV Fully Compatible with Existing Standards," IEEE Trans. **COM-32** (8), 948–953 (1984)

A.6   T. Nishizawa et al.: "HDTV and ADTV Transmission Systems: MUSE and its Family," Unpublished oral presentation, Natl. Assn. of Broadcasters, Las Vegas, April 11 (1988); Y. Tanaka, K. Enami, H. Okuda: "Compatible MUSE Systems," NHK Laboratories Note 375 (January 1990)

A.7   M.L. Dertouzos, R.K. Lester, R.M. Solow: *Made in America* (MIT Press, Cambridge, MA 1989)

A.8   M. Isnardi et al.: "A Single-Channel Compatible Widescreen EDTV System," 3rd HDTV Colloquium, Canadian Dept. of Communication, Ottawa (1987)

A.9   "Hierarchical High Definition Television System Compatible with the NTSC Environment," North American Philips Corp., Briarcliff Manor, NY (1987)

A.10 W.F. Schreiber, A.B. Lippman: "Reliable EDTV/HDTV Transmission in Low-Quality Analog Channels," J. SMPTE **98** (7), 496–503 (1989)

A.11 *Spectrum Compatible HDTV System* (Zenith Electronic Corp., Glenview, IL 1988)

A.12 Y. Faroudja, J. Roizen: "Improving NTSC to get Near-RGB Performance," J. SMPTE **96**, 750–761 (1987)

A.13 "Digicipher HDTV System Description," (General Instrument Corp, San Diego, CA 1991)

A.14 "Channel Compatible DigiCipher HDTV System," (Massachusetts Institute of Technology, 1992)
      "Digital Spectrum Compatible,"(Zenith/AT&T, 1991)
      "Advanced Digital TV," (ATRC, 1992)

A.15 M.D. Windram, G.J. Tonge, R.C. Hills: "The D-MAC/Packet Transmission System for Satellite Broadcasting in the United Kingdom," EBU Tech. Rev., No. 227 (1988)

A.16 R.J. Iredale: "A Proposal for a New High Definition NTSC Broadcast Protocol," IEEE Trans. **CE-33** (1), 14–27 (1987)

A.17 E. Chalom: B.S. Thesis, Massachusetts Institute of Technology, Electrical Engineering and Computer Science Department (1987)

A.18 F.W.P. Vreeswijk, M.R. Haghiri: "HDMAC Coding for MAC Compatible Broadcasting of HDTV Signals," Third Intl. Workshop on HDTV, Torino, Sept. 1989

A.19 R. Storey: "Motion-Compensated DATV Bandwidth Compression for HDTV," IBC-88, Brighton, England (1988)

A.20 Commission of the European Communities: "Proposal for Council Decision on an Action Plan for the Introduction of Advanced Television Services in Europe," COM **92** (154 final) Brussels (1992)

A.21 G. Theile, G. Stoll, M. Link: "An Introduction to the MASCAM System," EBU Review Technical, No. 230 (8) 158–181 (1988)

A.22 "Description of the COFDM System," CCETT, Groupement D'Interêt Economique, Cesson Sévigné, France, (1990)
      "Principles of Modulation and Channel Coding for Digital Broadcasting to Mobile Receivers," EBU Review Technical, (8) pp. 168–190 (1987)
      D. Pommier, P.A. Ratliff, E. Meier-Engelen: "A Hybrid Satellite/Terrestrial Approach for Digital Audio Broadcasting with Mobile and Portable Receivers," Proceedings, NAB Engineering Conference, pp. 304–312 (1990)

A.23 A.G. Mason, G.M. Drury, N.K. Lodge: "Digital Television to the Home - When Will It Come?," Proceedings International Broadcasting Convention, IEE Conf. 327 pp.51–57 (1990)

A.24 "Digital Terrestrial Television Development in the SPECTRE Project," A.G. Mason, (National Transcommunications Ltd, UK, 1992), N.K. Lodge, (Independent Television Commission, UK, 1992)

A.25 "Design of an HDTV Codec for Terrestrial Transmission at 25 Mbit/s," P. Weiss, B. Christensson, J. Arvidsson, H. Anderson, (HD-Divine, Sweden), G. Petrides, C. Foster, (Vistek Electronics, UK), IBC, Amsterdam (1992) E. Stare: "Development of a Prototype System for Digital Terrestrial HDTV," Tele. English Ed. **No. 1** (2) (1992)

A.26 R. Monnier, J.B. Rault, T. deCouasnon: "Digital Television Broadcasting with High Spectral Efficiency," Thomson-CSF/LER, France, IBC, Amsterdam (1992)

A.27 P. Bernard: "STERNE: The CCETT Proposal for Digital Television Broadcasting, CCETT, France IBC, Amsterdam (1992)

A.28 W.F. Schreiber: "All-Digital HDTV Terrestrial Broadcasting in the US: Some Problems and Possible Solutions," Symposium International-Europe/USA, Paris (1991)

A.29 P.G. deBot: "Multiresolution Transmission over the AWGN Channel," Nederlandse Philips Bedrijven B.V., (1992) (in press)

A.26 ...

A.27 ...

A.28 ...

A.29 ...

Bibliography

Chapter 2

C.E.K. Mees, T.H. James: *Theory of the Photographic Process* (Macmillan, New York 1977)
A. Rose: *Vision: Human and Electronic* (Plenum, New York 1973)
W. Thomas (ed.): *Handbook of Photographic Science and Engineering*, SPSE (Wiley-Interscience, New York 1973)

Chapter 5

H.S. Hou: *Digital Document Processing* (Wiley, New York 1983)
T.S. Huang, O.J. Tretiak: *Picture Bandwidth Compression* (Gordon and Breach, New York 1972). Proceedings of a conference held at MIT in April 1969. A good collection of papers on the then state of the art
T.S. Huang, W. Schreiber, O. Tretiak: "Image Processing," Proc. IEEE **59** (11), 1586–1608 (1971). Extensive references
International Conference on Communication, ICC-79, Vols. 1, 2, Boston (IEEE, New York 1979). This is a yearly conference, and subsequent issues contain many useful articles on image coding
N.S. Jayant (ed.): *Waveform Quantization and Coding* (IEEE, New York 1976)
A.N. Netravali, B.G. Haskell: Digital Pictures, Representation and Compression (Plenum, New York 1988)
A.N. Netravali, J.O. Limb: "Picture Coding, a Review," Proc. IEEE **68** (3), 366–406 (1980). This paper contains an excellent bibliography
W.K. Pratt: *Digital Image Processing* (Wiley, New York 1978). Extensive references
W.K. Pratt (ed.): *Image Transmission Techniques* (Academic, New York 1979). This book contains a very perceptive article on transform coding by A.G. Tescher
W.F. Schreiber: "Picture Coding," Proc. IEEE **55**, 320 (1967)
Special Issues:
    Special Issue on Redundancy Reduction, Proc. IEEE **55** (3) (1967)
    Special Issue on Digital Communications, IEEE Trans. **COM-19** (6), Part I (1971)
    Special Issue on Digital Picture Processing, Proc. IEEE **60** (7) (1972)
    Special Issue on Two-Dimensional Signal Processing, IEEE Trans. **C-21** (7) (1972)
    Special Issue on Digital Image Processing, IEEE Trans. **C-7** (5) (1974)
    Special Issue on Digital Signal Processing, Proc. IEEE **63** (4) (1975)
    Special Issue on Image Bandwidth Compression, IEEE Trans. **COM-25** (11) (1977)
    Special Issue on Digital Encoding of Graphics, Proc. IEEE **68** (7) (1980)
    Special Issue on Picture Communication Systems, IEEE Trans. **COM-29** (12) (1981)
    Special Issue on Visual Communication Systems, Proc. IEEE **73** (4) (1985)
J.C. Stoffel (ed.): *Graphical and Binary Image Processing and Applications* (Artech House, London 1982). A good collection of articles

Chapter 6

W.H. Banks (ed.): *Halftone Printing* (Pergamon, New York 1964)
Eastman Kodak Co.: "Kodak Handbook of Newspaper Technique," Q165. (There are many other useful Kodak Data Books in the Q Series. Some are available in large camera stores.)
R.L. Gregory: *The Intelligent Eye* (McGraw-Hill, New York 1970)
R.W.G. Hunt: *The Reproduction of Color* (Fountain, Tolworth, England 1987)
W.M. Ivins, Jr.: *Notes on Prints* (MIT Press, Metropolitan Museum of Art, New York 1930)
W.M. Ivins, Jr.: *How Prints Look* (Beacon, Boston 1958)
W.M. Ivins, Jr.: *Prints and Visual Communication* (MIT Press, Cambridge, MA 1965)
Journal of the Optical Society of America (a continuing series)
C.E.K. Mees, T.H. James: *The Theory of the Photographic Proces* (Macmillan, New York 1977)
A.H. Phillips: *Computer Peripherals and Typesetting* (Her Majesty's Stationery Office, London 1970)
Pocket Pal: International Paper Co. (Available in many graphic arts and artists' supply stores)
Proceedings, Technical Association of the Graphic Arts (Rochester, New York) (a continuing series)
TAPPI: The Journal of the Technical Association of the Pulp and Paper Industry (a continuing series)
J.A.C. Yule: *Principles of Color Reproduction* (Wiley, New York 1967). (Excellent references)

Chapter 7

F.W. Billmeyer, Jr.: "Survey of Color Order Systems," Color Res. Appl. **12** (4), 173–186 (1987)
Color as Seen and Photographed, Kodak Publication E-74; *Kodak Filters for Scientific and Technical Uses*, Kodak Publication B-3; there are many other valuable Kodak publications in this field
Committee on Colorimetry, *The Science of Color* (Optical Society of America, Woodbury, NY 1963)
R.M. Evans: *An Introduction to Color* (Wiley, New York 1948)
R.M. Evans, W.T. Hanson, W.L. Brewer: *Principles of Color Photography* (Wiley, New York 1953)
A.C. Hardy: *Handbook of Colorimetry* (MIT Press, Cambridge, MA 1936)
R.W.G. Hunt: *The Reproduction of Colour in Photography, Printing, and Television* (Fountain, Tolworth, England 1988) (distributed by Van Nostrand-Reinhold, NY)
R.S. Hunter: *The Measurement of Appearance* (Wiley, New York 1975)
F.J. In der Smitten: "Data-Reducing Source Encoding of Color Picture Signals Based on Optical Chromaticity Classes," Nachrichtentech. Z. **27**, 176–181 (1974)
D.B. Judd: "Basic Correlates of the Visual Stimulus," in *Handbook of Experimental Psychology*, ed. by S.S. Stevens (Wiley, New York 1951) p. 811
D.B. Judd, G.S. Wyszecki: *Color in Business, Science, and Industry* (Wiley-Interscience, New York 1975)
J. Limb, C. Rubinstein, J. Thompson: "Digital Coding of Color Video Signals – A Review," IEEE Trans. **COM-25** (11), 1349–1385 (1977)
D.L. MacAdam: *Sources of Color Science* (MIT Press, Cambridge, MA 1970)
D.L. MacAdam: *Color Measurement* (Springer, Berlin, Heidelberg 1985)
S. Newhall, D. Nickerson, D.B. Judd: "Final Report of the O.S.A. Subcommittee on the Spacing of the Munsell Colors," J. Opt. Soc. Am. **33** (7), 385 (1943)
M. Southworth: *Pocket Guide to Color Reproduction* (Graphic Arts, Livonia, NY 1979)
M. Southworth: *Color Separation Technique* (Graphic Arts, Livonia, NY 1979)
L. Stenger: "Quantization of TV Chrominance Signals Considering the Visibility of Small Color Differences," IEEE Trans. **COM-25** (11) 1393–1406 (1977)
G. Wyszecki, W.S. Stiles: *Color Science* (Wiley, New York 1967)
J.A.C. Yule: *Principles of Color Reproduction* (Wiley, New York 1967)

Chapter 8

W.F. Schreiber: "Improved Television Systems: NTSC and Beyond," J. SMPTE **93** (8), 734–744 (1984)

W.F. Schreiber: "Advanced Television Systems for the United States: Getting There from Here," J. SMPTE **97** (10), 847–851 (1988)

W.F. Schreiber: "Considerations in the Design of HDTV Systems for Terrestrial Broadcasting," J. SMPTE **100** (9), 668–677 (1991)

W.F. Schreiber: "Spread-Spectrum Television Broadcasting," J. SMPTE **101** (8) 538–549 (1992)

Appendix

J. Freeman: "A Cross-Referenced, Comprehensive Bibliography on High Definition and Advanced Television Systems, 1971–1988," J. SMPTE **99**, 909–933 (1990)

Subject Index